The Earth's Surface Temperature:
Basic Science Including Carbon Dioxide

Why Human Efforts to Control Climate are Futile
And Reasonable Use of Fossil Fuels is Essential

Robin Smart

Disclaimer and Terms of Use Agreement

The author and publisher of this ebook and the accompanying materials have used their best efforts in preparing this ebook. The author and publisher make no representation or warranties with respect to the accuracy, applicability, fitness or completeness of the contents of this ebook. The information contained in this ebook is strictly for educational purposes. Therefore, if you wish to apply the ideas contained in this ebook you are taking full responsibility for your actions. The author and publisher disclaim any warranties (express or implied), merchantability, or fitness for any particular purpose. The author and publisher shall in no event be held liable to any party for any direct, indirect, punitive, special, incidental, or other consequential damages arising directly or indirectly from any use of this material, which is provided "as is" and without warranties. All references are for informational purposes only and are not warranted for content, accuracy or any other implied or explicit purpose. The author and publisher are not in any way associated with Google. Many different persons and their different opinions are described in this ebook. The author and publisher believe they all individually sincerely and honestly hold their opinions which differ from those of others described in this ebook and reject any assertion that any individual described in this ebook is not honest or sincere in holding their view. This ebook is © copyrighted by Dr Robin Smart 2023 under all applicable international, federal, state, and local laws with all rights reserved. No part of this may be copied or changed in any format, sold or used in any way other than what is outlined in this ebook under any circumstances without express permission from Dr Robin Smart, except by a reviewer who may quote brief passages in a review. This ebook has been prepared for publication and submitted to Amazon Kindle Books by Windshadow Pty Ltd, Australian Company Number 621 425 107.

Introduction

The idea that human use of fossil fuels is causing a dangerous rise in world temperatures with consequent severe effects on our climate including rising seas, severe storms, flooding and wildfires, has been promoted vigorously for over 40 years. Rising levels of Carbon Dioxide in the atmosphere are blamed due to the "Greenhouse Effect" of that gas. This view has been dominant now for over 20 years, including by virtually all governments, the United Nations and many other international organizations. Massive efforts are being made to move to renewable energy sources and reduce human Carbon Dioxide production.

Like most I followed this huge shift casually through the mass media largely for many years. But about a decade ago I became concerned about the consequences of this change on our advanced civilization, especially the practicalities of supply of affordable electricity to run our homes, schools, hospitals, industries, workplaces; and of energy supply for transport, the lifeblood of society. I also remembered reading "Ice: A Chilling Scientific Forecast of a New Ice Age" by Sir Fred Hoyle FRS, eminent astrophysicist, in the 1980's. I recalled Sir Fred emphatically dismissing in less than a page on scientific grounds the arguments of those stating carbon dioxide was a dangerous greenhouse gas.

Seven years ago I started researching for my own interest the basic science behind this matter. The modern internet makes this very possible. I gradually realised that many scientific disciplines are involved in determining the factors behind the surface temperature of our Earth. And, while there is a great deal of publicity concerning the shift to renewables and the threat of carbon dioxide, there is very little detailed scientific analysis available for the general public, including children and young people. Rather there are a lot of dogmatic very confident statements expertly presented.

I also found there is a large number of scientists in this field who do not consider carbon dioxide to be a threat, and some of them are discussed here as well as many who do.

This book aims to present to the general reader, including those with no scientific education, as well as children and young people, the basic science behind the Earth's surface temperature, especially the place of carbon dioxide, in a form which hopefully can be generally understood. As the project went on my own opinion clarified, and the latter part of the book therefore attempts to deal with the many consequences of the shift.

But you, the reader, must judge for yourself the result. And it is vital that the general population becomes familiar with the basic science behind this matter so that people everywhere can make up their own minds about the wisdom or folly of these massive steps being taken by their governments. It is insufficient to simply trust one set of "experts".

About the Author:

Robin Smart

MB ChB (Otago) FRCS (Eng) FRCSEd FRACS FUSANZ FNZMA

Robin Smart was born in Wellington, New Zealand, in 1945. In 1962 he won the maths and science prize at Scots College there, and in 1963 was Proxime Accessit to the Dux. In 1964 he gained a place at the University of Otago Medical School with all A's, including an A+ for physics, in the entry examinations. After graduating in 1969 he specialized in Urology which he has practiced for over 40 years. This dynamic surgical specialty has changed greatly in that time with advances in endoscopy, imaging and technology.

Robin has a lifelong interest in sailing and has owned cruising yachts for over 40 years. This has led to an intense interest in weather. He has completed five major ocean crossings, three as skipper/owner and one in 1983 in the pre-GPS era as navigator using astronavigation. Two were of the Tasman Sea, the most recent in 2017. He continues to work and sail.

Dedication

To the memory of my father, Fred Smart MSc FISE FICE FNZIE,
who taught me mathematics and much else.

Contents

Introduction iii
About the Author: iv
Chapter 1: Heat Energy 1
Chapter 2: The Earth's Atmosphere: 15
Chapter 3: The Earth's Surface Temperature 27
Chapter 4: Water 53
Chapter 5: The Oceans 77
Chapter 6: Ice and Snow 87
Chapter 7: Sea Level 97
Chapter 8: Carbon Dioxide 103
Chapter 9: Carbon Dioxide Heating of the Atmosphere 139
Chapter 10: Methane 175
Chapter 11: The Earth's Surface Temperature 185
Chapter 12: Conclusions 223
Chapter 13: Alternatives to Fossil Fuels 237
Chapter 14: Gas and Coal – a Modern View 307
Chapter 15: Some Consequences of the Forced Shift to Renewables 325
Detailed Summary of Contents and List of Figures 335

Chapter 1

Heat Energy

Temperature:

Things at the same temperature do not transfer heat energy between them. But if an object has a higher temperature than another heat energy will be transferred from it to the lower temperature object.

Temperature is measured by referring to certain changes which occur as the heat energy in an object changes. The most common object used is water; when water freezes to ice this is referred to as 0 degrees centigrade. When the heat energy in water increases to the point that it boils to steam (at sea level) this is 100 degrees centigrade.

As temperature, that is the contained heat energy in an object, increases changes occur. These are used in instruments to measure temperature. A common one used is the metal mercury. This is liquid from minus 40 degrees centigrade up, so is liquid at common temperatures we experience at the Earth's surface. As the heat energy, temperature, in metals increases they increase in size. This occurs in mercury too, so it gets bigger as its temperature increases. Mercury thermometers use this to measure temperature. They consist of a glass lower bulb and a thin glass tube extending up from this. As the temperature, heat energy, in the liquid mercury in the bulb increases it expands and pushes up into the thin tube. How far up it goes depends on the temperature of the mercury. The freezing point of water is marked as 0 degrees centigrade, the boiling point of water at sea level 100 degrees centigrade. And in between the scale is marked from 1 to 100.

Alcohol also expands as its temperature increases and freezes at minus 114 degrees centigrade, so is liquid at common temperatures on the Earth's surface, like Mercury. It is very commonly used in simple thermometers. The alcohol is coloured blue or red to make it visible.

More sophisticated modern temperature measuring instruments use electronic methods not mercury. These include **thermocouples, resistance temperature detectors,** and **radiative temperature detectors.** These all use differences which occur as temperature changes. Thermocouples use voltage changes, resistance detectors use electrical resistance changes, and radiative detectors use changes in heat (infrared) radiation.

An old mechanical method still used commonly is the **bimetallic bar.** In this two bars made of different metals are strapped together. As the temperature changes the bars expand (if hotter) or contract (if colder) differently so the bar bends. This can be connected to a dial to measure temperature. This method is used in, for example, barbecue ovens, air conditioning units.

Water boils to steam at a temperature, heat energy level, which is marked as 100 degrees centigrade. This is at sea level as said. At higher altitudes the air is thinner, air pressure is lower, and water boils at a lower temperature. This because water turns to vapour, gas, we call steam, more easily when the gas pressure at the water surface is lower. In space, where there is no gas pressure, water evaporates to gas at all temperatures, even well below 0 degrees centigrade. So the centigrade scale we use refers to sea level atmospheric pressure to measure 100 degrees centigrade when water boils.

Centigrade is the same scale as Celsius. There are other scales: Fahrenheit is used in the USA especially. It is historically older than Centigrade, and was devised in 1724 by a physicist Daniel Fahrenheit. He invented the mercury thermometer also, and his temperature scale, degrees Fahrenheit, was the first. He set zero at the freezing point of a salty brine solution, and the temperature of the human body as 90 degrees (later corrected to 96), the freezing temperature of water is 32 degrees Fahrenheit. Water boils at sea level at 212 degrees Fahrenheit. A degree Fahrenheit is smaller than a degree centigrade, 5/9 ths the size.

Another important scale in the scientific world is the Kelvin scale. This was named after an eminent physicist, William Thomson, who in the 1880's determined that the lowest temperature possible, when objects have no heat energy is minus 273.15 degrees centigrade. Zero on the Kelvin scale is at minus 273 degrees centigrade, meaning that zero centigrade is 273 degrees Kelvin. One degree Kelvin is the same size as one degree centigrade. He was made Lord Kelvin as an honour recognizing his achievements as a physicist, but also because he opposed home rule for Ireland. He was Irish.

As said at first temperature measures the amount of heat energy in gases, liquids and solids. If these are at different temperatures heat energy moves from the object with more heat energy (hotter) to that with less (colder).

The scientific symbol for degrees of temperature is °. So 100 degrees centigrade is 100° Centigrade, often called 100° C.

Heat Energy:

All matter – gases, liquids, solids – is made of atoms. The atoms are linked to make molecules. While the molecules in many solids may be large with millions of atoms each, those in gases are much smaller. Our atmosphere is mostly Nitrogen, 78%, and two nitrogen atoms make each molecule (N2). 21% is oxygen, also two atoms per molecule, O2. Water vapour makes up from 2 to 4%, and has two atoms of hydrogen and one of oxygen per molecule (H 2 O). Argon gas makes up 0.9%, one atom per molecule. All other gases make up just 0.1% (one molecule in 1000 molecules of air) of the atmosphere.

These include Carbon Dioxide, each molecule has one carbon atom and two oxygen atoms (CO2), which makes up 0.04% of the atmosphere, one molecule every 2500. And Methane, which has one Carbon atom and four hydrogen atoms per molecule (CH4). This makes up 0.00017%, 1.7 molecules per million in the atmosphere.

Heat energy is the vibration of molecules intensely and rapidly. The greater the amount of heat energy, temperature, the more vigorous the vibration of the molecules. At zero degrees Kelvin, minus 273 degrees centigrade, the vibration of molecules stops, they are still.

Moving Heat Energy:

Heat energy moves from objects – gases, liquids, solids – with more of it to those with less of it. This means that the vibration of molecules spreads from objects with it more intensely to those with it less vigorously.

Conduction

In solids it spreads directly from one molecule to the next, and the next, and the next, on and on, until it has spread through the solid object evenly. This is called *conduction* of heat energy. It does this because in solids the molecules are very close to each other, making the solid denser and heavier than liquids or gases.

Convection

In liquids the heat energy vibration also spreads directly to the next molecule as in solids. But the vibration pushes the molecules apart, which can happen in liquids because the molecules are not tightly bound to each other as they in solids. Hence liquids flow. Solids don't. When the molecules are further apart that part of the liquid is lighter compared to colder parts with less heat energy, lower temperature. So the warmer parts float higher in the liquid, moving up. This results in a current of warmer liquid moving up. This warmer liquid current in turn warms the part of the liquid it moves into, shifting the heat energy there. This is called *convection*. If you heat water on a stove in a pot you can see this happening. It is the most important way heat energy is moved in liquids.

In gases the same thing happens. Initially the hotter molecules, with more heat energy, vibration of molecules, transfer their greater vibration to nearby molecules directly. Then the warmer molecules push apart, making the warmer gas lighter, less dense. This gas then moves up relative to nearby cooler gas starting a *convection* current. This results in movement of heat energy within the gas, upwards. This is the most important method of heat movement in gases, like liquids.

These are all variations of heat energy, the vibration of molecules, being directly passed on to the adjacent molecules. There is another method of moving heat energy called *radiation.*

Radiation

We are all familiar with radiation. The sun heats us and our world by radiation. When objects – solids, liquids, gases – have heat energy within them they naturally emit radiation, a pure form of energy. It is referred to scientifically as *electromagnetic radiation.* It is in the form of waves of pure energy, like waves in the sea, or sound waves in the air. Only it moves in empty space extremely fast as it has no weight or mass. It moves at the speed of light, indeed light is a form of this radiation energy. The speed is 1080 million kilometers per hour, or 300,000 kilometers per second, or 186,000 miles per second. In gases, and in some liquids and solids, this radiation moves very slightly slower.

As said the radiation is wave like, and the length of the waves can be measured, the *wavelength*. Radiation with longer waves has fewer waves per second, because the speed of the radiation is the same for all radiation, light speed. Radiation with shorter waves has more waves per second. The number of waves per second is called the *frequency* of the radiation.

Clearly the distance the radiation travels in one second is the wavelength times the number of waves per second, frequency. This speed is always the speed of light, 300,000 kilometers per second. The absolutely huge speed means that there are a massive number of these tiny waves per second in a radiation beam.

Figure 1 shows the wavelength and frequency for the tiny section of electromagnetic radiation which we can see with our eyes, the **Visible Light Spectrum.** As shown the wavelength is totally minute at 0.4 millionths of a meter for blue light, and slightly longer for red light at 0.7 millionths of a meter. In between are other colours, orange, yellow and green. The remaining two colours, indigo and violet, are slightly shorter wavelengths than blue. "Roy G Biv" is a popular way of remembering this.

A millionth of a meter is a ***micron.*** There are a thousand microns in a millimetre, it is tiny.

The vertical line on the left shows the frequency, number of tiny waves per second, for each colour. For red light there are 4 hundred thousand billion waves per second. For blue, 8 hundred thousand billion. The other colours are in between.

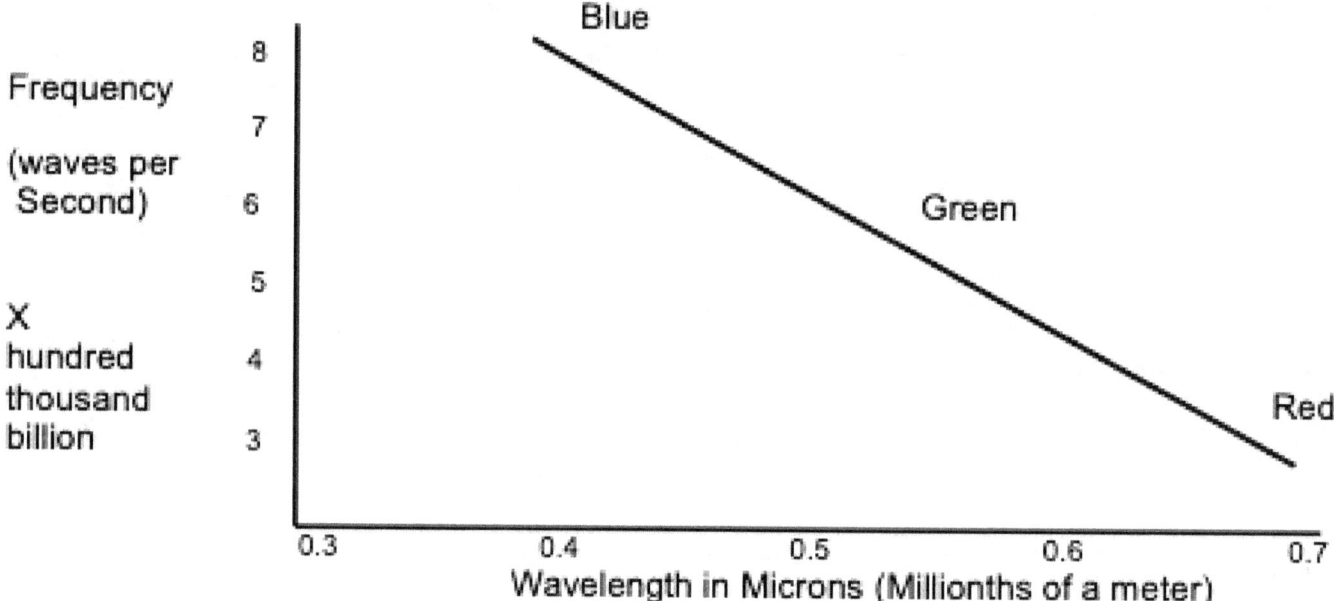

Figure 1: Visible Light

To clarify *Figure 2* shows 4 waves of red light. The wavelength is shown as 0.7 microns, and the 4 waves take 100,000 billionths of a second to pass by at the speed of light, 300,000 kilometers per second.

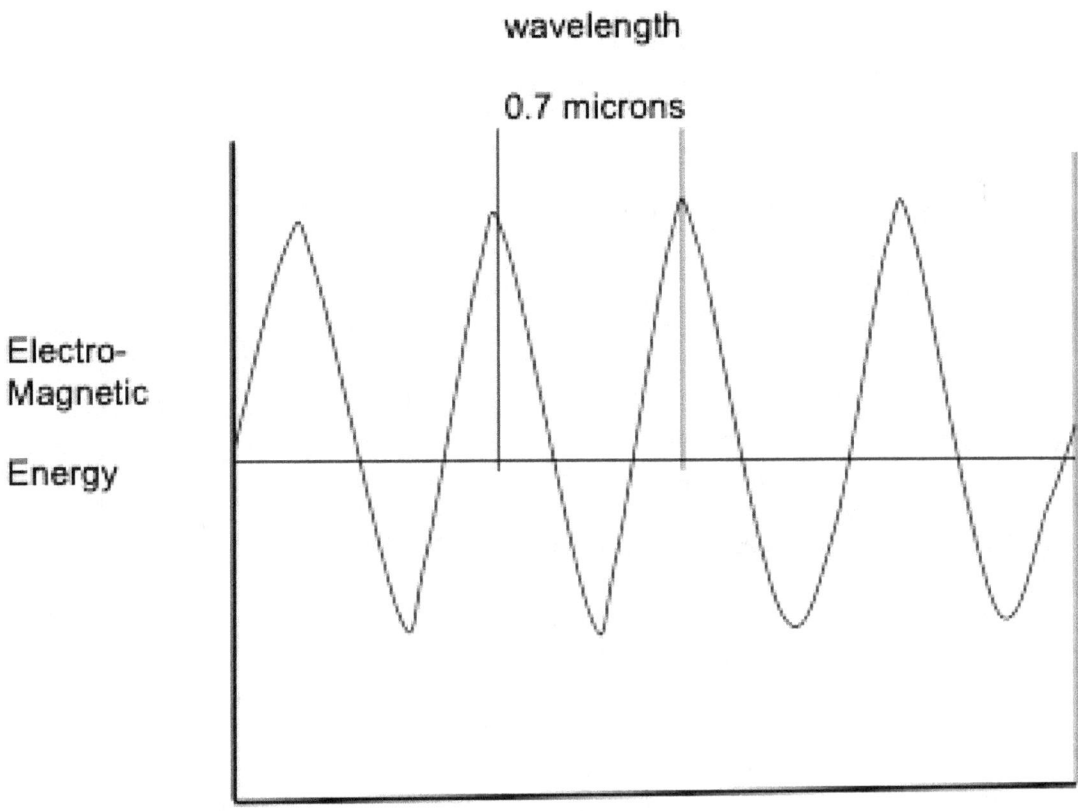

Figure 2: Diagram of Red Light Waves. 4 waves occur in one hundred thousand billionth of a second. Each wave is 0.7 microns long. Speed is 300,000 kilometers per second.

To write these as numbers for frequency (number of waves per second) of these colours it is 400,000,000,000,000 for red, and 800,000,000,000,000 for blue, waves per second.

Clearly it is very long winded and clumsy to talk about these huge numbers in this way. Doing any calculations on them like this is not possible. So scientists and mathematicians use a simple method to discuss huge or tiny numbers. This system is shown in *Figure 3.*

In Figure 3 the actual number is to the right of the vertical line, and to the left is the scientific code for that number.

In this system for large numbers instead of writing all the zeros, the number of zeros is given as the small number above and to the right of 10. For example, 100 is referred to as 10^2, 1000 is 10^3 – all the way to 10^{14} for one hundred thousand billion as shown.

For smaller numbers than one the small number is a minus number. 0.1 becomes 10^{-1}, 0.01 is 10^{-2}, all the way to 0.000000001 which is 10^{-9}.

Power	Actual Number
10^{14}	100,000,000,000,000
10^{13}	10,000,000,000,000
10^{12}	1,000,000,000,000
10^{11}	100,000,000,000
10^{10}	10,000,000,000
10^{9}	1,000,000,000
10^{8}	100,000,000
10^{7}	10,000,000
10^{6}	1,000,000
10^{5}	100,000
10^{4}	10,000
10^{3}	1000
10^{2}	100
10	10
1	
10^{-1}	0.1
10^{-2}	0.01
10^{-3}	0.001
10^{-4}	0.0001
10^{-5}	0.00001
10^{-6}	0.000001
10^{-7}	0.0000001
10^{-8}	0.00000001

Figure 3: Actual Number for each 10^{power}

The system is referred to scientifically as the ***Logarithmic System***. The little number is referred to as ***Power***. For example 100 is ***10 to the power 2***.

Another big advantage of this system apart from concisely writing huge numbers is that ***multiplication*** and ***division*** are much easier. For example, to multiply 100 X100, which is 10^2 X 10^2, just add the small numbers together and that is the answer - 10^4, which is 10,000. As another example, say, 400,000 X 2,000,000, which is four hundred thousand multiplied by two million, that is 4 X 10^5 X 2 X 10^6, the answer is 8 X 10^{11}. This is 800,000,000,000 (eleven zeros) or eight hundred billion.

For division the small numbers are subtracted. For example 1000 divided by 100 is $10^3 \div 10^2$, which 10^1, just plain 10. an example of a bigger number, say 4 X 10^9 (four billion) divided by 2 X 10^6 (two million) is 2 x 10^3 (two thousand).

To return to Radiation:

There are a huge variety of Radiation types depending on the wavelength of the Radiation waves. They are shown in *Figure 4.* Starting at the top are very short wavelength radiation waves, the wavelength being shorter than 0.001 microns (millionths of a meter) called ***gamma rays.*** Because their wavelength is so short there has to be a large number of them passing in one second to be at the speed of light, 300,000 kilometers per second, which all radiation moves at. So gamma rays have very high frequency. They are very high energy waves also, and would pass right through our bodies easily, damaging us. They are mostly found in stars and the Sun's radiation. They are largely blocked from reaching the Earth's surface by the very strong magnetic field which surrounds the Earth, because it's centre is molten iron and magnetic, and deflects harmful radiation away. This protection shield is called the ***Van Allen Belt.***

X – Rays, used in medicine, have slightly longer wavelengths than gamma rays from 0.1 microns to 10^{-6} (0.000001) microns. These short high energy waves also shoot through the soft tissues of our bodies and many objects, and are used to produce pictures of our bodies' internal structures as a result in X–rays and CT scanning. They also can damage us, and are also produced by the Sun's and stars' radiation. They are thankfully also blocked from the Earth's surface by the magnetic field of the Van Allen Belt. The dose we are given for X–rays and CT scans is regulated to safe levels.

Next are ***Ultraviolet Waves*** from 0.4 microns to 10^{-3} (0.001) microns wavelength. While not visible to our eyes this radiation does cause our skin to go browner, because melanin pigment producing cells in the skin respond to it, and they cause sunburn. Ozone in the Earth's upper atmosphere blocks most ultraviolet radiation from reaching the surface, but some does get through. Again the Sun and stars produce them.

Then we come to the tiny part of radiation which we can see with our eyes, the ***Visible Spectrum*** with wavelengths from 0.4 microns to 0.7 microns, shown in Figure 1.

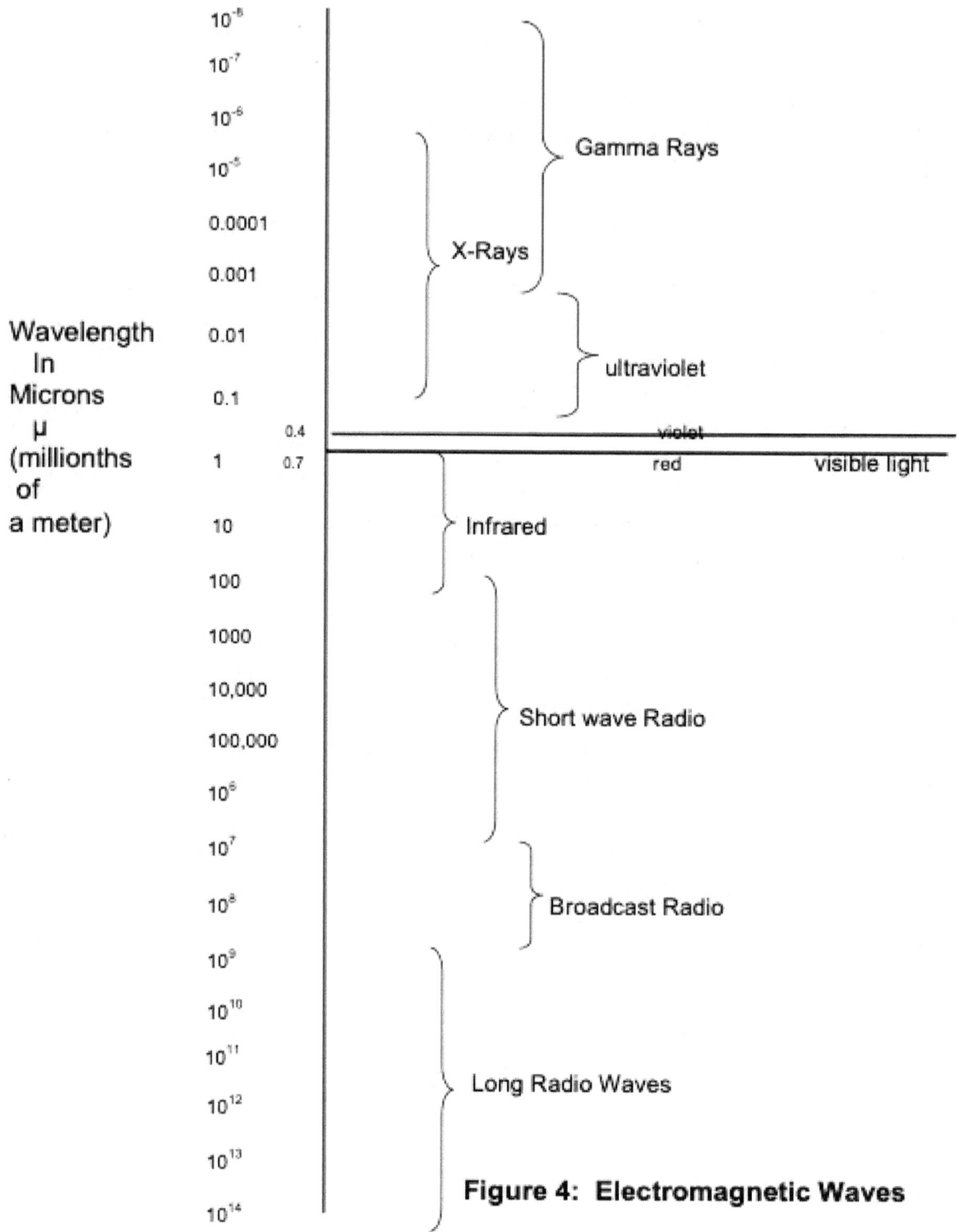

Figure 4: Electromagnetic Waves

Figure 1 shows that that the colour with the longest wavelength is red. So the next section of radiation is called **Infrared** - meaning **below red**. Its waves range from 0.7 microns to 350 microns. They cannot be seen with our eyes but we feel them as heat waves. Many objects we regard as hot such as heaters, engines, fires, cooking elements produce large amounts in infrared radiation which feels hot to us.

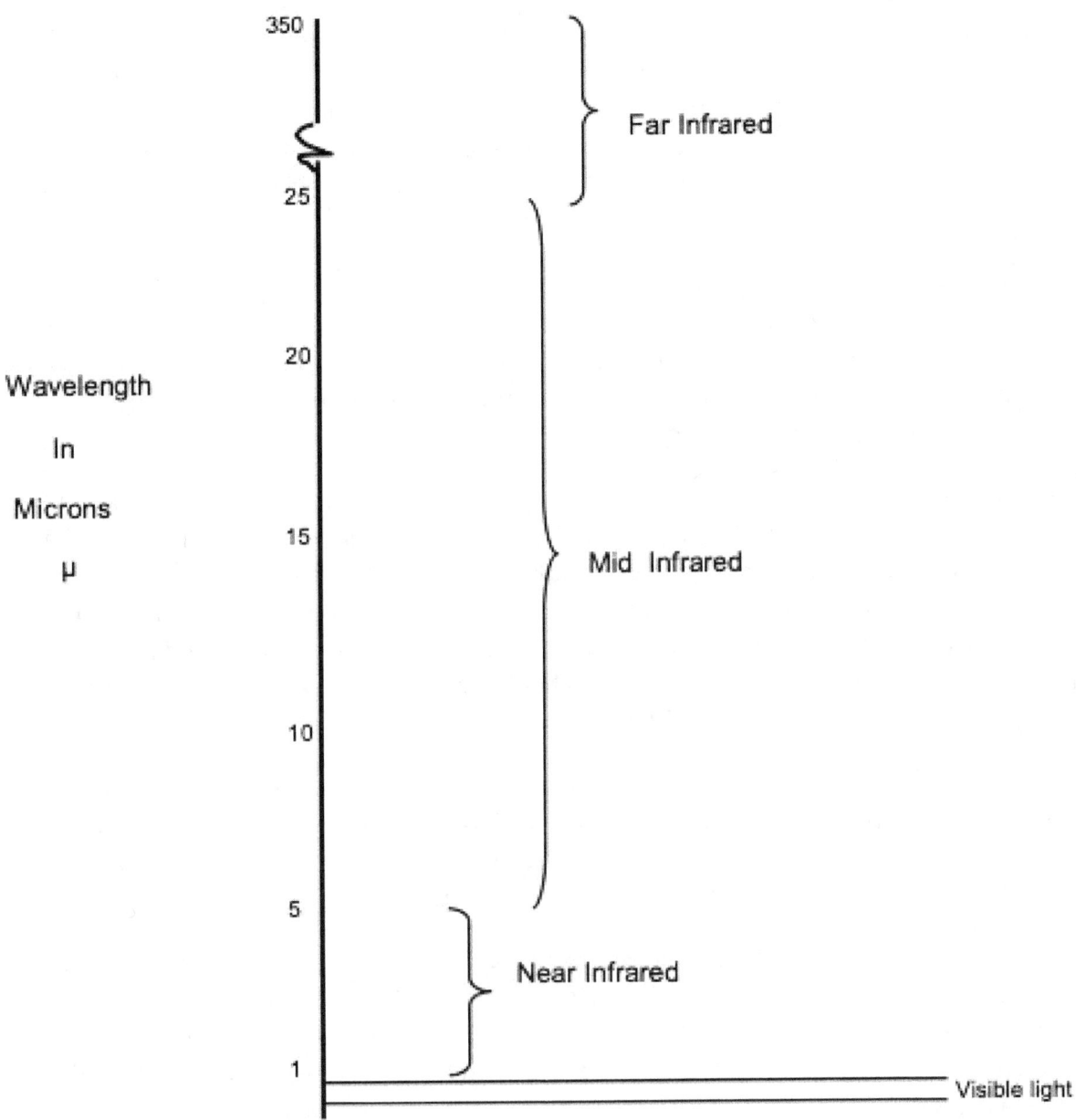

Figure 5: Heat Waves -- Infrared

1 Micron μ = 1 millionth of a meter

The Infrared radiation is divided into **Near Infrared** from 0.7 microns to 5 microns, **Mid Infrared** from 5 to 25 microns, and **Far Infrared** from 25 to 350 microns wavelength, as shown in **Figure 5**.

Longer radiation waves are used in radio and television transmission, also produced by stars and the Sun. They are shown in Figure 4 also. The longest waves are very very long, 10^{14} microns, which is 100,000 kilometers! It follows they are very low frequency.

In general very short high frequency waves have much more energy than longer lower frequency waves.

Earlier we discussed that objects – solids, liquids, gases – with heat energy produce radiation. All objects with a temperature more than absolute zero, 0° Kelvin, or minus 273° Centigrade, produce radiation. This is called **Blackbody Radiation** scientifically. The wavelength of that radiation depends on how much heat energy the object has, that is its temperature, how hot it is. Objects with higher temperatures, more heat energy, produce more waves of radiation per second, that is higher frequency, and these waves have shorter wavelengths. Cooler objects produce fewer waves per second, lower frequencies, and these have longer wavelengths.

We see this commonly when cooking. When the stove element is just warm we can feel the heat coming from it but cannot see this radiation because it is in the **Infrared** part of radiation, its waves are too long for us to see, longer than 0.7 microns. If the element gets hotter it glows red. This is because the heat radiation waves get shorter as the element temperature increase, and the radiation moves into the red part of the visible spectrum we can see. This is **Red Hot.**

The Sun is an example of this too. The temperature of the Sun's surface is 6000° Centigrade, and so most of the sunlight which reaches us on the Earth's surface, shown in *Figure 6,* after harmful radiation like gamma rays, X – Rays and Ultraviolet light has been blocked by the Van Allen magnetic belts and Ozone as mentioned before, is able to be seen by us in the Visible Spectrum from 0.4 to 0.7 microns wavelength. Some of it has longer wavelengths in the near infrared up to 3.2 microns, as shown in Figure 6. We cannot see that part but feel it as heat. Because the Sun is so hot its radiation is of shorter wavelength than cooler objects, and we can see that radiation as *light* to us.

Another example of an object producing radiation because of its heat energy is the Earth itself. The average temperature of the Earth is about 10° Centigrade to 15° Centigrade. There are clearly parts which are much hotter, such as deserts on a hot day, and parts which are much colder, such as Antarctica in the long dark winter. But 10° to 15° Centigrade is an average across the Earth.

The wavelength of the Earth's radiation is therefore much longer than the Sun's, which is at 6000° Centigrade. The Earth's radiation wavelength ranges from 4 microns to 90 microns as shown in *Figure 7.* The most intense radiation has a wavelength of about 10 microns, a lot of it occurs at wavelengths which are longer as shown – at least 40% is at greater wavelengths than 15 microns. We cannot see this radiation but can feel it as heat.

Visible light near infrared

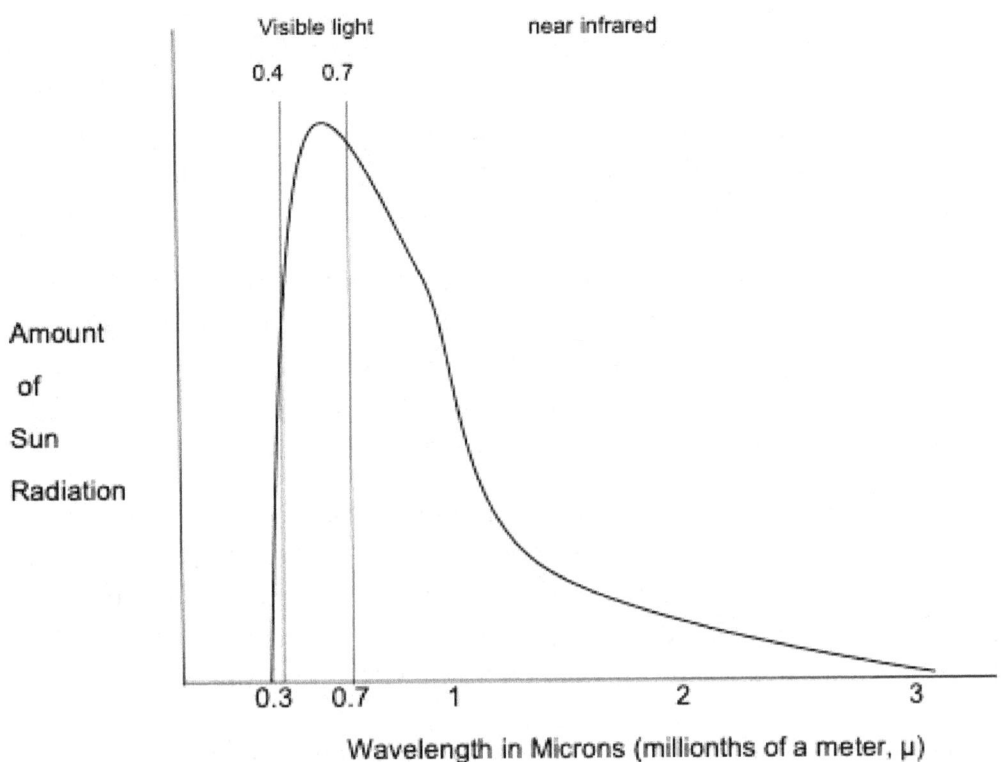

Figure 6: showing the wavelengths of the Sun's radiation which reach the Earth's surface, most intensely in the *visible light* range from 0.4 to 0.7 microns, but also in the *near infrared* range from 0.7 to 3.2 microns.

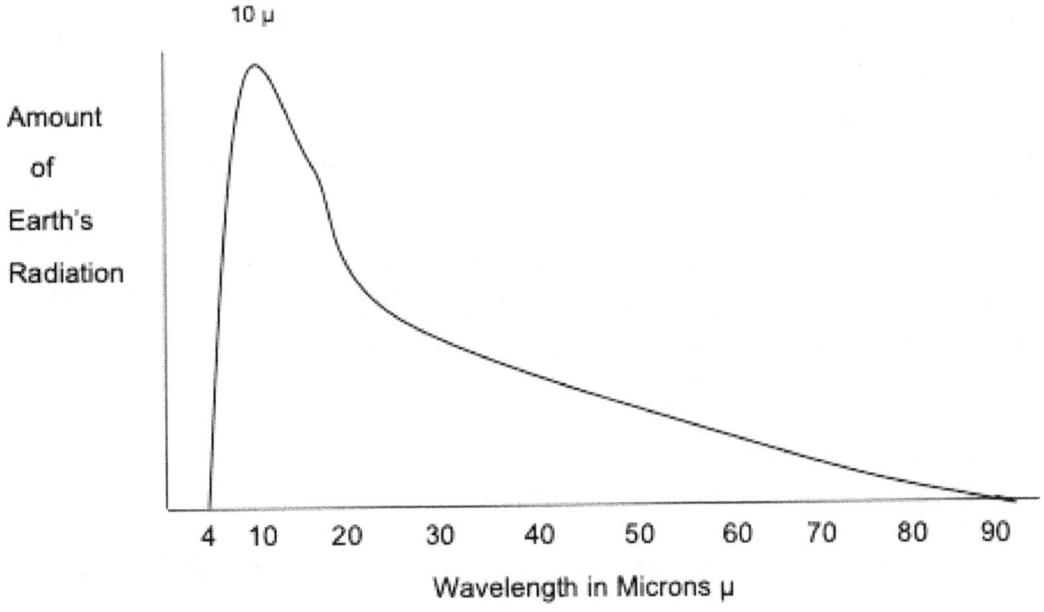

Figure 7: Earth's Energy (Heat) Radiation in the Infrared.

We earlier discussed that heat is the extremely rapid vibration of the tiny molecules in an object. When radiation strikes an object it causes the molecules to move, becoming hotter, the temperature of that part of the object increases. The heat then spreads to nearby parts of the object, either by directly vibrating adjacent molecules, called *conduction,* or, in liquids and gases, the hotter molecules float upwards because the heat vibration has forced the molecules further apart, so that the hotter part is lighter, floats upwards, shifting the heat up in the liquid or gas by *convection.* This is the most important way heat shifted in liquids and gases.

Every tiny molecule in an object produces radiation if it is at a temperature above absolute zero, minus 273° Centigrade, and therefore vibrating with heat energy. And the wavelength of that radiation depends on the temperature of the molecule just as for large objects like the Sun and the Earth we discussed above. So molecules at the surface of the Sun at 6000° Centigrade produce radiation in the visible light wavelength. And molecules on the surface of the Earth produce radiation in the infrared spectrum.

The hotter an object – solid, liquid or gas – is the more *Energy* there is in its radiation, as you would expect. The amount of energy in the radiation goes up very rapidly as the heat energy, temperature, of the object increases. It goes up to *the power 4 of the Kelvin temperature.* So an object at 10° Centigrade, which is 283° Kelvin (0° Kelvin is minus 273° Centigrade) has radiation energy proportional to 283 X 283 X 283 X 283, or 283^4. This is 6,414,247,921 worked out, say 6 billion, or 6×10^9 in the logarithmic scale discussed in Figure 3.

In contrast the Sun, at 6000° Centigrade, which is 6000° Kelvin, (ignoring the small 283° C difference for this calculation), has radiation energy proportional to 6000 X 6000 X 6000 X 6000, which worked out is 1.2×10^{15} or 1200 trillion It means the Sun puts out about 500,000 times the radiation energy *per square meter of its surface* the object at 10° Centigrade, for example the Earth does. This relationship between the radiation energy output and temperature of an object is described by the *Stefan-Boltzmann Law,* worked out by Josef Stefan in 1879. He used data measured by John Tyndal a few years before on radiation energy from objects at different temperatures.

To complete the comparison between the energy put out by the Sun at 6000°C and the Earth at 10°C allowance must be made for the much bigger surface area of the Sun, which is 6×10^{18} square meters compared to the Earth which has 5.1×10^{14} square meters. This means the Sun has 1.2×10^4, or twelve thousand, times the surface area of the Earth. So the Sun's energy output is 500,000 X 12,000 times that of the Earth, which is 6×10^9, or six billion times.

The heat radiation made by molecules in solids and liquids quickly strikes nearby molecules, because they are so close together in solids and liquids, and is absorbed by those molecules, making them hotter. Thus the heat energy spread by *radiation* in solids and liquids is similar to conduction and convection discussed before – the radiation heat energy cannot get far in these dense objects.

At the *surface* of solids and liquids the radiation heat energy is free to move into what is next, which, in our environment on Earth, is usually the gas of our atmosphere.

What happens to heat radiation energy traveling through a gas depends on what the gas is made of and the wavelength of the heat radiation. For many gases heat radiation just passes straight through and is not absorbed by its molecules, which are not affected by the heat passing through and their temperature, heat content, is not altered. To understand why this happens it helps to appreciate that the solid nucleus at the centre of each atom in the molecules is truly minute compared to the space of the atom, which is at least 100,000 X as big as the nucleus. This space is occupied by tiny *electrons* whizzing around. Electrons are minute compared to the nucleus, less than 0.0005 the mass, or 5 ten - thousandths. This means that the vast amount of the space of a molecule is empty space, not solid. In solids and liquids the molecules are close together, and heat radiation will hit another molecule soon and cannot travel far as said above. But in gases the molecules are far apart, so heat radiation can travel very long distances without being absorbed in some gases.

But some gas molecules do absorb radiation, *but only for certain wavelengths.* As a general rule gas molecules which are made from only two atoms and are simple in shape, tend to not absorb heat radiation and let it pass straight through, are not heated as a result. Whereas gas molecules which are more complicated, made of 3 or more atoms, are not symmetrical molecules, tend to absorb heat radiation, but only in certain wavelengths. This is because when the heat radiation strikes these more complicated molecules it sets up heat vibration *within the molecule only if it is a certain wavelength for that molecule.* So only heat radiation of the right wavelength can heat those more complicated gases. Heat radiation which is not the right wavelength will just pass straight through those gases and not heat them, similar to the simpler gases. This limitation of the wavelength of heat radiation which a gas molecule can absorb is called its ***fingerprint absorption.***

If heat radiation is of the right wavelength for that gas, in its fingerprint range, then it will heat the gas. If it is not, it won't.

When the gas molecule has been heated by heat radiation of the right wavelength for it, it very quickly starts to radiate heat radiation away itself, because it is now hotter. This heat radiation produced by the molecule has wavelengths depending on its temperature according to the *blackbody* radiation discussion above, where the radiation of the Sun, the Earth, and other objects was discussed. It has a broad swathe of wavelengths, not just a very limited number, as in fingerprint radiation absorption. This was shown in Figure 6 and figure 7. The molecule produces this blackbody heat radiation in *all directions*, up, down, sideways, not just up or down.

References and Further Reading for Chapter 1 are given at the end of Chapter 2.

Chapter 2

The Earth's Atmosphere:

To look now at our atmosphere and how the different gases in it absorb the Sun's radiation, and the Earth's heat radiation: as said above the atmosphere is made up of nitrogen 78%; oxygen 21%; water vapour 2 – 4%; argon 0.9%; all other gases less than 0.1%; these including carbon dioxide 0.04%; methane 0.00017% (1.7 parts per million).

Nitrogen, making up 78% of the atmosphere, is a simple symmetrical molecule with 2 atoms, called N2. It does not absorb any radiation of any type, including heat radiation from the Earth which is from 4 to 90 microns wavelength as explained in Figure 7. So heat radiation from the Earth's surface passes straight through nitrogen gas in the atmosphere without heating it.

Argon, making up 0.9% of the atmosphere, is similar. There is only one atom per molecule, it is symmetrical, heat radiation just passes straight through it without heating it.

Oxygen makes up 21% of the atmosphere and there are two oxygen atoms in each molecule, O2. it is a comparatively simple symmetrical molecule but it does a little bit of radiation absorption. This is shown in *Figure 8.*

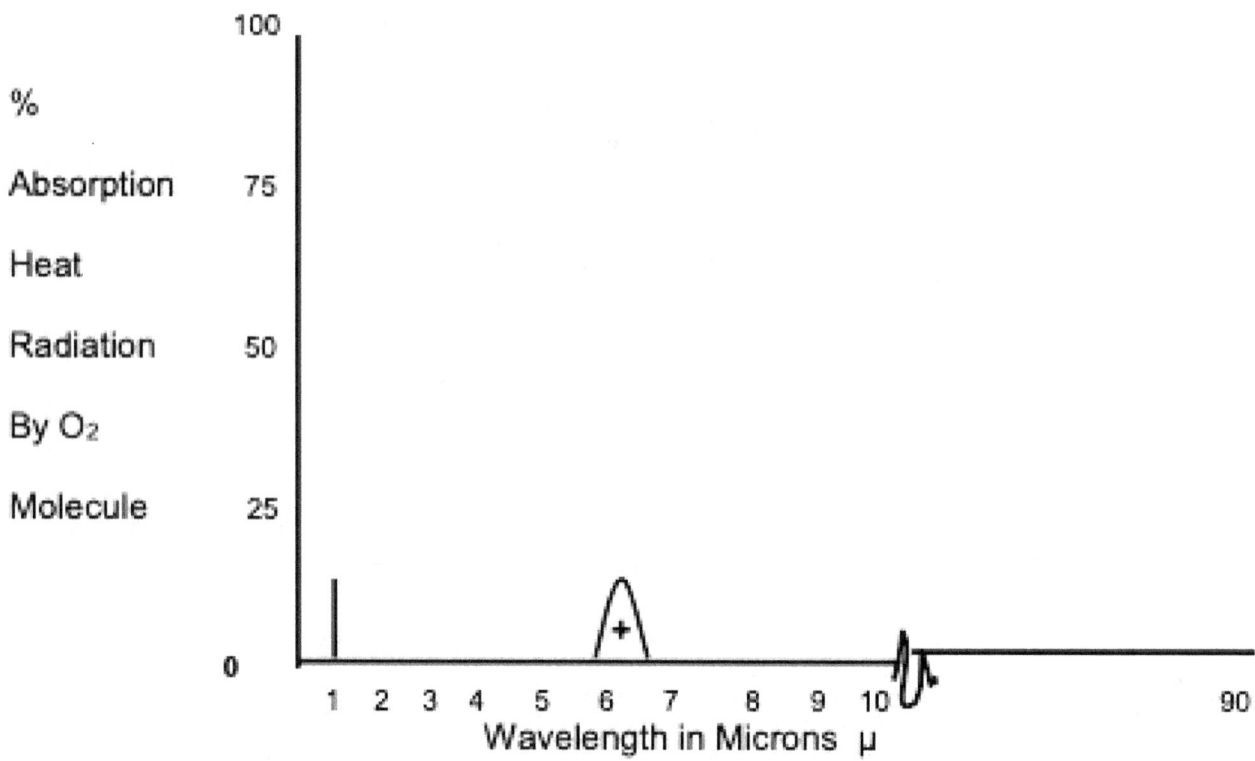

Figure 8: Radiation Absorption by Oxygen Molecules O2

As shown Oxygen absorbs heat radiation in a very limited fashion. There is a tiny absorption at wavelengths less than 1 micron (in fact 3 little peaks at 0.63, 0.69 and 0.76 microns wavelength), then a small amount at 6.2 microns, with a little spread of this from 5.8 to 6.8 microns. All other wavelengths Oxygen lets pass unaffected.

The squiggle in the lower horizontal line is showing a change in scale of this line to allow for wavelengths up to 90 microns to be shown. To the left of the squiggle the scale is larger to show the wavelengths Oxygen absorbs better.

To assess the effect, if any, of Oxygen in the atmosphere on heat radiation passing through it we shall first consider any effect on the Sun's radiation on Earth. Figure 6 shows this to be of wavelengths from 0.4 microns to 3.2 microns, the visible light section and near infrared as shown in Figure 5. Oxygen has only a tiny effect on the visible section. It has no effect on the Near Infrared section. So it has practically no effect on the Sun's radiation at the Earth's surface.

Does it have any effect on the Earth's heat radiation from the surface through the atmosphere to space? Figure 7 shows that the Earth's heat radiation is from 4 microns to 90 microns wavelength. Oxygen absorbs a tiny amount at 6.2 microns as mentioned and shown in Figure 8, in the Mid Infrared section. This has really no effect on heat radiation passing from the Earth's surface to space. The heat radiation just passes straight through Oxygen apart from the tiny amount absorbed a little bit at 6.2 microns wavelength.

Water Vapour, H2O, makes up 2 – 4 % of the atmosphere. Water is a comparatively complex asymmetrical molecule compared to the ones we have discussed. It has two Hydrogen atoms and one Oxygen atom. And its heat radiation absorption is much greater. In fact it is by far the most important heat radiation absorber in our atmosphere.

This is shown in Figure 9.

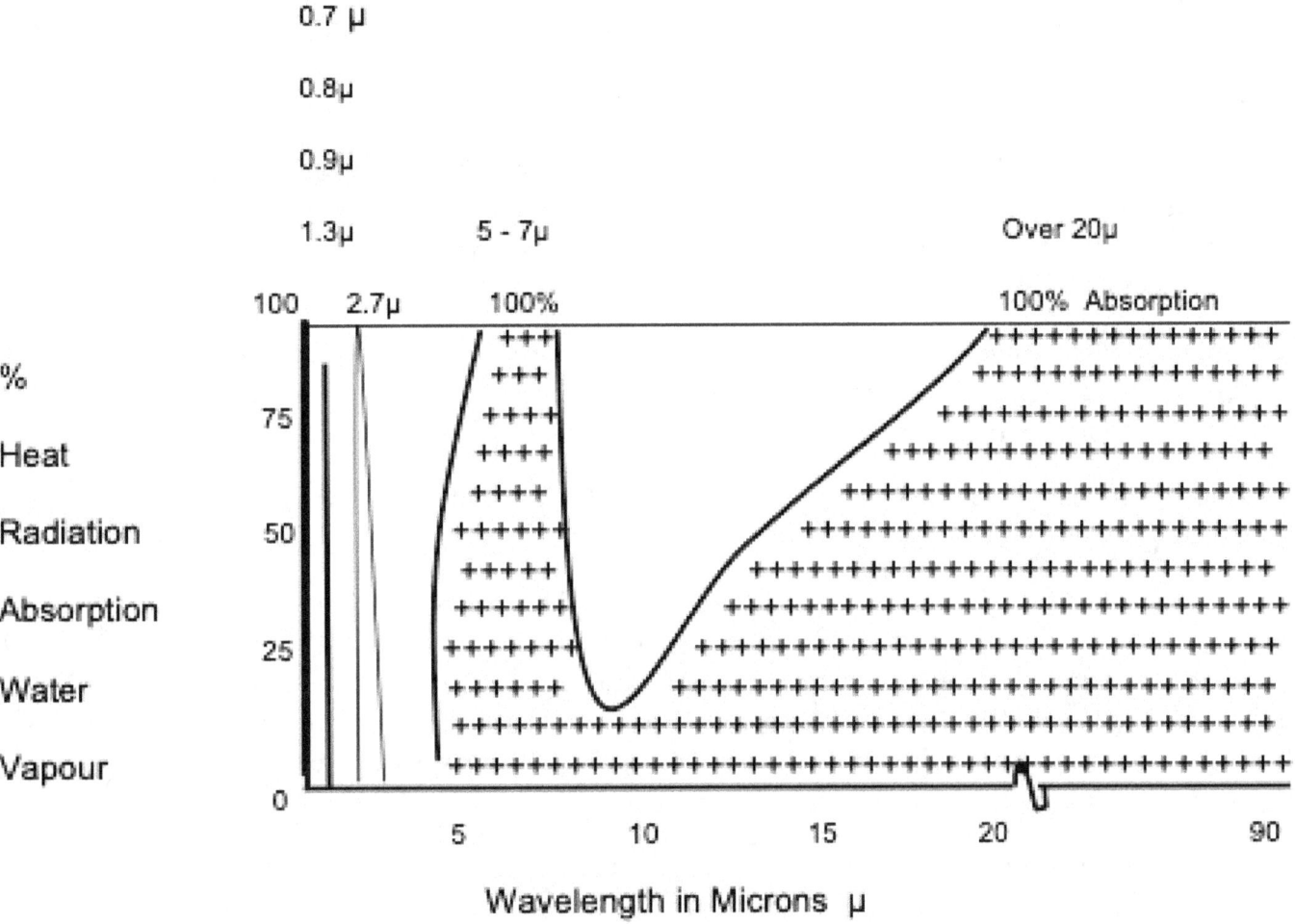

Figure 9: Radiation Absorption by Water Vapour Molecules H_2O

Water vapour absorbs heat radiation intensely for a number of wavelengths, indeed the absorption is 100% for many, meaning that *all* the heat radiation for those wavelengths is absorbed by the water vapour in the atmosphere.

To start with the effect on the Sun's radiation at the Earth's surface, wavelengths 0.4 to 3.2 microns as shown in Figure 6. There are a series of small spike like absorption bands at 0.7, 0.8, 0.9, 1.3 microns as shown by the figures above the left end of the vertical line, and all symbolized by the vertical line at about 1 micron on the graph. It was not possible to fit these all in this graph with this scale, so they are shown that way. Then there is a slightly broader 100% absorption spike from 2.5 to 3 microns, peaking at 2.7 microns. The presence of water vapour in the atmosphere would have some effect on reducing the amount of the Sun's radiation getting to the Earth's surface, resulting in it being cooler. We intuitively know this is true.

Moving onto the effect on the Earth's heat radiation as shown in Figure 7, from 4 microns to 90 microns. It can be seen that there is broad absorption by water vapour from 4.5 microns to 7 microns, reaching 100% from 5 to 7 microns as shown. Then there is some reduction in absorption to its lowest amount at 9 microns as shown in Figure 9. Then the absorption gradually climbs back up to 100% at 20 microns heat radiation wavelength. And, importantly, it remains at 100% beyond 20 microns through 90 microns.

This means that *all* the heat radiation leaving the Earth's surface with wavelengths from 20 to 90 microns will be absorbed by water vapour in the atmosphere, and also heat radiation with wavelengths from 5 to 7 microns will be too, with a lesser effect on heat radiation from the Earth's surface with wavelengths between 7 and 20 microns.

If there is no water vapour in the atmosphere, as might occur over large desert areas, then the heat radiation in these wavelengths might escape into space directly from the Earth's surface. But that is very uncommon, usually there is, as said before, between 2 and 4% water vapour in the atmosphere.

So atmospheric water vapor is a very powerful preventer of heat radiation escaping from the Earth's surface directly into space.

Carbon Dioxide, CO2 : This makes up 0.04% of the atmosphere, for every molecule of CO2 there are 2500 other molecules of other gases especially nitrogen, oxygen and water vapour. The amount of CO2 in the atmosphere is often described in *parts per million,* meaning the number of molecules of CO2 for every million molecules of all gases in the atmosphere. It is around 400 parts per million which equals 0.04%. Using the same measure nitrogen in the atmosphere is 780,000 parts per million, 78%. So CO2 is one of the trace gases, less than 1%, in the atmosphere.

CO2 is a molecule made up of three atoms, one carbon and two oxygen. It is a complex shape like water vapour molecules so does have some heat radiation absorption properties. These are shown in Figure 10.

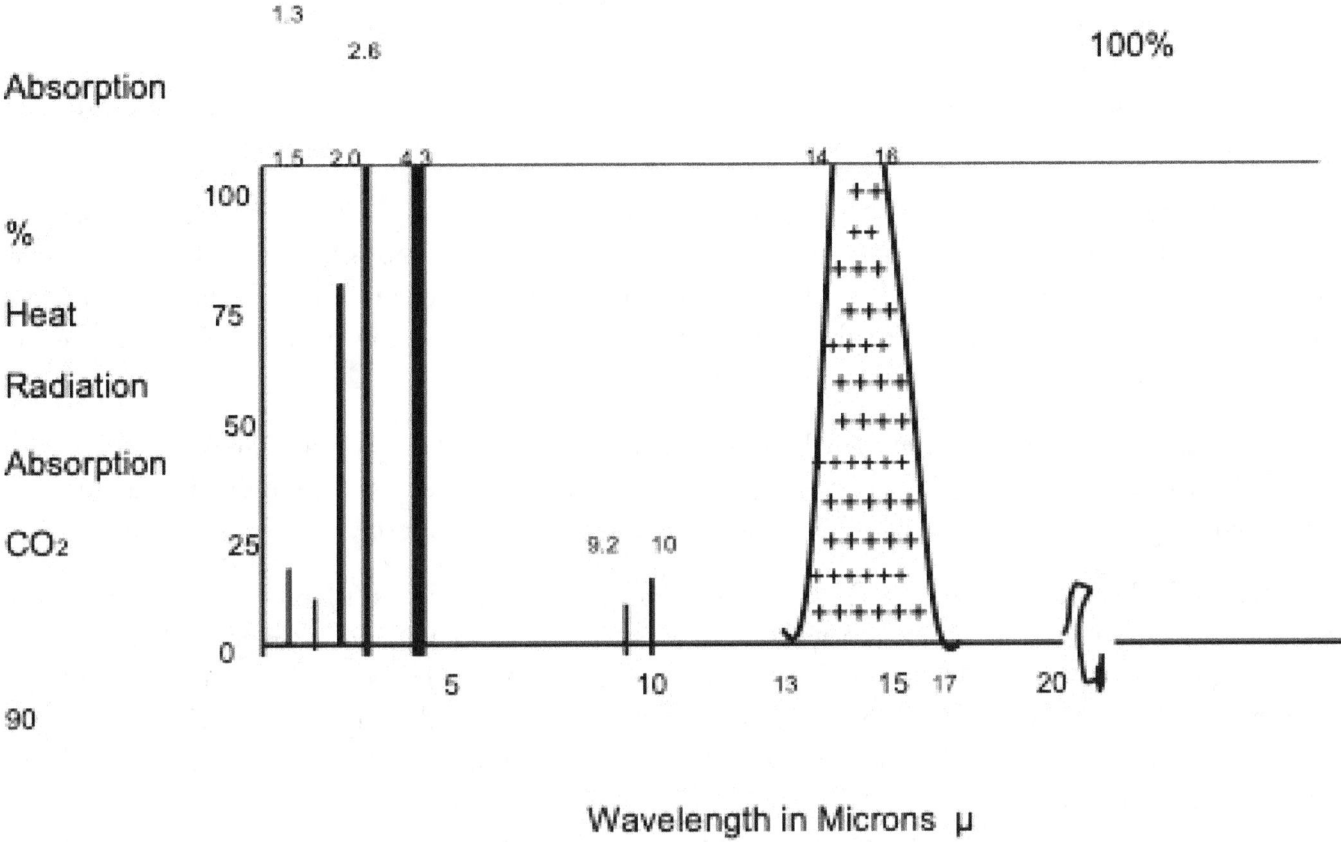

Figure 10: Radiation Absorption by Carbon Dioxide Molecules CO₂

Looking first at the effect on the Sun's radiation from 0.4 microns to 3.2 microns as shown in Figure 6. There are tiny bands of weak absorption at 1.3 and 1.5 microns, then stronger but narrow ones at 2 and 2.6 microns. The one at 2.6 microns is a little wider, about 2 microns wide from 2.5 to 2.7 microns. It is clear that CO2 in the atmosphere has only a slight effect on sunlight passing to the Earth's surface.

Regarding the effect on the Earth's heat radiation from 4 microns to 90 microns as shown in Figure 7 there are a number of wavelengths affecting this as shown, especially up to 17 microns. In order to show this more clearly again the same change of scale at 20 microns is used as in Figure 9 for water vapour: the squiggle marks the change in scale from 20 to 90 microns.

Carbon Dioxide has no absorption effect above 17 microns, unlike water vapour which powerfully absorbs radiation above 20 microns. So heat radiation with longer wavelengths the 17 microns passes through carbon dioxide unaffected.

For wavelength from 4 to 17 microns there are some areas of absorption. There is a narrow but 100% area at 4.3 microns, this goes from 4 to 4.4 microns. To assess the importance of this on Earth's heat radiation

Figure 7 shows there is not much to affect for that wavelength, so it is of little importance. There are weak areas at 9.2 and 10 microns as shown, again of little importance.

Then there is an area on intense absorption from 13 to 17 microns, it reaches 100% absorption from 14 to 16 microns.

In the 1970s and 1980s when research on the effect of carbon dioxide on heat radiation from the Earth's was earlier many of the instruments used, for example those on the American NASA Nimbus 4 satellite, could not measure heat radiation greater than 25 microns. So only the band from 4 to 25 microns was considered, and it was concluded that carbon dioxide affected about 25% of that wavelength band, hence reduced the heat radiation from the Earth by about 25%, it was thought.

But later research has shown, as mentioned above, that at least 40% of the heat radiation of the Earth is in the wavelengths from 15 to 90 microns, most above 17 microns not affected by carbon dioxide, so it is now considered that carbon dioxide affects about 8% of the heat radiation from the Earth's surface, not 25%.

Carbon dioxide does have an effect on the Earth's heat radiation, but much less than water vapour which as discussed absorbs intensely radiation greater than 20 microns. But an interesting feature of carbon dioxide's absorption in the band from 14 to 16 microns, shown to be 100% in Figure 10, is that it is very intense absorption. Indeed, at sea level, even though carbon dioxide is only 0.04% of the air there, *all* of the Earth's heat radiation in that wavelength band from 14 to 16 microns is completely absorbed by CO_2 molecules in just 10 meters of air – even though only one in 2500 molecules is a CO_2 one! This is discussed in detail in the later section on carbon dioxide's effect on the Earth's surface temperature. But a consequence of this intense absorption in this narrow wavelength band is that increasing the concentration of CO_2 in the air *does not have any more significant effect on the heat radiation from the Earth's surface because all the radiation in that narrow band which CO_2 affects has already been absorbed at lower concentrations of CO_2.* It is like a car going flat out – it cannot go faster. But more of that in the later section.

Methane, CH_4, makes up just 0.00017% of the atmosphere, 1.7 molecules per million, a very tiny amount. It is a complex molecule with 5 atoms, one carbon and 4 hydrogen, but it has only a minor radiation absorption effect as shown in Figure 11:

Again looking first at the effect, if any, on the Sun's radiation to the Earth's surface from 0.4 microns to 3.2 microns wavelength: There are small weak absorption bands at 1.5 and 2.2 microns of little importance, and another more intense band at 3.2 microns, right at the edge of the Sun's radiation wavelengths. Methane has only a slight effect on the Sun's radiation striking the Earth's surface.

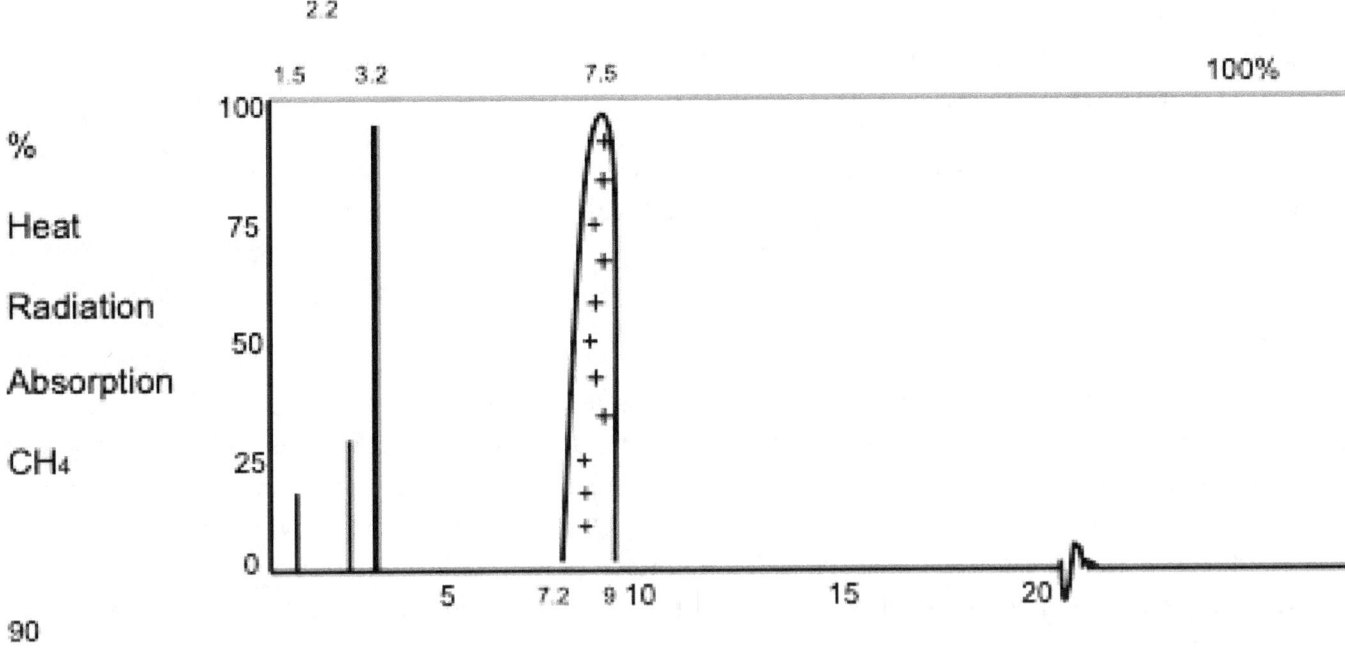

Figure 11: Radiation Absorption by Methane Molecules CH₄

Concerning Methane's effect on the Earth's heat radiation leaving its surface there is a band of intense absorption from 7.2 microns to 9 microns, peaking at 100% absorption at 7.5 microns. There are no other absorption bands. Looking back at Figure 7 which shows the heat radiation wavelengths of the Earth's radiation it can be seen that methane would have only a small overall effect, perhaps about 1%, blocking the radiation. When the tiny amount of methane in the atmosphere, 1.7 molecules per million, is considered the effect becomes very tiny.

Ozone, O3 :

This is a very important trace gas in the high atmosphere, the *Stratosphere.* The reason it is important is that Ozone absorbs harmful radiation from the Sun in the high energy wavelengths from 0.1 to 0.3 microns in the ultraviolet radiation band. If these wavelengths could pass to the Earth's surface they would imperil many living organisms by destroying complex cell structures such as chromosomes. Figure 6 shows that the Sun's radiation reaching the Earth's ranges from 0.4 to 3.2 microns. The Ozone Layer was discovered by Charles Fabry and Henri Buisson, French physicists, in 1913. They realised the Sun must be emitting radiation wavelengths shorter than 0.4 microns because of it's surface temperature of 6000 °C, which meant that it's *Blackbody radiation* included shorter wavelengths than 0.4 microns. They looked and found Ozone is the stratosphere, and realised this was absorbing the shorter wavelengths and preventing them reaching the Earth's surface.

The Layers of the Atmosphere

The bottom layer of the atmosphere is called the **Tropopshere.** This goes from the surface to about 8 kilometers up at the poles, and 18 kilometers at the equator. This layer contains 80% of the mass of the atmosphere, and virtually all of the water vapour. The water vapour causes it to be a turbulent restless layer, the source of storms, rain, weather of all kinds. The top of the troposphere is called the **Tropopause.**

There is virtually no water vapour above the Tropopause because it is too cold. The temperature there is minus 50 °C, all water vapour has turned to ice well and truly, the ice falls, cannot go higher, so above that the atmosphere is very dry, without water. As water vapour is essentially the source of the turbulence, storms and wind in the troposphere without it the higher atmosphere is very calm, still and dry. This **Stratosphere** goes up to 50 kilometers above the surface.

A factor in the turbulence of the Troposphere is that the **temperature** decreases markedly as one goes higher. It is around 15° C at the surface, but minus 50° C at its top, the tropopause 8 to 18 kilometers up. This makes **convection currents** in the air much stronger because the warmer thinner air from the surface rises faster and faster in the colder denser air higher up. This is most obvious to us looking at thunderstorms approaching. The **thunderheads** are upward moving columns of warmer air rushing up to the tropopause. As they rise they cool quickly, the water vapour in them turns to water and ice, with heavy rain forming and hail.

In the **stratosphere** by contrast there is no water vapour. The gases in the stratosphere are consequently much more transparent to heat radiation than those in the troposphere below. So most heat radiation is free to pass into space without causing powerful convection turbulence as in the troposphere. This explains why it is a still layer. We experience this when we fly as most airplanes fly in the lower stratosphere in clear air. Looking out the window often we see the thick cloud layers made of water far below.

There is, however, some absorption of radiation from the Sun in the stratosphere. This occurs in the **ultraviolet** part of the Sun's radiation wavelengths shorter than 0.4 microns. **Oxygen molecules,** O_2, absorb ultraviolet radiation with a wavelength of about 0.3 microns (remember that visible light is from 0.4 to 0.7 microns). This causes the O_2 molecules to break up to O single atom ones, then form **O_3** ones called **Ozone.** These molecules absorb virtually all the ultraviolet radiation from the Sun with wavelengths from 0.2 to 0.35 microns in the stratosphere. Most of the ultraviolet radiation from the Sun with wavelengths shorter than 0.2 microns is blocked by Nitrogen and Oxygen molecules, and free atoms and other particles higher in the outer atmosphere, the **Thermosphere,** which absorb these very short wavelengths. As said above Ozone completes this process of protecting life Earth from these harmful wavelengths which would severely damage living organisms.

But the ultraviolet radiation from the Sun also causes Ozone to in turn break down to single atom O molecules, which quickly join to form the common O_2 form. This process is sped up greatly by some chemicals which get into the stratosphere, principally oxides of Nitrogen (molecules of Nitrogen and Oxygen, NO). Some chemicals produced by man also speed up greatly the destruction of Ozone, the most well-known being **fluorocarbons,** complex molecules of Fluorine and Hydrocarbons used, for example, as fluids in fridges. These caused the **Ozone hole** to form over Antarctica, causing great concern that life on Earth would be threatened by uncontrolled ultraviolet radiation, finally resulting in international agreement to ban man-made Fluorocarbons by the *Montreal Protocol* in 1987. Since then Ozone has recovered partially.

The ongoing destruction and creation of Ozone results in it being present only in very small amounts, 10 molecules per million virtually all in the stratosphere, and mostly at around 25 kilometers height. The air is thin there, and if all the Ozone in the stratosphere was compressed to the thicker density of air at sea level it would be only 3 mm thick. Yet the absorption of ultraviolet radiation from the Sun by these Ozone molecules results in warming of the stratosphere air, which rises smoothly to a higher level, resulting in the temperature at the top of the stratosphere being warmer, minus 3°C, than its bottom, minus 50°C. This results in convection currents being weaker and the stratosphere calmer than the troposphere as said.

Concerning the temperature of the surface of the Earth the stratosphere has virtually no influence slowing the passage of heat from the surface to frigid space beyond the atmosphere. This is because the stratosphere gases, largely Nitrogen and Oxygen, are practically transparent to the great majority of the blackbody heat radiation of the Earth as shown in Figure 7 with wavelengths from 4 to 90 microns. And because the layer allows warmer gases to smoothly rise and lose their heat by radiation to space, which is very cold indeed at minus 270° C, freely.

The Atmosphere beyond the Stratosphere:

The next layer up is the *Mesosphere* from 50 to 85 kilometers above the surface. The air is very thin, , and the temperature decreases with increased height from minus 3° C to minus 90° C at its top, the coldest part of the atmosphere. Most meteors burn up and are seen as *shooting stars* in the Mesosphere.

Above this is the *Thermosphere* from 85 to about 500 kilometers up. The air is extremely thin here, so thin that only a hundred thousandth of the mass of the atmosphere is here and there are big distances between molecules. The Sun heats these markedly and their temperature increases with height. It can get to 2000° C in daytime. But if we were to go there it would feel very cold indeed because there is little heat energy from the widely separated gas molecules.

The *edge of space* is in this layer at 100 kilometers high, called the *Karman Line.* Above this it is not possible for aircraft to use lift from their wings to fly.

Many satellites including the International Space Station orbit in this layer, the air is so very thin. The Sun's radiation causes gas molecules here to become electrically charged, and sometimes to break up into single atoms, a process called *photo disassociation.* The charged particles cause radio waves to bounce around the Earth, meaning that we could hear radios on the opposite side of the Earth in the days before satellites. High in the thermosphere the atoms are all very light, just Helium, oxygen atoms, nitrogen atoms and Hydrogen. The oxygen and nitrogen atoms absorb the Sun's very short wavelength radiation less than 0.2 microns wavelength helping to protect life on Earth along with ozone as discussed above.

A curious feature of the Thermosphere is a band of *Sodium atoms* 10 kilometers thick above 90 kilometers. The sodium is supplied by meteors.

Above this again is the *Exosphere.* This starts at about 500 kilometers up. It ends where the Earth's gravity no longer pulls gas molecules back to Earth, about half way to the Moon. There are tiny amounts of

Hydrogen, atomic oxygen, Helium, Carbon Dioxide here in the lower parts, and almost nothing in the upper part, after which it is true outer space.

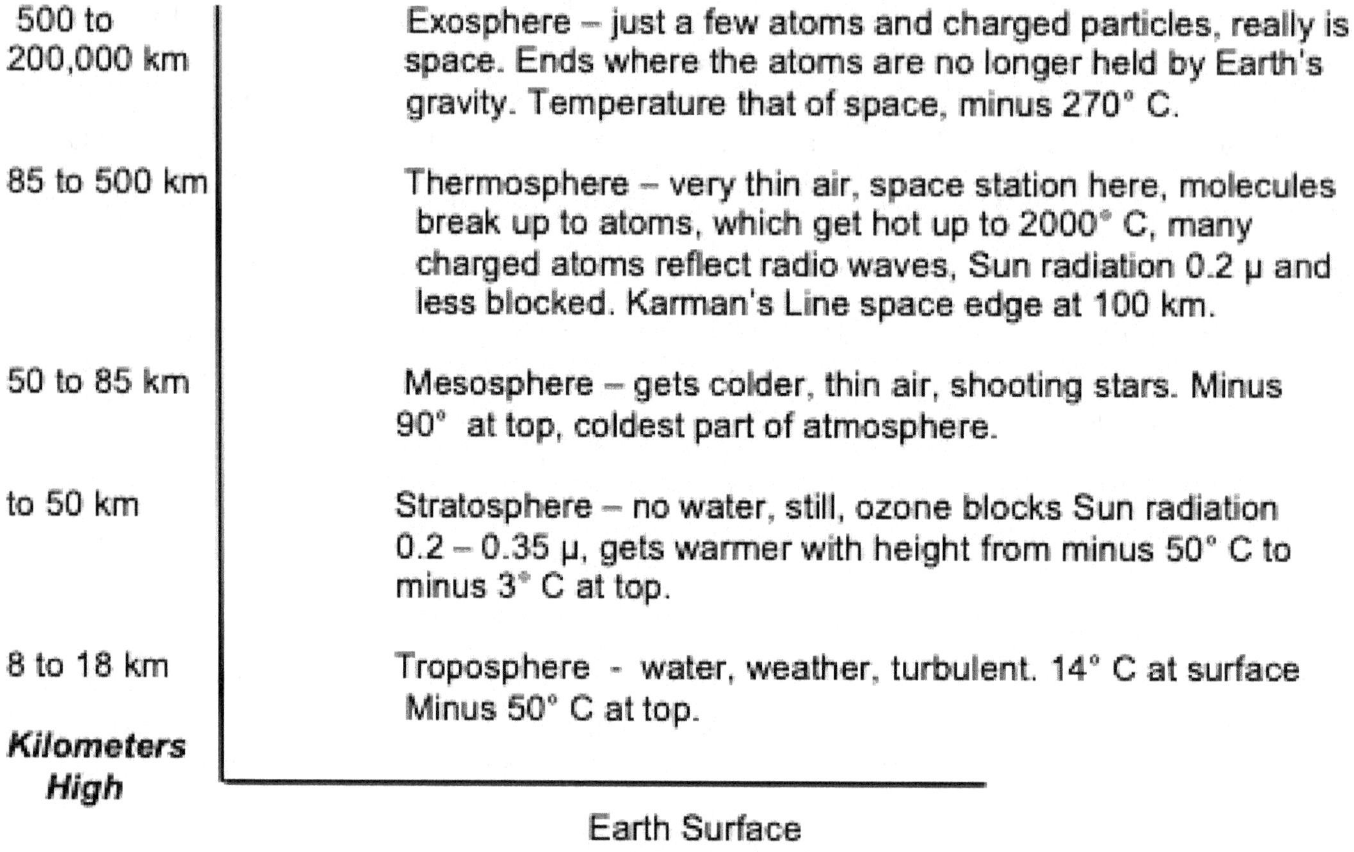

500 to 200,000 km	Exosphere – just a few atoms and charged particles, really is space. Ends where the atoms are no longer held by Earth's gravity. Temperature that of space, minus 270° C.
85 to 500 km	Thermosphere – very thin air, space station here, molecules break up to atoms, which get hot up to 2000° C, many charged atoms reflect radio waves, Sun radiation 0.2 μ and less blocked. Karman's Line space edge at 100 km.
50 to 85 km	Mesosphere – gets colder, thin air, shooting stars. Minus 90° at top, coldest part of atmosphere.
to 50 km	Stratosphere – no water, still, ozone blocks Sun radiation 0.2 – 0.35 μ, gets warmer with height from minus 50° C to minus 3° C at top.
8 to 18 km	Troposphere - water, weather, turbulent. 14° C at surface Minus 50° C at top.

Kilometers High

Earth Surface

Figure 12: the Layers of the Atmosphere.

These outer layers, the Mesosphere, the Thermosphere and the Exosphere have no influence on the surface temperature of the Earth as they are too thin and so high that any heat they have is radiated away into space.

Other Atmospheric Gases:

There are a number of other trace gases in the atmosphere, but they are of little importance if any concerning the temperature of the Earth's surface, and so are not detailed here.

References and further reading for Chapters 1 and 2:

If a reference does not work a computer search will usually reveal the article.

www.irina.eas.gatech.edu/EAS8803_Fall2009/lec6.pdf (Sun radiation).

www.irina.eas.gatech.edu

www.ceadservl.nku.edu

www.webbook.nist.gov/cigi/cbook.cgi

www.journals.ametsoc.org/view/journals/bams/96/9/bams-d-13-00286.1.xml Turner and Mlawer (2010) (Earth blackbody radiation Antarctic ground based study).

Patchett, L. et al. Far-infrared radiative properties of water vapour and clouds in Antarctica. Bulletin of American Meteorological Society, Vol. 96, Issue 9, pp 1505-1518. DOI: https://doi.org/10.1175/BAMS-D-13-00286.1 (Earth Blackbody radiation).

Clough. B.J. et al. Atmospheric radiative transfer modelling: a summary of the AER codes. J. Quant. Spectrosc. Radiat. Transfer, 91, 233-244. DOI: 10.1016/j.jqsrt.2004.05.058. (Earth Blackbody radiation).

https://pressbooks-dev-oer.hawaii.edu/atmo/chaptrer/chapter-2-solar-and-infrared-radiation/ (Earth infrared radiation).

Sir Fred Hoyle FRS; book: "Ice - A Chilling Forecast of a New Ice Age" page 130.

Early scientists studying especially CO2 radiation absorption:

Svante Arrhenius 1896. Swedish scientist. (Postulated that more CO2 would result in more absorption and an increase in atmospheric temperature, and less CO2 in an ice age).

Knut Angstrom 1900. Swedish scientist. (Described some absorption wavelengths of CO2 and the intensity of absorption meaning that further absorption could not occur with more CO2. Disagreed with Arrhenius).

Langley 1898.

Rubens and Aschkinoss 1905 (First to describe absorption wavelengths at 15μ of CO2.)

Schaefer and Philips 1926.

Martin and Baker, University of Michigan, 1932.

McElroy and Fugel. Atmosphere – Ocean; 46:1, 15-37.

www.reseach.vu.ni/en/publications/three-dimensions.amd-ozone-distribution-based-02-assimilation-of-nad-JacobsvanPeet

Ya, Virolaina et al, Atmospheric and Oceanic Physics, 47, 480. (2011)

www.chem.libretexts.org. Search IR Spectroscopy. California State Uni.

www.agupubs.onlinelibrary.wiley.com/doi/pdf/10.1029/97JDO3285

Soleman et al, J Geophysics Research, Vol 103, No. D4, 3847-58, Feb 27, 1998.

www.C:/Users/reception/Downloads/PS-20190224.pdf

www.esa.int/Applications/Observing-the-Earth/Altius/Measuring_ozone.

www.earthobservatory.nasa.gov/features/SOLSE

www.sciencedirect.com/topics/earth_and_planetary_sciences/stratosphere

www,en.wikipedia.org/wiki/Van_Allen_radiation_belt

www.albany.edu/faculty/rgk/atm101/structur.htm

www.encyclopedia.com/science/encyclopedias-almanacs-transcripts-and-maps/charles-fabry

Chapter 3

The Earth's Surface Temperature

If the temperature of an object, including the surface of the Earth, is steady, not changing, the heat energy entering it must be equal to the heat energy leaving it.

If the heat energy entering it is greater than that leaving it then the temperature will increase.

If the heat energy leaving it is greater than the heat energy entering it then the temperature will decrease.

This is just plain common sense and is called scientifically the *First Law of Thermodynamics.*

Sources of Heat Energy entering the Earth's Surface:

1: From the Sun.

2: From the hot molten iron centre of the Earth.

3: Friction effects due to the rotation of the Earth, for example sea tides.

The Sun:

Important Factors affecting the Sun's Energy Arriving on the Earth:

- The energy output of the Sun.
- The distance of the Earth from the Sun.
- The Rotation of the Earth.
- Reflection of the Sun's Energy back into Space by the Earth.

The Energy Output of the Sun:

The Sun, with a surface temperature of 6000° C, produces by nuclear reaction largely converting Hydrogen to Helium, 4×10^{26} watts of energy. By contrast human made energy totals 17×10^{12} watts (14 trillion). So the Sun produces 23×10^{12} times as much energy as humans do, 23 million million times as much (23 trillion).

The key factor regarding the surface temperature of the Earth is the energy output of the Sun. This used to considered completely constant but with knowledge it is now realised there are changes.

4.5 billion years ago, when the Earth formed and the Sun was young, it produced 30% less energy than it now does.

Studies of substances affected by the Sun's radiation including Carbon (C^{14}), Beryllium (Be^{10}) suggest that there is a variation over very long time periods. To summarise if the Sun's energy output rises these substances decrease on Earth, and if it falls they increase. This is because when the Sun is more active these substances are swept away from Earth. They can be found in cores of ancient ice from deep in Antarctica, and also in tree remains.

After the telescope was invented (possibly by the Dutch eyeglass maker Ilans Lippershey in 1608) studies of the Sun's sunspots, which decrease in number when the Sun cools, showed a correlation with cooler times on Earth. In 1894 Edward Maunder, an English astronomer, reported that very few sunspots had occurred between 1645 and 1715. This coincided with the coldest part of the "Little Ice Age" which went from 1300 to 1850 when European temperatures dropped causing many rivers to freeze, crops to fail with resulting famine, glaciers to advance. Modern studies of this time suggest global temperature was probably 0.9° C cooler than now. Sunspot studies are still very useful today.

The *Schwabe Cycle* is a cycle where the number of sunspots increases, then decreases every eleven years. It was discovered by Samuel Schwabe, a German astronomer, in 1843. The cycles are numbered from one in 1755. We have just finished cycle 24 in 2020 and are now in cycle 25. There are other Sun activity cycles recognized such as the *Hale Cycle* every 22 years, the *Gleissberg Cycle* every 88 years, and *Suess Cycles* every 200 years.

Other studies have looked at the pattern of the Aurora in the far North including those by John Dalton over 50 years in the 1800s, after whom the *Dalton Minimum,* a very cold period in 1816 was named. As well as decreased aurora, which are increased when the Sun is more active, he noted there were very few sunspots. There was no summer and European temperature dropped by 1° C.

Tree ring studies also have suggested cooler and warmer periods. Ancient Beetle studies have also helped.

Studies of the Sun's magnetic field cycle by *Valentina Zharkova* from Northumbria University of Newcastle upon Tyne in the UK, using data from the Wilcox Solar Observatory in California have suggested that the Sun has a regular magnetic field cycle of about 11 years waxing and waning in energy output, and that about every 350 – 400 years there is a "Grand Solar Cycle" where the Sun's output decreases resulting in temperatures on Earth decreasing. It is suggested that if the Sun's output decreases 0.22% the Earth's surface temperature decreases by about 1° C.

Modern satellite studies of the Sun's activity over 20 years have shown a decrease of about 0.1% every eleven years, and it is considered this could decrease the Earth's troposphere atmosphere temperature by 1° C.

There are also many historical records of warmer and cooler periods in the recent past recorded by man.

During man's recorded history there have been the following interesting cooler periods:

- Sumerian cold period: 3370 to 3300 BC.

- Egyptian cold period: 1410 – 1370 BC.

- Homeric (Greek) cold period: 810 – 720 BC.

- Greek minimum cold period: 390 – 330 BC

- Middle Ages cold period: 1010 – 1050.

- Wolf minimum, Middle Ages: 1280 – 1350 AD.

- Maunder minimum: 1645 – 1715.

- Dalton minimum: 1823 – 1833.

- The Little Ice Age from 1300 - 1850 AD includes the last three above.

By contrast there have been warmer periods also in man's recorded history. These include the Medieval Warm Period from 950 to 1250 AD, before the Little Ice Age. This was when the Vikings lived in Greenland happily. And it has been shown that the change was world-wide from Pacific Ocean studies and New Zealand cave stalagmite studies. The temperatures then may have been warmer than today by 1 to 1.5° C.

There was also another warm period in Roman times (0 to 100 AD). Obviously between the cooler periods noted above there were warmer periods.

In the more recent past there was a cool period from 1880 to 1915 world-wide, 0.9° C colder than now. Then a warmer period from 1915 to 1940. Then a cooler period from 1945 to 1977 when the average temperature dropped 0.5° C. At that time there was concern the current Ice Age was becoming dominant. Since 1977 there has been warming.

What is interesting about Valentina Zhakarov's work is that is fits in with all these cycles generally. And it predicts a cooler period with more severe winters from now until 2050, part of the "Grand Solar Cycle" mentioned above.

What is important is that the Sun is the huge force concerning the Earth's surface temperature. Tiny variations in its output have a big - to us - effect on the surface temperature. And as scientific knowledge develops awareness of the variation in the Sun's output grows. The so-called Solar Constant, which stated that the Sun had a constant output of energy which never varied looks uncertain at least.

Earlier opinions that the Sun's output was constant were limited in value by technical limitations in instruments used to measure it, including early satellite measurements. It now seems certain that there are variations which may account for some at least of the surface temperature changes, given that the Sun is the "Big Boss".

Ice Ages:

The bigger picture is that the Earth has had many ice ages over its history. For the last 700,000 years the approximate cycle has been an ice age lasting 100,000 years, then a warm period lasting about 20,000 years. We are in a warm period in the ice age, called the *Quaternary Ice Age,* which commenced 2.5 million years ago. Our warm period started 11,000 years ago and is called the *Holocene*. Scientific opinions on how long it will last vary from 50,000 years to very much shorter.

The reasons for ice ages remain uncertain. As well as variations in the Sun's output theories concerning the Earth's orbit around the Sun and its axis of rotation tilt (Croll and Milankovitch), collisions with asteroids and comet tails, (Sir Fred Hoyle) are prominent. Other theories include cycles in major ocean currents. One certainty is that there must be a large land mass near a Pole, such as Antarctica at present, enabling polar ice sheets to form and fall into the ocean and cool it. Antarctica arrived at the South Pole and started icing up 40 million years ago. The movement of tectonic plates means that there are long periods without a large polar land mass covered with ice, which falls into the sea cooling it and Earth. The major ice ages since the Earth was formed 4.5 billion years ago are:

- Huronian ice age – 2.4 to 2.1 billion years ago, lasted 300 million years.

- Cryogenian ice age – 850 to 635 million years ago, lasted 215 million years. Ice almost reached the equator, so called "Snowball Earth".

- Ordovician ice age – 460 to 430 million years ago, lasted 30 million years.

- Permo-Carboniferous ice age – 360 to 260 million years ago, lasted 100 million years. Parts of Africa then at the South Pole.

- Jurassic – 170 to 110 million years ago. Lasted 60 million years. But this was not a full ice age, there was no Polar land mass covered with ice sheets.

- Quaternary ice age – started 2.5 million years ago. Antarctica at the South Pole, Greenland close to the North Pole.

In between ice ages there have been much warmer periods than the present day. During the Cretaceous period 60 to 100 million years ago the average Earth temperature was 9° C to 14° C higher than now, up to 28° C average across the Earth. There have been 3 other such warm periods in the last 550 million years, since complex life with multiple cells evolved.

Over the last 50 million years the Earth's average surface temperature has steadily decreased to a low 20,000 years ago of about 8° C in the depths of the ice age. Since then it has risen to present levels of about 14° C where it has been for the last 10,000 years.

There is a very good graph of this at : www.en.wikipedia.org/wiki/Geologic_temperature_record .

Distance from the Sun:

This is a basic factor controlling the amount of energy from the Sun the Earth receives, and the Earth's temperature.

The average distance is 149.6 million kilometers. The Earth's *orbit,* or path around the Sun is almost a perfect circle but not quite, so the distance varies a little as the Earth goes round the Sun. At the closest, which occurs in January each year, it is 147 million kilometers away. At the farthest 152 million kilometers in July.

To illustrate how important small changes in distance from the Sun are this change of 5 million kilometers makes a 7% difference in the amount of Sun radiation received by the Earth.

Other planets in the Solar system illustrate this too. Venus is 108 million kilometers from the Sun and has an average temperature of 480° C. Mars is 228 million kilometers from the Sun and has average temperature of minus 60° C.

Rotation of the Earth:

The Earth rotates taking 24 hours to complete one rotation, so each place on Earth has about 12 hours receiving radiation from the Sun and about 12 hours at night when heat radiation is lost from the Earth into space (as an annual average, it varies with the seasons as will be explained). A person standing at the Equator is moving at 1600 kilometers per hour because of the rotation.

If the Earth rotated slower, say it took 48 hours to make a complete rotation, each place would have longer to receive radiation from the Sun in daytime and would get hotter. And at night it would have longer to loose heat and would get colder. The change from daytime temperature to nighttime temperature would be greater.

If it rotated faster, say 12 hours for a rotation, the change would be less.

600 million years ago, when complex life first started on Earth, a complete rotation took only 21 hours. Because of the energy used up in tides of the ocean and in the molten liquid centre of the Earth, caused by the gravity of the Moon and the Sun, the rotation is slowing at about 2 thousandths of a second each century.

Obviously we are completely used to the 24 hour cycle, which affects the change in temperature from day to night.

Concerning other nearby planets Mars has a similar length day to us, 24 hours and 37 minutes. Its daytime temperature rises to 20° C in some places and nighttime drops to minus 110° C, lower to minus 170° C at the poles, a range of about 130° C.

Mercury has a very long day, taking 58 of our days for one complete rotation. As a result it gets very cold at night, which lasts for 29 of our days and the Sun, although very close, does not heat that part of Mercury. The temperature drops to minus 170° C. In the Mercury daytime the reverse happens, the temperature gets to 470° C, so the range is about 640° C.

Axis Tilt:

This is a major factor controlling the Earth's temperature. If there was no tilt it would mean we would have no weather seasons and the Sun would always be making the same path in the sky all year. Every day everywhere would be 12 hours daylight, 12 hours night time.

The opposite would be if the tilt was 90°, a right angle. The planet Uranus has this situation with a tilt of 98°. This means that places on it are dark for half the year, then daylight for the other half. A year on Uranus is 84 years, so there is 42 years of day and 42 years of night in some polar places.

The reason for this is that the planets go round the Sun in their orbits and the *axis* of their spin, or rotation, stays the same relative to wider space. The planets in our Solar system go around the Sun as though they were all on one huge flat disc, like dust on a record. Obviously each planet is at a different distance from the Sun, Mercury the closest, Neptune the farthest. Their *axes of rotation*, although different for each planet, stay the same, with some variations due to wobbling, compared to this huge flat disc which is called the *Plane of the Solar Elliptic*.

The *Axis* of anything rotating is the straight line about which the object turns. For example for a wheel rotating the *axis* goes through the axel of the wheel.

It is thought that Earth started out 4.5 billion years ago with no axial tilt, the axis of its rotation a right angle to the Plane of the Solar Elliptic. Then it may have been struck by a large planet and the axis knocked over. Currently the tilt is 23.5 degrees. There are 90 degrees in a right angle, 360 degrees in a circle. The same symbol, ° , is used for degrees of *angle* as for degrees of temperature.

The tilt means that as the Earth goes round the Sun each year to us the Sun rises higher in the sky in daytime for half the year, and the days get longer, it gets warmer. Then the reverse happens the other half of the year.

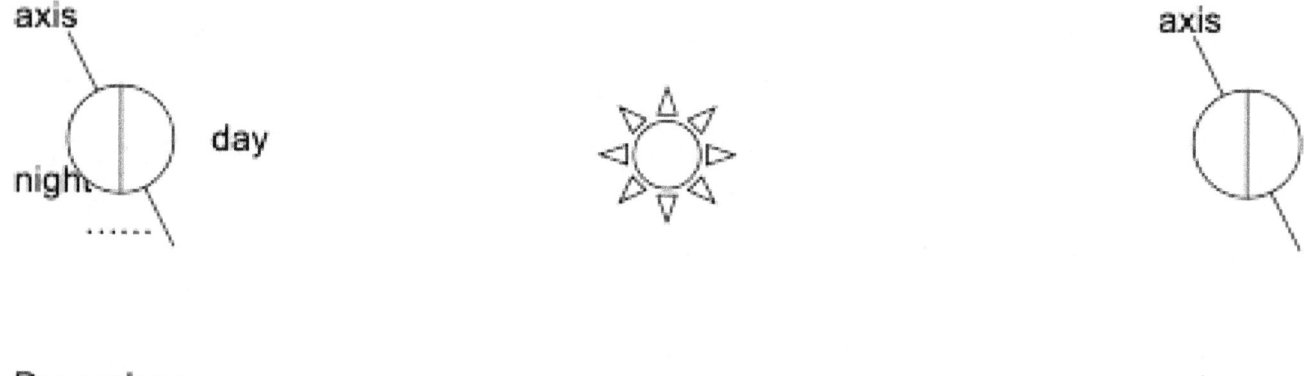

Figure 13: Showing the Earth in December and June, at opposite sides of its orbit around the Sun. The axis of the Earth's daily rotation stays the same relative to outer space. This results in northern places getting summer in June, and Southern places in December.

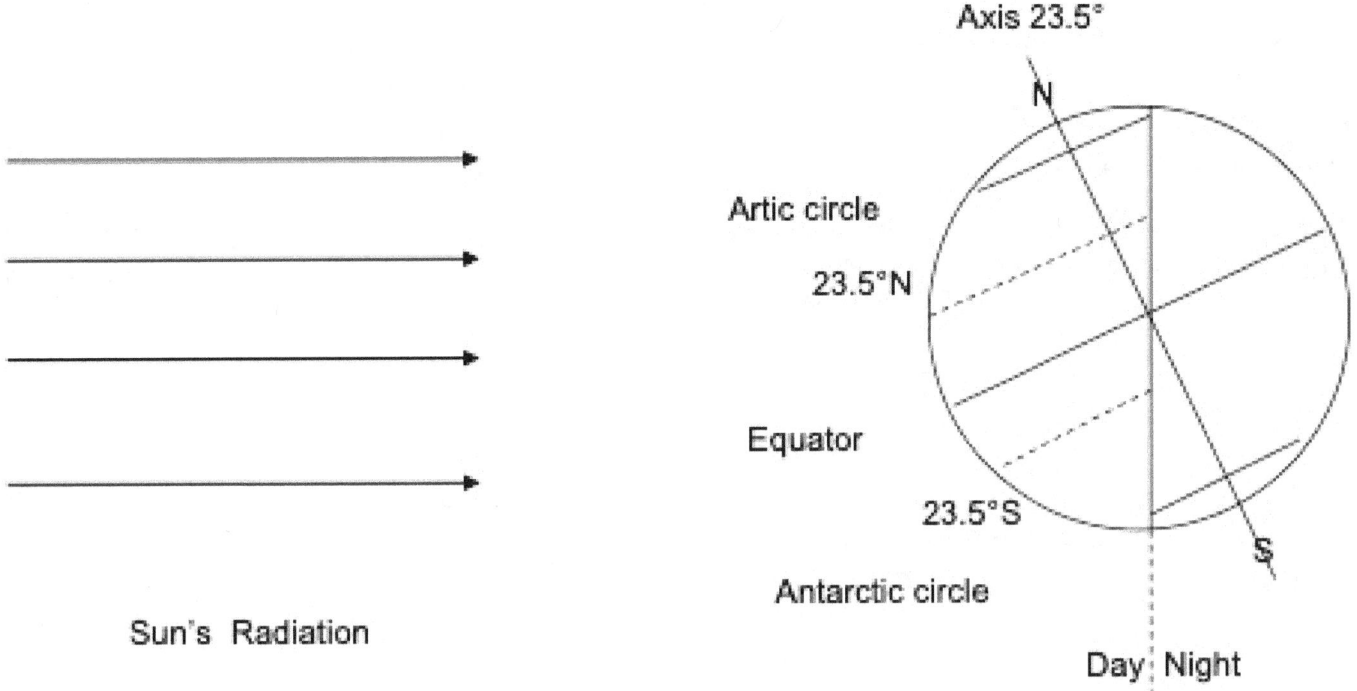

Figure 14: Showing the Earth's situation in June when it is summer in the northern places and winter in southern. N is the North Pole, S the South Pole, marking the axis of the Earth's rotation. The vertical line is the day/night boundary. The Equator is shown. The dotted lines show the length of daylight at positions in the North and the South. The daylight length in the South is only half that of the equivalent place in the North, the nights are twice as long in the South as North. In addition the Sun's rays are much more overhead in the North, and much lower in the South at this time, making them weaker in the South. Hence it is much colder in the South in June.

The opposite is the situation in December.

Notice that at the Equator it is always a 12 hour day, and 12 hour night, right through the year. The tilt of the axis has no effect there, there is no winter or summer.

Notice also that close to the North Pole there is an area where there is no night in June. And close to the South Pole an area where there is no daytime, just night. In December these places have the reverse situation. The boundary where this happens is called the **Artic Circle** in the north and the **Antarctic Circle** in the south. The **Latitude** of these is 66.5° North and 66.5° South. Latitude of a place is the size of the angle from the Equator. Places on the Equator have a latitude of 0°. whereas the North Pole is at 90° North, and the South Pole at 90° South.

The Artic and Antarctic circles are at 66.5° latitude because this is 23.5° less than 90°, where the Poles are. This because the tilt of the Earth is 23.5° as said above.

Another consequence of the tilt being 23.5° is that the Sun is directly straight above different places during the year. When the Sun is straight above a place it is called being at its **Zenith.** In June, because of the tilt of 23.5°, places with a latitude of 23.5° North have the Sun straight above in the middle of the day. In December places with a latitude of 23.5° South have the Sun straight above at midday. All places north of 23.5° North and south of 23.5° South never have the Sun straight above. These two lines, or circles as they go right around the Earth, are called the *Tropic of Cancer* and the *Tropic of Capricorn.* They are named after Constellations of stars above them.

Figure 14 shows the situation in June and December when the axial tilt cause mid summer and mid winter. The situation when the Earth is half way between these points in its orbit around the Sun is shown in Figure 15.

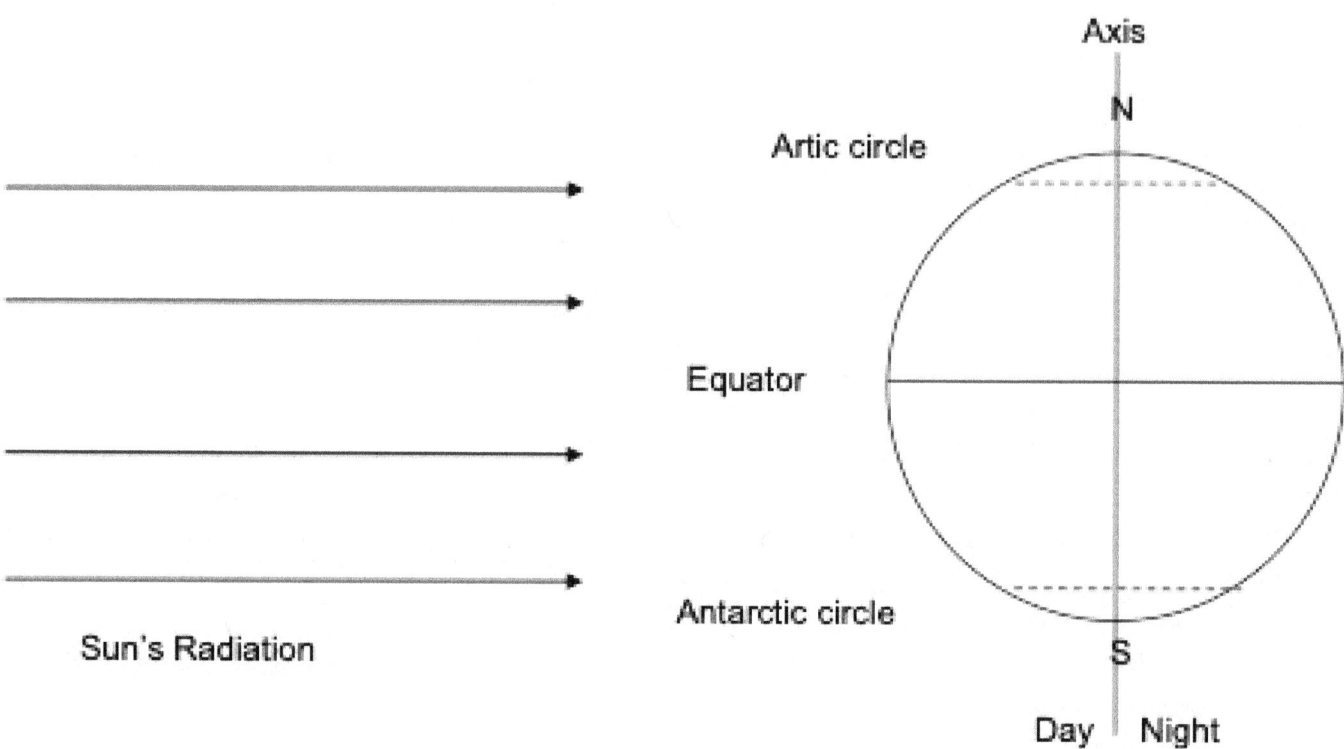

Figure 15: showing the situation in March and September when the Earth is halfway around its orbit around the Sun. The tilt of the axis of the Earth's daily rotation is now not tipping the Poles towards or away from the Sun as in June and December resulting in the days and nights everywhere being 12 hours long. this time is called the *Equinox,* meaning equal days and nights everywhere.

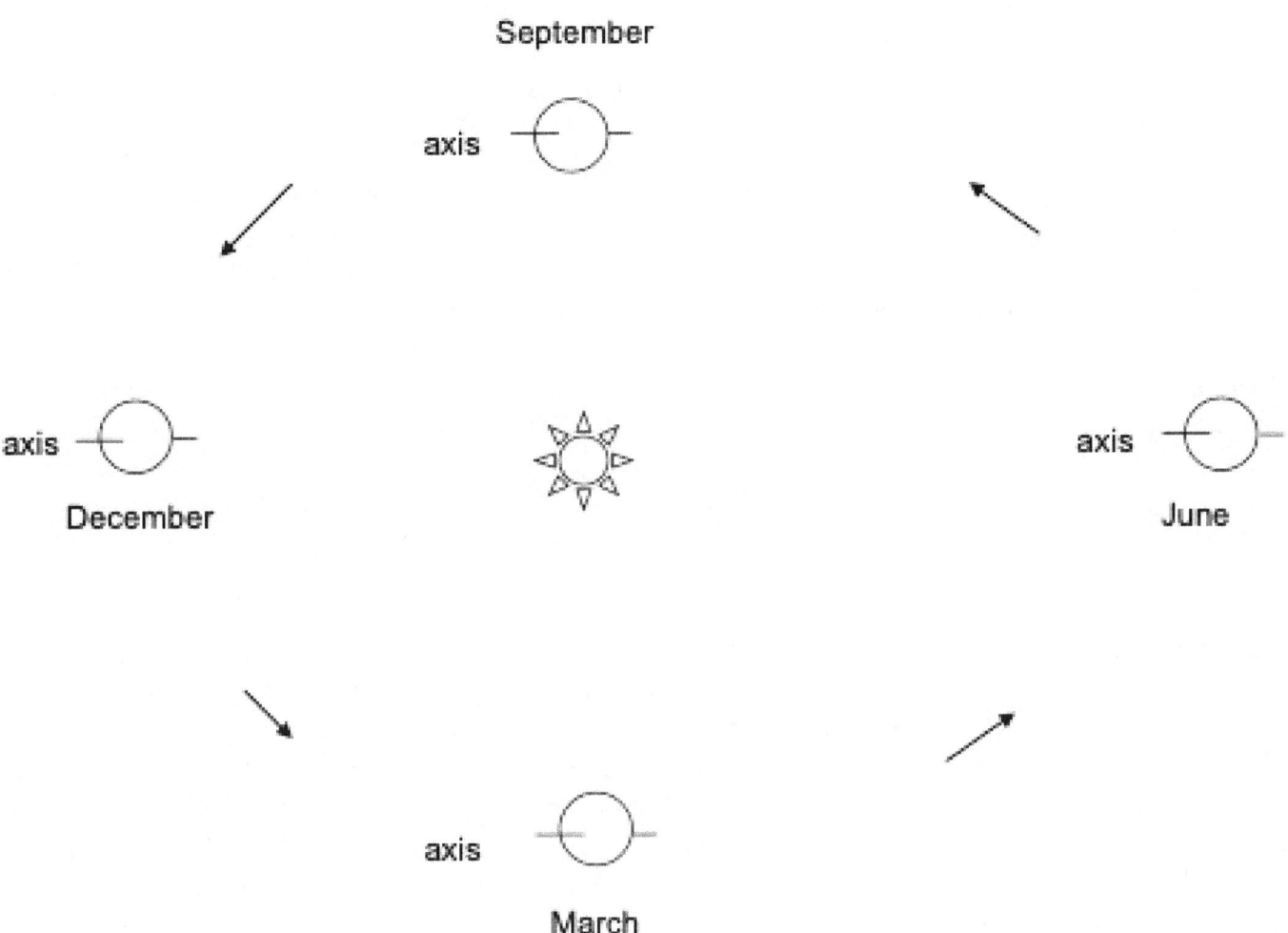

Figure 16: showing a view looking down on the Earth's orbit around the Sun for the year, North Pole upper most. It shows that in March and September, the Equinoxes, the tilt is at right angles across the Sun's radiation direction so days and nights are 12 hours everywhere.

Over very long periods, many thousands of years, the tilt varies. This has been studied by many observers for hundreds of years but a prominent scientist in our era is Milutin Milankovitch, a Serb who in the 1920s studied this intensely and concluded variations in the tilt were the cause of Ice Ages. The tilt varies from 22° to 24.5° over a cycle lasting 41,000 years. In the present cycle it is decreasing, and will be at its minimum in 12,000 years, 22°. It was at its maximum about 9000 years ago.

When the tilt is less summers are cooler because the daylight hours in summer are shorter, the Sun does not move as far from the Equator in summer. This makes the temperature on Earth generally cooler. The amount of Sun energy arriving on Earth is about 6% less when the tilt is only 22° compared to 24.5°.

Another interesting variation is the wobble of the Earth's axis of rotation, just like a child's top rotating and wobbling. Only it takes a long time to wobble, 26,000 years. This is the time for one wobble of the axis of rotation. The angle of the wobble is also about 23°. It means the angle of the Sun's radiation changes, and when it is less the Earth is cooler because summers are cooler. The wobble is called ***Precession of the Earth's axis of rotation.*** This makes about a 3% difference to the amount of Sun radiation reaching Earth. In addition it means that every 13,000 years the axis tilt points the opposite way as the wobble has gone through half a circle. Which means that at that position in the Earth's orbit around the Sun, when it used to be the southern hemisphere summer, it is now winter.

A further interesting variation is that the Earth's orbit around the Sun, as said above, is not a perfect circle. Sometimes the Earth is closer to the Sun, currently in January, the summer in the southern hemisphere. That makes Australian summers warmer, and British winters warmer, due to about 7% more Sun radiation than when the Earth is further away, currently in July. But the distorted circle orbit is also slowly rotating, every 112,000 years it goes around the Sun once relative to wider outer space. In addition the shape of the orbit varies from almost a perfect circle to a more distorted ellipse (egg shape) in a cycle lasting about 100,000 years. Currently it is moving back towards a perfect circle. When it is at its most distorted egg shape the distance from the Sun to the Earth varies more during the year, resulting in 23% less Sun radiation reaching Earth when it is furthest away compared to 6 months later when it is closest. Currently this difference, as explained above, is 7% because of a 5 million kilometer difference.

All of these very long term minor variations make no difference to us in our daily lives obviously. But they may be part of the explanation for ice ages and warmer periods occurring over millions of years, although not the complete explanation. When all the above variations are coinciding to cool Earth, in cycles varying from every 41,000 years to 100,000 plus years they are major factors in an ice age developing. Other factors include a large icy continent close to a Pole shedding icebergs to cool the oceans as discussed above, and large bodies colliding with Earth creating global dust clouds which block the Sun radiation resulting in sudden cooling. An example of that is the 170 kilometer diameter crater at Chicxulub in Mexico considered to be the result of a 10 kilometer wide asteroid colliding at 60,000 kilometers per hour 65 million years ago. The resulting cold period is considered to have wiped out many species including the dinosaurs. Other scientists consider that huge volcanic eruptions resulting in Sun blocking dust have been factors. Yet others point to changes in ocean deep cold water cycles.

What is important to us is that the Sun is the "Big Boss". Our days, seasons are consequence of the Earth's daily rotation and annual journey around the Sun. The nature of our planet is governed by its distance from the Sun, which means that water can exist in solid, liquid and gas form here, in marked contrast to all other planets in the Solar System.

The Sun's Radiation Energy reaching the Earth:

Measuring Energy:

Radiation Heat energy is measured in ***kilowatt hours,*** just like electrical energy being used in our houses. To explain further the unit of energy is called the ***joule,*** named after an English physicist, James Joule, who studied energy in the 19th century.

Joules are used to measure other kinds of energy too, such as mechanical energy. but the more common measure we are familiar with is the ***Watt.***

If one joule of energy is used or made in one second it is called one watt. Watts measure the *rate* of energy being used or made. The *rate* of energy is known as **Power.** A **kilowatt** is 1000 watts. It is 1000 joules of energy used or created in one second.

As there are 60 seconds in a minute and 60 minutes in an hour, it follows there are 3600 (60 X 60) seconds in an hour. So in a **kilowatt hour,** or **kWh,** there are 3600 X 1000, or 3.6 million joules of energy. A joule of energy is a very small amount, so usually we use kilowatt hours in measuring quantities of energy.

As discussed above the Sun puts out radiation energy at a rate of 4×10^{26} watts. As the Sun has an area of 6×10^{18} square meters, it follows that each square meter on the Sun puts out 6.5×10^7 watts, which is 65 million watts.

Watts per square meter are abbreviated as w/m^2. A square meter is an area one meter by one meter square.

By the time this energy has travelled 150 million kilometers to Earth it has spread out markedly and decreased to 1366 w/m^2. This amount of energy strikes the top of the atmosphere.

Not all this energy passes through the atmosphere to reach the Earth's surface. What happens to it in the atmosphere is shown in Figure 17. This is on a clear day with no clouds, and the Sun directly above the ground, at its *zenith* as discussed before.

Figure 17 shows the effect of the atmosphere on incoming Sun radiation energy on a clear day with the Sun directly overhead at its zenith. 20% is absorbed in the atmosphere, 10% is diffused and scattered in the atmosphere by collision with molecules and particles, dust. Of this some escapes into space, some finally gets to the surface. 77% of the incoming Sun radiation finally arrives at the surface.

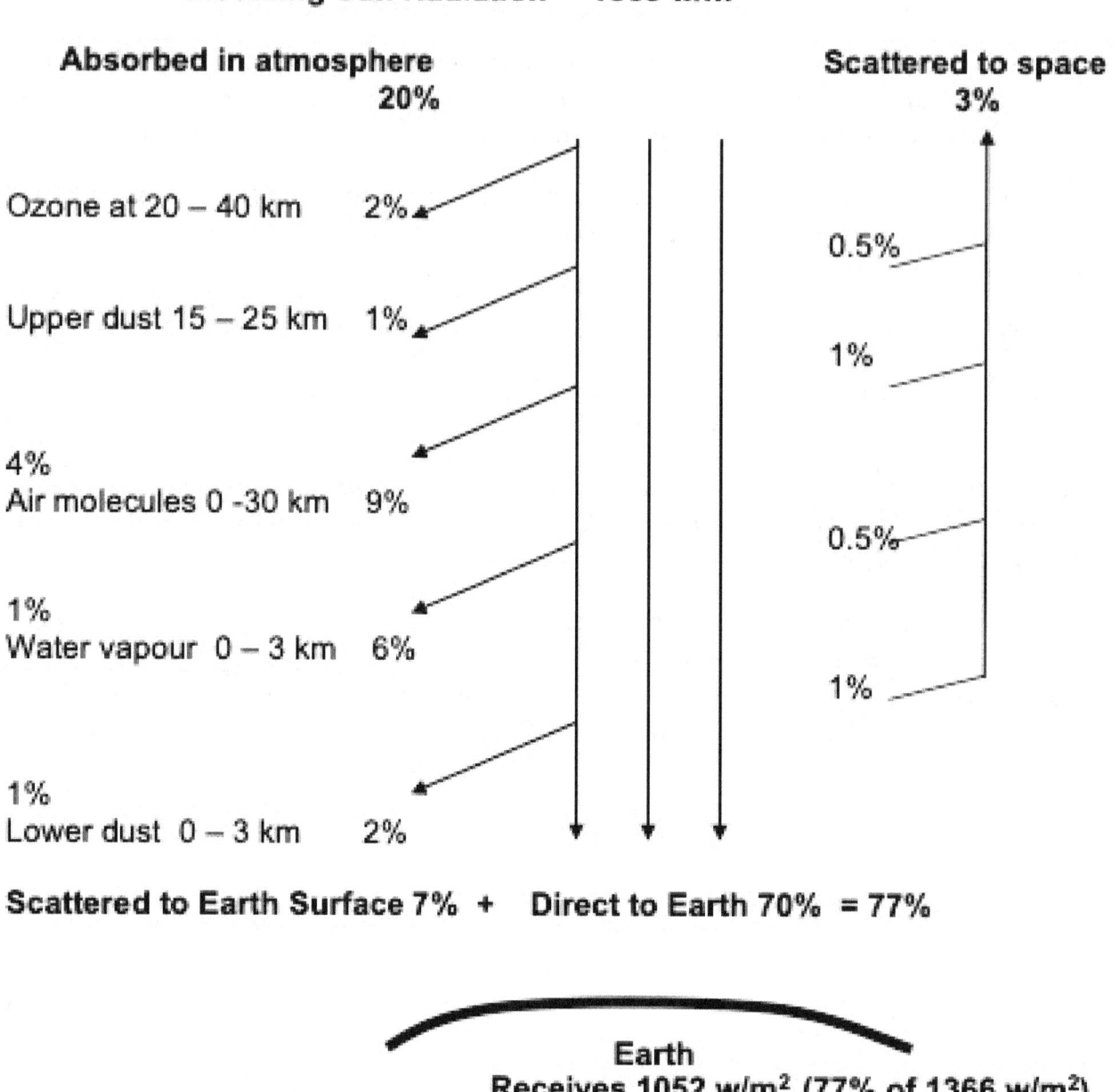

Figure 17: showing the atmosphere's effect on the Sun's radiation passing through to the Earth's surface.

On a **Cloudy Day** it is very different. Clouds are very effective at reflecting energy back upwards into the atmosphere, and at absorbing Sun radiation energy. The amount of radiation getting through clouds to the surface can drop to only 20% in thick cloud cover.

Once at the surface what happens to the radiation next depends on the nature of the surface. If the surface is very absorbent of Sun radiation, remembering that it is in the wavelengths 0.4 to 3.2 microns as discussed above in Figure 6 then most of it will be soaked up in the surface and heat it. If the surface is very *reflective* then some or all of it will be reflected back into the nearby atmosphere. The proportion that is reflected varies depending on the nature of the surface, and is called the *Albedo* of the surface.

Some examples of Albedo are:

- Snow 90%
- Ground 10 - 40%
- Trees 15 - 18 %
- Flat Water: if the Sun is overhead, 10%; if low, close to the horizon, 90% - because Sunlight reflects off water when the Sun is low.
- The Ocean 6%. Waves increase the absorption of sun radiation.
- Ice 50 - 70%, clean ice 90%.
- Man made structures, such roads, paved areas, buildings vary depending on the colour and texture of the surface. In general the closer to white the colour is the higher the albedo, reflectivity. The ultimate high albedo surface is a mirror. The opposite, very low albedo, occurs the closer to black the colour is and the rougher the surface. Such surfaces absorb the Sun's heat radiation very well.

The Sun's heat radiation, which includes visible light as discussed above, striking the ocean is largely absorbed into the first few meters of surface water. The warmer water is then distributed more widely in the ocean by wave action and currents, some of which are very large. For example the Gulf Stream takes huge amounts of warm ocean water from the tropics in the Caribbean to Europe.

When the Sun's radiation strikes land it warms only the top few centimeters because land transmits heat deeply poorly, it is a good *insulator*. So the surface of land heats up quickly in the Sun, compared to the ocean, because the heat energy does not travel deeper or more widely.

Clean snow reflects most of the Sun radiation, about 90%. So Sun radiation does not melt clean snow, it is melted by warm air passing over it.

Clean ice is different. Sun radiation penetrates deeply into it before being absorbed, up to 10 meters. This means the resulting heat is deep within the ice, for example in Antarctica and Greenland, so melting does not occur until a huge amount of heat energy has been delivered to thick ice, which in the Antarctic averages over 2 kilometers deep and in parts is 4 km deep.

These considerations mean that the oceans absorb the biggest amount by far of the Sun's radiation energy. Two thirds of the Earth's surface is ocean, more in the Southern Hemisphere. The ice and snow towards the Poles reflects most of the energy, which is much weaker there as explained next.

When the Sun's radiation arrives at the surface of the Earth it is not striking a flat Earth, but a sphere. This means that the radiation is spread over a greater area towards the Poles than near the Equator. This happens because the Sun is lower in the sky so its radiation spreads out, just like a shadow which is longer when the

Sun is lower, and absent when the Sun is directly above. So the amount of Sun heat radiation per square meter is less towards the Poles than near the Equator. This is shown in Figures 18 and 19.

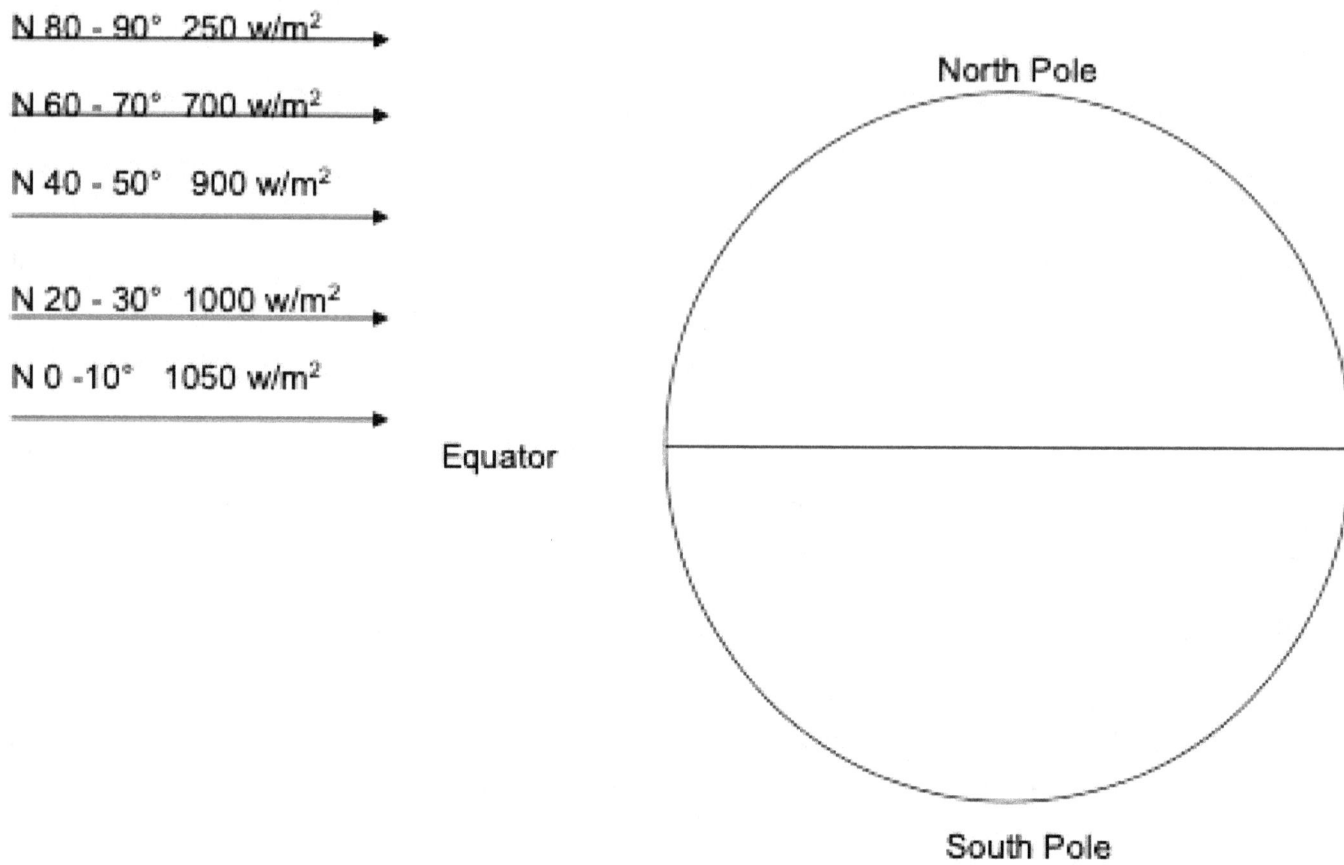

Figure 18: showing the Sun's radiation power reaching the Earth's surface at various latitudes at midday on a clear day at the Equinox when the Sun is directly above the Equator. As one goes away from the Equator and closer to the Pole the Sun appears lower in the sky at midday, shadows are longer, and the Sun's radiation is spread over a bigger area and is less powerful.

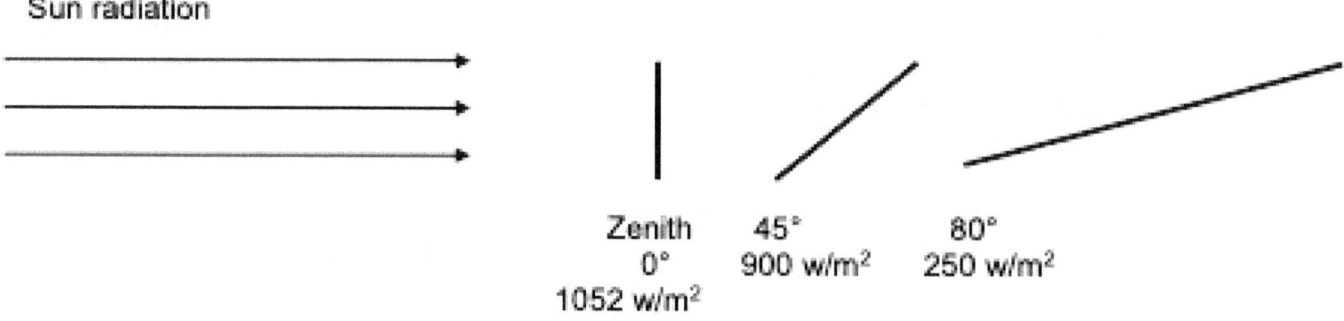

Figure 19: showing the effect of shifting away from where the Sun is directly overhead (at its Zenith) towards the Poles. The Sun is lower in the sky, its radiation is spread out more when it strikes the surface and is weaker.

Although not shown the situation in the Southern Hemisphere is the same.

At other times of the year when the Sun is directly above latitudes away from the Equator the situation is the same but tilted accordingly.

A similar weakening of the Sun's radiation occurs every day, due to it becoming lower in the sky after midday during the afternoon. And in the morning it is initially very low, then rises, its radiation strengthens becoming greatest at midday.

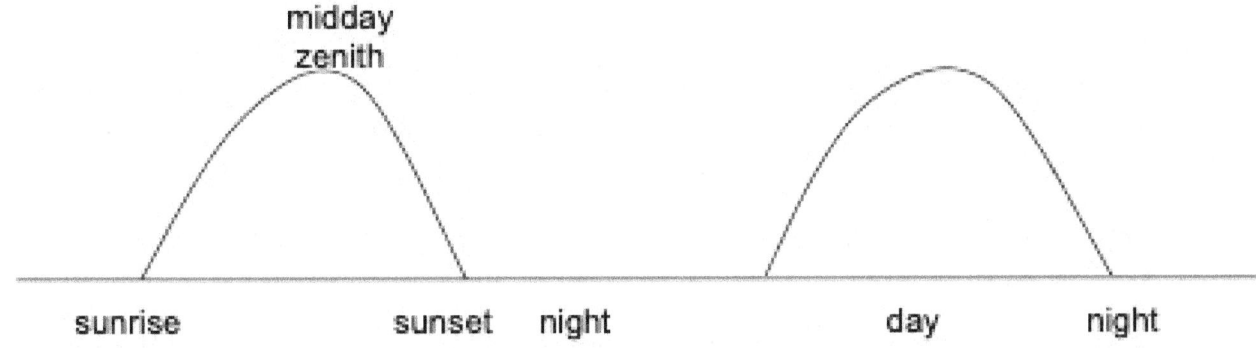

Figure 20: showing the effect of the Earth's daily rotation on the rate of Sun radiant energy striking the surface. There is no Sun radiation at night obviously. Then after sunrise the Sun is very low in the sky, the radiation is spread out as in Figure 21. The radiation gets more concentrated into a smaller area as the Sun rises. In places where the Sun is directly above at midday (at its zenith) it gets to about 1000 w/m² on a clear day as explained above.

For places where the Sun does not get directly above because they are north or south of the latitude where the Sun is directly above at that time of the year the peak the radiation peak at midday follows the guide in Figure 20 for clear days.

Then as the Sun gets lower in the afternoon the radiation heat energy gets spread out again so the amount per square meter decreases, and stops for that place after sunset obviously.

The amount of Sun radiant energy striking a place during a whole day depends on the latitude of the place, the time of the year, for example summer, winter, and if it is clear or cloudy. Cloud makes a very big difference, if thick the Sun radiation reaching the surface can be decreased by up to 80%.

In clear sky situations most of the Sun's radiation reaching the surface arrives directly, but about 10% of it is scattered by dust and water or ice particles in the atmosphere as shown in Figure 17. On cloudy days this *Diffuse Radiation,* scattered Sun radiation, makes up more of the total sun radiation reaching the Earth's surface, as much as 30% or greater. The *Diffuse Radiation* still is of the same wavelength band as *Direct Radiation* from the Sun, 0.4 to 3.2 micron wavelength, as explained in Figure 6.

The Total Amount of Sun Radiation Energy Arriving on Earth Per Day:

So far the Sun's radiant energy has been discussed regarding its amount in **watts per square meter**, w/m^2. as explained before a watt of energy is one **joule of energy per second.** So watts measure the **rate energy is delivered or used per second.** We are familiar with this, for example electric heaters, light bulbs are labelled with how many watts, or kilowatts (1000 watts in a kilowatt) they use.

We also know that the amount of energy heaters, lights, use depend on how long they are on for. The total energy used depends on the watts and the time.

It is common to measure the total amount the Sun's radiant energy striking a place in the same way. The measure is done in **kilowatt hours, kWh.** 1 kilowatt hour is energy of 1 kilowatt going for 1 hour. The same measure is used for our electricity meters measuring how much electrical energy we have used in our houses.

A kilowatt hour is quite a lot of energy. It is 1000 watts for one hour. That is 1000 joules per second for one hour, which has 3600 seconds. So a kilowatt hour has 3,600,000, or 3.6 million joules, as discussed above. This is approximately the amount of heat energy required to heat three buckets of water (30 litres) from cold (20° C) to hot (50° C).

The Sun's radiant energy gets spread over the Earth's surface, in contrast to an electrical appliance like a water heater where the energy just goes into the device. So the measure is done for each square meter and is in **kilowatt hours per square meter, kWh/m^2**.

It is useful to know how much energy arrives from the Sun at each place on the surface. That is measured in **kilowatt hours per square meter per day, kWh/m^2/day.** This varies depending on the latitude and season of the place as discussed above. Figure 21 below details this.

Northern Midsummer

90° N ⟶	12.6 kWh/m² /day
40° N ⟶	11.6
0° ⟶	9.2
40° S ⟶	3.5
90° S ⟶	0

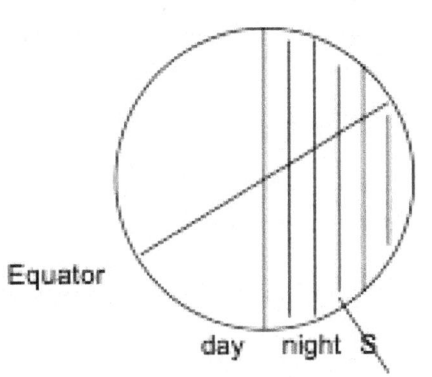

Equinox – March and September

90° N ⟶	0
40° N ⟶	7.9
0° ⟶	10.3
40° S ⟶	7.9
90° S ⟶	0

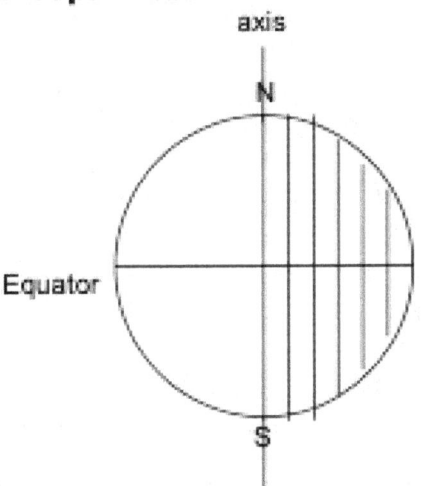

Southern Midsummer

90° N ⟶	0
40° N ⟶	3.7
0° ⟶	9.8
40° S ⟶	12.4
90° S ⟶	13.5

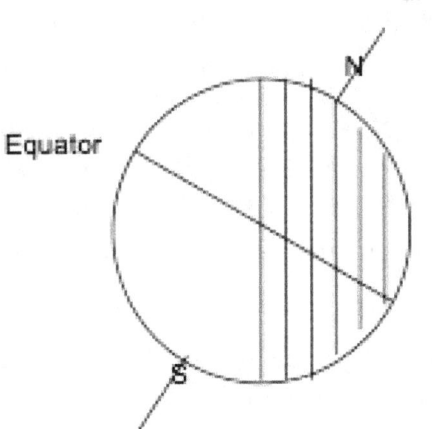

Figure 21: showing the total Sun radiation energy received each day per square meter at northern midsummer, the equinoxes in March and September, and southern midsummer for the Poles, 40° North and 40° South, and the Equator. The Equator receives a steady high amount throughout the year. The Poles receive very high amounts at midsummer, but nothing from equinox to equinox, 6 months of the year. The South Pole receives a little more than the North Pole at midsummer because the Earth is 5 million kilometers closer to the Sun then as discussed earlier. The mid-latitudes receive a varying amount from high amounts at midsummer to low amounts at midwinter.

The *average* received each day over the whole Earth is about 6 kWh/m². For the whole year the *average* amount of Sun radiation energy for the different latitudes in Figure 23 is described below:

- North Pole, 90° N: 3.1 kWh/m²
- 40° N: 7.8
- Equator: 9.9
- 40° S 8.8
- South Pole, 90° S: 3.4

So the Equator region receives the most over the year as we expect. And the Polar regions the least because of the long dark winters.

The Polar regions also reflect the majority of the Sun energy they receive because large areas are covered with ice and snow, with their high **albedo** as discussed earlier. This clearly makes the surface colder. In contrast the equatorial regions and mid latitudes have large areas of ocean, which reflects far less Sun radiation and are warmer.

The big picture is that the Earth receives a massive amount of Sun radiation. More energy arrives on Earth from the Sun in two hours than humans make and use in a year.

Humans make and use about 14 **Terawatts** of energy, which 14 million million watts (14 X 10^{12} watts). By contrast the Sun's radiation energy arriving at Earth is about **175,000 Terawatts** (175 X 10^{15} watts), more than 9000 times greater. A **Terawatt** is 10^{12} watts, a million million watts, or a billion kilowatts. Remember that a watt measures energy *each second, the rate of energy moving.*

Expressed in terms of radiation energy received from the Sun each year, the Earth's surface receives about 1.1 X 10^{18} kilowatt hours per year (eleven hundred thousand million million kWh). This more usually called 1.1 X 10^9 **Terawatt Hours**, which is 1.1 billion Terawatt Hours. By contrast this is nine thousand times the annual energy humans make and use, which is about 1.2 X 10^{14} kilowatt hours per year, or 120,000 Terawatt Hours.

Earlier some of the variations in the Sun's energy output were discussed. These included the **Schwabe Cycle** every 11 years, and the **Grand Solar Cycle** every 300 – 400 years. The Sun's output varies 0.1% in the Schwabe cycle, which is about 89 Terawatts, more than six times human energy use. And for the Grand Solar Cycle the variation is about 200 Terawatts, more than fourteen times human use.

Other Sources of Heat Energy for the Earth:

Geothermal:

The Earth has a very hot solid iron-nickel **Inner Core**, an extraordinary idea for us living on the, at times, very cold surface. The temperature at the centre is about 6000° C. The inner core is about 1220 kilometers diameter, 5000 kilometers beneath the surface. The diameter of the Earth is 12,000 kilometers, it is 6000 km from the surface to the centre.

Next is the **Outer Core,** a molten liquid layer of iron and nickel 2400 kilometers thick at a temperature of about 3500° C. Its top layer lies about 2800 kilometers beneath the Earth's surface. It causes the magnetic field of the Earth.

Next is the **Mantle.** This is about 2800 kilometers thick and is made of rocks which are very hot and semi-molten, like glue. The rocks are principally iron and magnesium silicate rocks. Its temperature varies from 1000° C at its top to 3500° C in the deeper parts close to the outer core. The mantle makes up about 85% of the volume of the Earth.

The outermost layer on which we live is the **Crust.** This ranges in thickness from only 5 kilometers at the bottom of deep ocean trenches to 70 kilometers under mountain ranges. It averages 15 km thick. It makes up only 1% of the Earth. It is composed of **Tectonic Plates** which move about on the glue like semi-molten mantle beneath. The thickness of the crust on which we live is about the same proportion as the skin of an apple is to the apple – yet we feel secure ! Its temperature averages 15°C at the surface and increases to up to 600°C at its depth.

The temperature of the Crust gets hotter the deeper one goes. It increases by about 30° C per kilometer of depth. The cause of all this heat is considered to be radioactive decay of substances for about 60% of it, and the rest is left over heat from the formation of the Earth more than four billion years ago. Considering all this it is not surprising that considerable amounts of heat reach the surface by **conduction** from the very hot interior. The rate of heat doing this is about 44 Terawatts, three times Human energy usage.

The Earth contains a massive amount of heat energy. It has about 3×10^{24} kilowatt hours (three million billion billion kWh). This is more than ten billion times the annual Human use of energy.

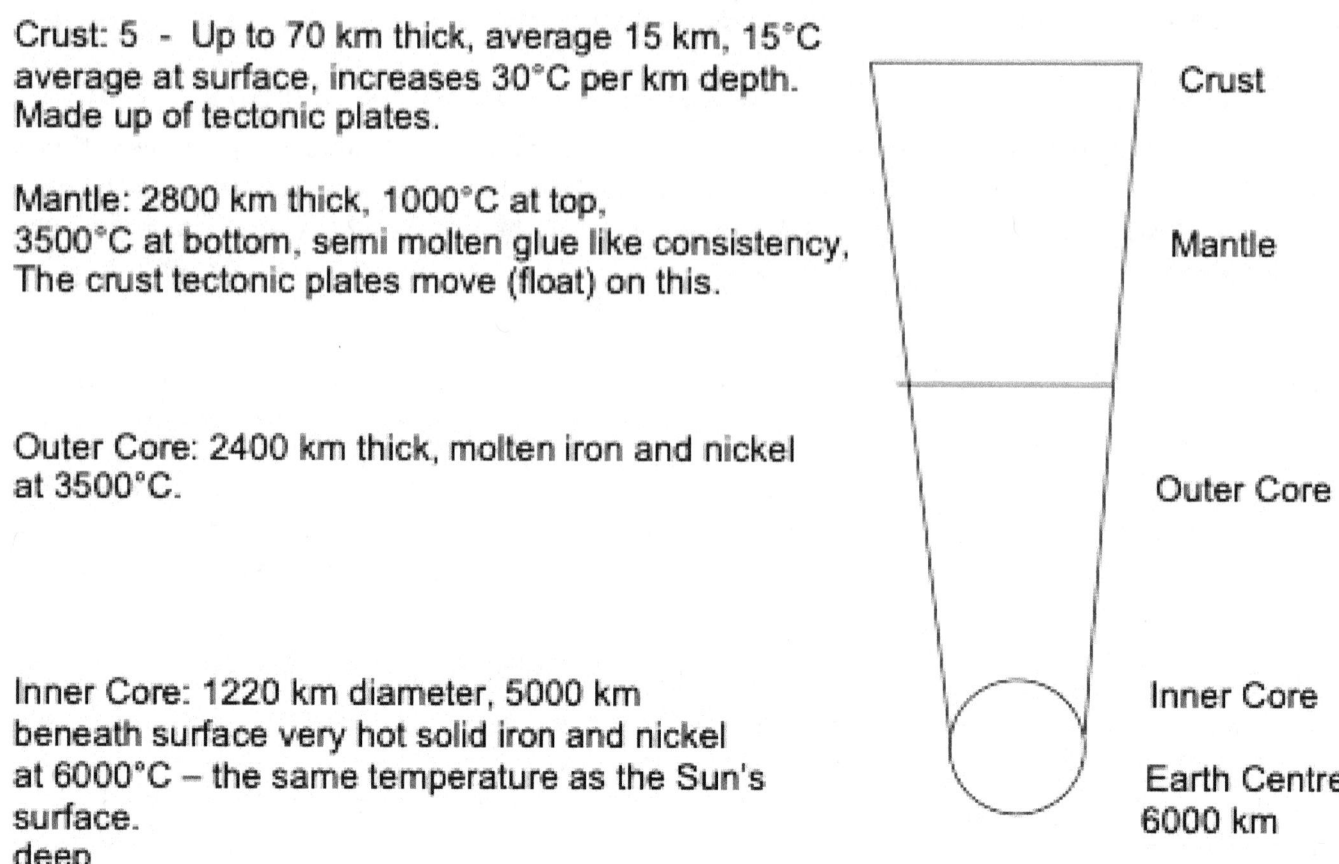

Crust: 5 - Up to 70 km thick, average 15 km, 15°C average at surface, increases 30°C per km depth. Made up of tectonic plates.

Mantle: 2800 km thick, 1000°C at top, 3500°C at bottom, semi molten glue like consistency, The crust tectonic plates move (float) on this.

Outer Core: 2400 km thick, molten iron and nickel at 3500°C.

Inner Core: 1220 km diameter, 5000 km beneath surface very hot solid iron and nickel at 6000°C – the same temperature as the Sun's surface.
deep

Figure 22: showing the layers of the Earth, their temperature and consistency.

Tidal Energy:

This is the last source of heat energy for the Earth's surface. As the Earth rotates a complete turn each day the gravitational forces of the Sun and Moon reduce the effect of the Earth's gravitation on the part of the Earth facing them. This results in liquids and gases, especially the oceans and atmosphere, rising a small amount – relative to the size of the Earth 12,000 km diameter – creating a bulge of the atmosphere and ocean familiar to us as the *Tide,* and similar distortions right through to the Earth's centre.

When the Sun and Moon are in line, such as when there is no Moon at night because it is in the daylight sky near the Sun, or near the *Full Moon* time the gravity of the Sun and Moon effectively combine and we see big tides, *Spring Tides.* When they are not in line we see a half Moon in the sky and the tides are weak, *Neap Tides.* This shown in **Figure 23**.

Because the Moon is much closer than the Sun its gravity has about twice the tidal effect on the Earth. The Sun, being so huge, has a large gravitation effect on the Earth, the principal effect is to hold the Earth in its orbit around the Sun. The Moon has a very wobbly orbit around the Earth going once around every 28 days (a *Lunar Month*). So tides occur on about a 25 hour cycle, or 6 ¼ hours between each tide.

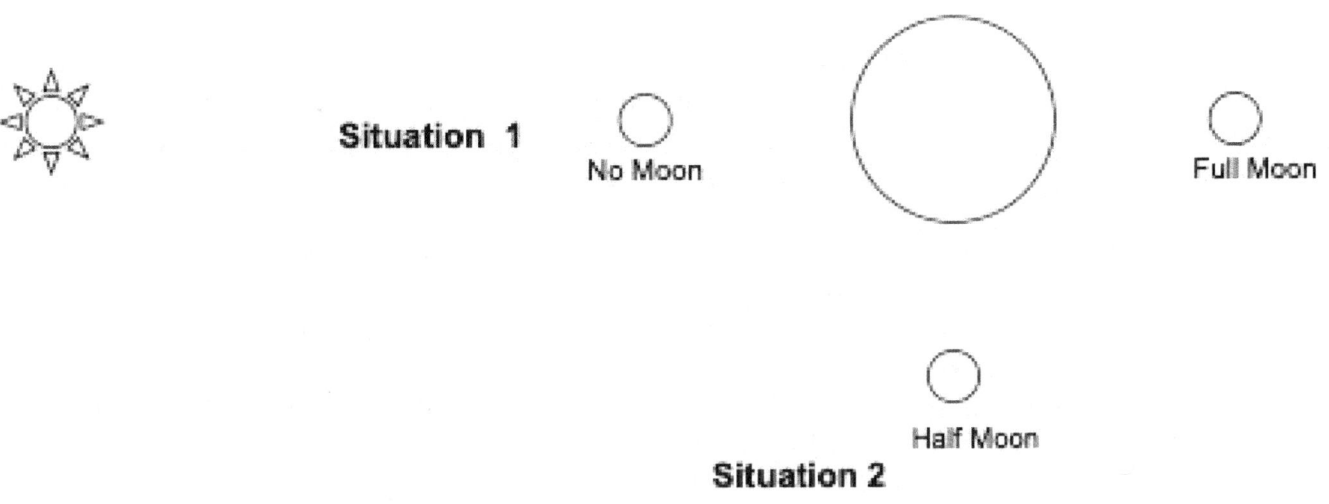

Figure 23: showing how the combined gravity of the Sun and Moon makes tides larger in Situation 1 when they are in line, and less in Situation 2 when they are not.

These tidal gravitational forces affect the whole of Earth, including the Mantel and Core, resulting in tidal distortion, although we only see the ocean effect. The forces result in ***friction and subsequent Heat generation*** as the Earth rotates. This totals about 2.4 Terawatts.

The friction is resulting in the rotation of the Earth gradually slowing, as discussed earlier. It slows about two thousandths of a second each century, so the Earth day has lengthened from 21 hours to 24 hours over the last 600 million years since complex multicellular life started here.

All other stars, planets and moons are subject also to tidal gravitational forces from neighbouring stars and planets. The forces tend to slow the rotation of these bodies until it stops altogether if enough time has passed. This what has happened to the Moon, which has stopped rotating in relation to the Earth, so we just see one side of it, never the far (called dark) side. It is also happening to the planet Mercury which rotates only one and a half times each Mercury year as opposed to Earth 364 times.

If the Earth had no Moon, and did not rotate in relation to the Sun there would be no tides. Many other aspects would be very different also, for example the side always facing the Sun would be extremely hot, and the dark side extremely cold. Although the Earth's rotation is gradually slowing it will take at least 4 billion and perhaps as long as 2 trillion (a trillion is a million million) years to stop!

Energy Loss from the Earth:

It is common sense that if heat energy keeps going into an object without equal heat loss, the objects temperature will keep increasing. We see this every time we boil the kettle. This is part of the ***First Law of Thermodynamics***.

As the temperature of an object increases it loses heat at a greater rate because it is now at a higher temperature than its surroundings, and heat energy flows from higher temperature – hotter – to lower temperature – colder areas. Heat energy flows by *conduction, convection, and/or radiation*, as discussed in chapter 1.

Eventually objects reach a balance, if they are not destroyed by burning, melting or vapourising away. At the balance heat energy going into the object is equalled by heat energy leaving, and the temperature of the object stays the same at that level. This is also part of the *First Law of Thermodynamics*.

We see this in everything around us that is staying at the same temperature. The heat energy going in is equalled by the heat energy going out. And it applies to the Earth as a whole as well. For the Earth to remain at a constant temperature the heat energy leaving must equal the heat energy arriving.

Considering the massive amount of Sun energy arriving there is an equal amount leaving for the temperature to stay the same. As about 175,000 terawatts arrives, 175,000 terawatts has to leave.

The Earth loses heat energy by radiating it away into surrounding space. The great bulk of energy leaving does so from the higher atmosphere, the *Stratosphere* and above layers. As only a tiny amount of the atmosphere, 0.1%, is in layers higher than the stratosphere, and that air is extremely thin, it follows that most heat radiation leaves from the stratosphere. Only a small amount, about 2% on average, leaves directly from the surface by radiation into space. This is because the *Troposphere* contains a large amount of water resulting in *convection* and *conduction* being the main form of heat transfer in this layer away from the Earth's surface. This is explained further in the next section on water.

Surrounding space is at a very low temperature, 3° Kelvin, or minus 270°C, so radiation readily flows away from the stratosphere and higher layers. It does so both in the daytime and also at nighttime.

The radiation energy leaving the Earth is in the *Infrared* part of the radiation spectrum discussed in chapter 1 and in Figure 5. The *wavelength* of this radiation follows the *Blackbody Radiation* guidelines discussed and shown for the Earth as a whole in Figure 7. The wavelength tends to be longer for cooler objects and shorter for hotter ones, including parts of the Earth. Everything on Earth, including the atmosphere layers, radiates heat radiation in the infrared spectrum of wavelengths, the wavelength varying according to that part's temperature.

Figure 12 earlier showed the temperature of the different layers of the atmosphere. As shown the temperature at the bottom of the stratosphere is minus 50°C, and at the top of it much warmer at minus 3°C. Heat radiation from this layer varies a little in wavelength because of this temperature variation, from a peak wavelength of about 10 microns at 3°C to a peak of about 13 microns at minus 50°C. This is shown in **Figure 24**. This also shows that much of the radiation escapes at other wavelengths from 4 to 90 microns, about 40% of it at longer wavelengths than 20 microns.

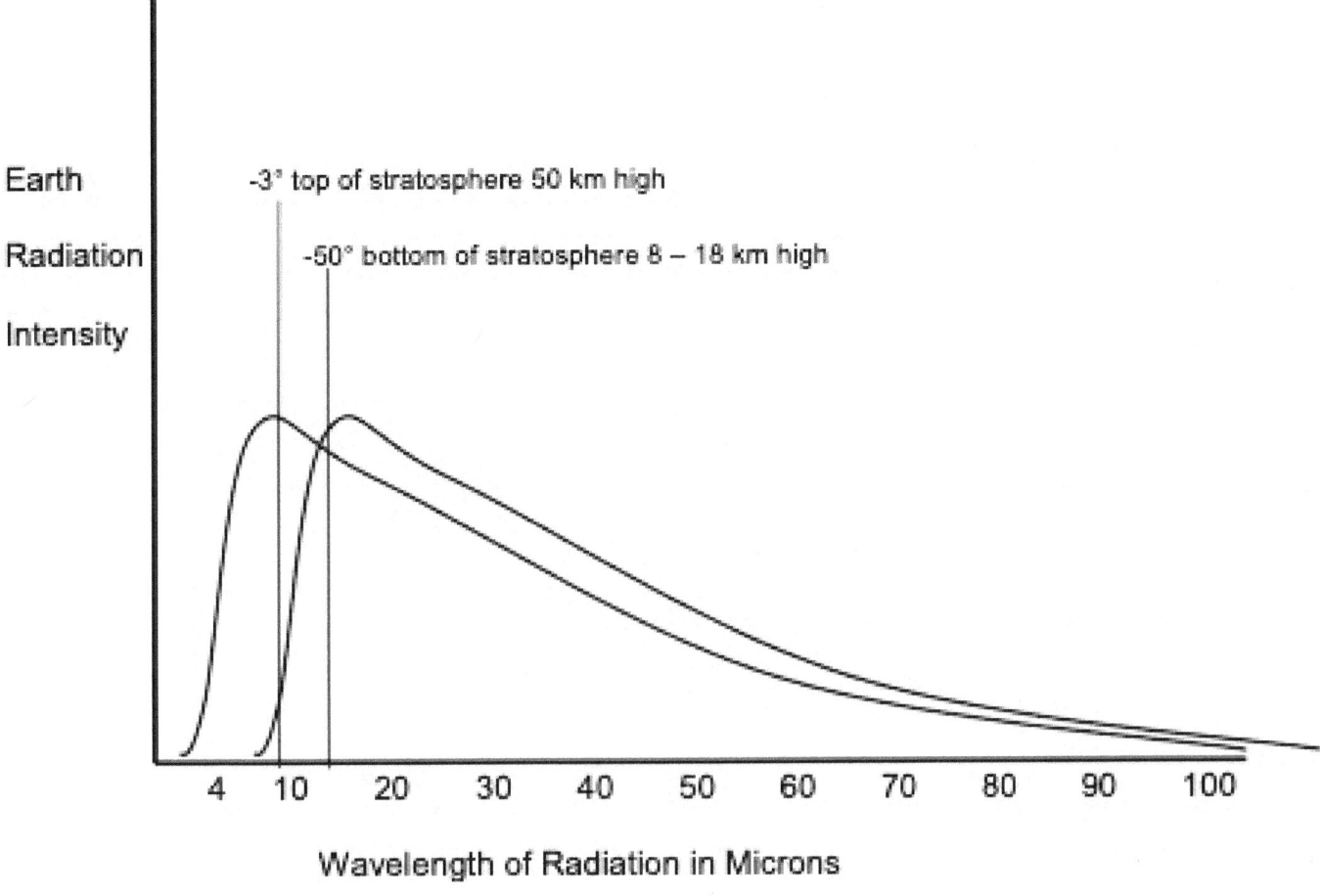

Figure 24: showing the wavelengths curve for heat radiation leaving the Earth from the stratosphere, the most important layer for Earth heat radiation loss. The stratosphere is minus 50°C at its bottom 8 – 18 km high and minus 3°C at its top 50 km high.

At sea level air has 1225 grams of air molecules per cubic meter. At the bottom of the stratosphere it has 100 grams per cubic meter, and at the top of the stratosphere 0.1 grams per cubic meter, one 10,000th of the sea level density. Above the stratosphere the air is so thin it carries only a tiny amount of heat energy to radiate into space. There is nothing here to stop heat radiation leaving Earth.

Over its very long history the Earth has reached a balanced temperature where outgoing heat radiation is the same as incoming, and the temperature overall stays the same. During the year parts of the Earth enter summer, then receive more heat radiation than they are losing, so the temperature increases, it gets warmer. Then those parts move into winter, they lose more heat radiation than they are getting, they get colder. Each year the same pattern repeats itself.

The big picture is that the Earth receives 175,000 terawatts from the Sun in energy radiation at high energy wavelengths from 0.4 microns to 3.2 microns. And puts out the same amount of energy radiation at lower energy wavelengths from 4 microns to 90 microns. Staying at the same temperature of 14°C average globally.

Watts measure the *Rate* of energy creation and use, one watt is one joule of energy used or created in a second. The *Amount* of energy is measured in joules, or, more commonly, kilowatt hours or terawatt hours as discussed above.

References and further reading for Chapter 3: The Earth's Surface Temperature.

If a link does not open do a google search and it should appear.

www.interestingengineering.com

www.pveducation.org

www.en.wikipedia.org

Wigley and Kelly, 1990.

www,swsc-journal.org

J .Space Weather Space Clim, 11 (2021) 17 html

www.ncbi.nim.nih.gov>articles>PMC7575229

Zharkova, Valentina. Temperature. 2020;7(3):217-222

https://pubs.usgs.gov

Rosenthal,Y.; Linsley, B.K.; Oppo, D.W.; (2013) ""Pacific Ocean Heat in the Last 10,000 Years." Science.342(6158); 617-621.

Cook, E.R. et al; "Evidence for a Middle Ages Warm Period in a 1100 year old tree ring reconstruction in past austral summer temperatures in New Zealand. "

Ice Ages Ancient and Modern. Wright, A.F.; Coope, G.R, et al. Seel House Press 1975.

www.pubs.usgs.gov/fs/fs-0095-/fs-0095-00.pdf

Crowley, Kim et al 1996.

www.en.wkipedia.org/wiki/Earth%27s_orbit

www.space.com>16903-mars-atmosphere-climate...

https://earthodservatory.nasa.gov

www.enwikipedia.org/wiki/milankovitch_cycles

Sir Fred Hoyle. Ice. Pp 60-65 and page 70.

https://climate.nasa.gov/news/2498/milankovitch-orbital-cycles-and-their-role-in-the-earths-climate/

www.history.com//topics/pre-history/why-did-the-dinosaurs-die-out

www.ecgllp.com/files/3514/0200/1304/2-Solar-Radiation.pdf

www.pveducation.org/pvcdrom/properties-of-sunlight/average-solar-radiation

www.nap.edu/read/4778/chapter/4

https://gpm.nasa.gov/education/videos/water-cycle-heating-ocean

www.nsf.gov>opp>antarct>science>icesheet

www.profhorn.meteor.wise.edu/wxwise/radiation/sunangle,html

www.applet-magic.com/insolation.htm T Watkins, San Jose State University.

www.en.wikipedia.org

www.news.mit.edu/2011/energy-scale-part3-1026

www.sws.bom.gov.au>educational

www.quora.com (sun energy).

www.esrf.fr>genera>Earth-Center-Hotter

www.en.wikipedia.org/wiki/Earth%27s_inner_core

https://phys.org/news/2015-12-earth-layers.html

Rybach, Ladislaus: Geo-Heat Quarterly Bulletin: 28; pp 2-7; 2007.

https://en.wikipedia.org/wiki/Earth_tide

Egbert and Ray, Nature 405, pages 775-8, June 2000

https://dummies.com/education/science/physics/flowing-from-hot-to-cold-the-second-law-of-thermodynamics/

https://en.wikipedia.org/wiki/Laws_of_thermodynamics

www.Nov79.com/gbwm/ntyg.html

www.nasa.gov/centers/dryden/pdf/87705main_H-352.pdf

E. Montoya and T Larsen: Stratospheric and Mesospheric Density-Height Profiles Obtained with the X15 Airplane. June 1964.

Chapter 4

Water

Because of Earth's location in the Solar System water exists in large quantities in its liquid state here. It is of course essential for life. It also is dominant concerning the Earth's surface temperature after the stage set by the conditions discussed before: the Sun's energy output, the distance from the Sun, the rotation of the Earth, and the tilt of its axis.

The Earth has a lot of liquid water, 1.3 billion cubic kilometers. We have a preconception that Earth has the most liquid water in the Solar system but it is now known that there is up to 50 times our volume of liquid water elsewhere:

- Europa, a moon of Saturn, has twice our volume beneath a solid ice crust 20 kilometers thick.

- Enceladus, another moon on Saturn, also has an ocean beneath a surface ice layers. Geysers of water were seen by the Cassini spacecraft in 2005.

- Ganymede, a moon of Jupiter, has a subsurface ocean 100 kilometers deep beneath a 150 kilometer ice layer.

- Ceres, the largest asteroid in the asteroid belt between Mars and Jupiter with a diameter of 1000 kilometers, has an ocean beneath an ice layer.

- Neptune and Uranus are made largely of ice but may have small quantities of liquid water. Pluto may have an ocean of liquid water beneath an ice layer.

- Mars has a lake of liquid very salty water 20 kilometers across 1.5 kilometers beneath its South Pole ice cap. The water ice cap is covered by a 1 meter thick layer of carbon dioxide ice, and the surface is at minus 70°C. There is no liquid water on the surface.

- Our Moon has water locked in soils, and some ice in the cold shadows of craters. It has about 1/100th of the soil water as the Sahara Desert.

Earth is the only place in our Solar System where large amounts of surface liquid water exist. Our oceans cover 71% of the Earth surface.

The reasons for this are the special properties of water, H2O, and the surface temperature of Earth determined by distance from the Sun and the Earth's rotation. Another very important factor is the *size* of the Earth with its diameter of 12,000 kilometers and large *mass* of 6×10^{24} kilograms. This means its *gravity* is strong enough to hold a dense atmosphere above the surface. If the Earth was significantly smaller, it would not hold an atmosphere as dense, and liquid water could not exist on the surface.

This is so because of water molecules, H2O, *escaping from the surface of liquid water.* This happens because liquid water molecules have heat energy. So some of the water molecules have enough energy to escape from the liquid and exist above the water surface. There they form a layer of *water vapour.* This called *evaporation.*

The reverse process where water vapour molecules above the liquid surface penetrate it and become part of the liquid is called *condensation.* Wherever there is liquid water both evaporation and condensation are going on.

If there was no atmosphere above the liquid water, these molecules would just escape into empty space. Eventually all the liquid water would disappear, all its molecules would have so escaped.

But on the Earth's surface there is an atmosphere because the Earth has enough gravity, because of its size and mass, to hold a dense one. The Moon, by contrast, is too small and has no atmosphere, and so has no liquid water. Mars also has no liquid water **because its atmosphere is too thin.**

The key to this is *atmospheric pressure. Pressure* is a force pressing on a surface. We are familiar with pressure when we dive into water – the further we go down the more *pressure* we feel on our bodies as the *weight* of water above us *presses* down on us. The same thing happens on the Earth's surface because of the atmosphere above us. The *weight* of the atmosphere above *presses* on us. Pressure is measured in *force per area of surface it is acting on.* A common system is *Pascals.* One *pascal* is one *newton* of force per *square meter* of area.

The atmospheric pressure at sea level on Earth is 101,300 Pascals. It is more common to talk about *Hectopascals,* there are 100 pascals in a hectopascal *hPa.* It follows there are 1013 hectopascals of atmospheric pressure at sea level.

Our household *barometers* measure atmospheric pressure. They usually measure it in *millibars.* One millibar is the same as one hectopascal.

The pressure of the atmosphere above us at sea level on Earth is 1013 hectopascals. By contrast on Mars the pressure is only 6 hectopascals. At this low pressure all the molecules in liquid water escape so there is no liquid water on Mars.

All the gas molecules in the air, above the Earth's surface add to the mass of the air there and to the pressure at the surface. All these molecules, be they water vapour, oxygen, nitrogen, argon, carbon dioxide, are at the same pressure in the air above the surface. But each contributes its own *partial pressure* to make up the total pressure.

For example, if the atmospheric pressure is 1013 hectopascals at sea level and main components of it – oxygen, nitrogen, argon, carbon dioxide – have the same proportions as discussed chapter one the partial pressures for each are shown below:

- Oxygen, 21% of atmosphere, partial pressure 21% of 1013 hPa – 121 hPa

- Nitrogen, 78% of atmosphere, partial pressure 78% of 1013 hPa – 790 hPa.
- Argon, 0.9% of atmosphere, partial pressure 0.9% of 1013 hPa – 9 hPa.
- Carbon Dioxide, 0.04%, partial pressure 0.4 hPa.

In this example the air is completely dry, there is no water vapour. But in practice there is always water vapour on Earth because there is so much liquid water. Typically there is between 2 and 4% water vapour. Obviously if, in the example above, there was 4% water vapor the partial pressure of it would be 40 hectopascals (4% of 1013 hPa), then all the other component gases would be a little less to make the total 100%.

On Earth the higher we go above the surface in the atmosphere the less the pressure gets because there is less above us, the **weight** of the atmosphere above us is less. At 45 kilometers above the Earth's surface our atmosphere has same pressure as the surface of Mars. The **temperature** also decreases in the troposphere, then increases in the stratosphere because of the Sun's radiation acting on ozone as discussed before chapter 2. The table below shows the pressure and air temperature at increasing heights above sea level on a day when it is 14° C at sea level:

Height in kilometers Pressure in hectopascals Temperature °C

In Troposphere:

- Sea level   1013   14
- 1 km   900   8
- 2 km   795   2
- 3   701   minus 4
- 4   616   minus 11
- 5   540   minus 17
- 6   471   minus 24
- 8 3   55   minus 37
- 10   264   minus 50
- 11   226   minus 55

In Stratosphere:

- 16   100   minus 50
- 32   10   minus 15
- 48   1   minus 3

In the Troposphere the temperature drops about 6.5° C per kilometer height. In the Stratosphere it stays at about minus 50° C up to 20 kilometers high, then warms more to reach close to 0° C at 50 kilometers high as discussed before.

The more heat energy liquid water has, that is the higher its *temperature,* the more water molecules escape from the surface and exist as vapour gas above it. This is common sense to us heating water for cooking and forming steam.

At each temperature of the liquid water a balance can form where the rate of water molecules leaving by evaporation equals the rate entering of water molecules entering by condensation. At that balance the amount, or volume, of liquid water stays the same. The *pressure* of the water vapour above the surface where this balance occurs at a given temperature of the liquid water is called the ***vapour pressure*** of the liquid water for that temperature.

The table below specifies what the vapour pressure is for liquid water at various temperatures:

Temperature ° C	Vapour Pressure hectopascals
0	6.1
10	12
20	23
40	73
60	198
80	471
100	1013 (sea level pressure)
120	1985
140	3611
200	15,527
374	220,640

The table shows that at 100° C the vapour pressure is the same as the atmospheric pressure at sea level. Clearly if the temperature of the liquid water goes higher than that because more heat is added the actual water vapour pressure cannot go higher because it just disperses into the air. So all the liquid water turns into vapour, steam, and the water boils. Many bubbles of vapour form throughout the liquid.

As another example at 6 kilometers up the atmospheric pressure is 471 hectopascals. So water would boil there at 80° C when the vapour pressure was that level.

Returning to the example of Mars with an atmospheric pressure of 6 hectopascals, liquid water would boil there when its vapour pressure was that level. But that is a lower pressure than the one at 0° C, 6.1 hectopascals, at which temperature *normal* water freezes to ice. So even liquid water just above freezing on Mars would boil away to vapour, hence there is no liquid water on Mars, but heaps on Earth.

The term *normal* water is used here because virtually all liquid water on Earth has impurities in it: dirt, bugs, little bits of all sorts of matter. The freezing temperature of normal water, at which it turns to ice is 0° C. But for ***very pure water*** the freezing temperature is lower, as low as minus 40° C. This called ***supercooled*** water. In nature supercooled water exists in clouds high in the atmosphere as will be explained. The reason for this is that in normal water the foreign matter particles act as a focus for ice formation, hence it occurs sooner when water is cooled,

But generally below 0° C water exists as ice, and above as liquid water with a layer of water vapour above it as explained.

Ice also has a layer of water vapour above it, meaning that some water molecules break out of the ice surface. This is called ***sublimation*** of the ice to water vapour. And if the air above the ice has water vapour in it, some molecules will join into the sold ice, this is called ***deposition*** of the vapor to ice. At no stage is liquid water formed at these temperatures below freezing.

As there is always a layer of water vapour associated with ice, there is a ***vapour pressure*** for the ice when equal amounts of water molecules are entering and leaving the ice surface, as for liquid water. This vapour pressure also increases with temperature, but is much lower than vapour pressures for liquid water. For example at minus 12° C the ice vapour pressure is only one hectopascal; and at minus 40° C, 0.1 hectopascal.

These changes from ice to liquid water to water vapour, and the reverse processes, are called ***phase changes.*** They are vital for the control of Earth's temperature because during them a huge amount of heat energy is either used or released.

The reason for this is that for water molecules to exist as a vapour, completely free to move in relation to other nearby molecules, they must have enough heat energy to resist being bound to the nearby molecules. So heat energy has to be added to water molecules to enable them to change from liquid water to water vapour. This heat energy is called the ***latent heat of evaporation.***

When the vapour water molecules go back to the liquid state, liquid water, they lose heat energy and release it into the surrounding air. They release the same amount of heat energy as was required to change them from liquid to vapour, that is the ***latent heat of evaporation*** again. But this is then called the ***latent heat of condensation,*** although it is the same amount of heat energy as discussed below.

One kilowatt hour of heat energy is used to turn 1.6 litres of water into vapour. And when this vapour turns back to 1.6 litres of liquid water, one kilowatt hour of heat energy is released, warming the surroundings.

When ice is melted to form liquid water one kilowatt hour of heat energy melts 12 litres of ice resulting in 11 litres of water, as water is denser than ice, hence ice floats on water. And when the water freezes to form ice the same heat energy is released warming the surroundings. These latent heats are the ***latent heat of melting*** and the ***latent heat of fusion*** respectively.

Water is unusual in that it's solid form, ice, is lighter than it's liquid, so as water gets colder and ice forms the ice moves to the surface and floats there. If ice were heavier than liquid water it would sink to the bottom and freezing would occur from the bottom up. The oceans, especially towards the poles, would be solid ice, unable to be heated and melted by the Sun. It is likely Earth would be a solid ice ball. But ice floats so it is not!

When ice turns directly into water vapour – **sublimation** – more heat energy is used. One kilowatt hour is needed to cause 1.4 litres of ice to turn into vapour. And when the vapour turns back to ice in **deposition** this heat energy is released into the surroundings.

To summarise these quantities of heat energy used or released in water phase changes:

- Latent heats of Evaporation and Condensation (liquid water to water vapour, water vapour to liquid water): one kilowatt hour every 1.6 litres of water.

- Latent Heat of Melting and Fusion (ice to liquid water, liquid water to ice): one kilowatt hour every 12 litres of ice.

- Latent Heat of Sublimation and Deposition (ice to water vapour, water vapour to ice): one kilowatt hour every 1.4 litres of ice.

The following steps explain why this is a mechanism for Earth losing a huge amount of the heat energy arriving from the Sun each day and remaining at a steady temperature:

1: Warm water on the Earth's surface heated by the Sun evaporates. As the oceans make up 71% of the Earth's surface and the great bulk of the Earth's water is there (98.5%) the greatest amount of evaporation occurs from the ocean surface. A much smaller amount occurs from the land's surface because once the surface layer of land has dried out no further evaporation occurs, the land just continues to heat up due to the Sun. Only 0.001% of the Earth's water is in soil.

For every 1.6 litres of water evaporating into vapour in the air above the surface one kilowatt hour of heat energy is lost from the water – so the water cools. The heat in the water is shifted to the water vapour in the air above.

2: The warm air containing water vapour rises from the surface because it is lighter than the surrounding air, **convection** occurs as explained chapter one. As it does so cooler air from nearby flows in to fill the space left by the warm air rising. This cooler air in turn is warmed by the warm ocean surface and receives water vapour from more evaporation, so the process continues. The warm moist air rises further. The higher it goes the less the **atmospheric pressure** is as explained earlier. So the rising air expands, gets thinner, occupies a bigger volume, its pressure decreases to match the surrounding air at that height, ***and its temperature drops,*** it gets colder.

This is the same process which occurs in refrigeration. In fridges a gas expands and gets colder, it takes heat from the surroundings and cools them. The same thing happens as warm air rises in the atmosphere. It expands because of lower pressure higher up and cools. This is part of **Boyles Law** concerning the relationship between pressure, volume, and temperature for gases.

3: Eventually the moist air originally from the surface reaches a height where the temperature is so low that the water vapour within it is *saturated*. It has reached the *vapour pressure* for water vapour at that temperature as explained previously. So then the water vapour turns into liquid water until the air is no longer saturated with water vapour. As explained before this process of *condensation,* the reverse of evaporation, releases one kilowatt hour of heat energy into the air for every 1.6 litres of liquid water formed from the water vapour, warming the surrounding air.

So now the heat energy originally from the water at the surface has been transferred to the air high above. This warm air being lighter than surrounding rises further and gets colder until the water vapour within it is again *saturated,* at *vapour pressure,* because of the colder temperature at greater height. So the process repeats, the water vapour *condenses* to form liquid water, heat is released into the air at one kilowatt hour per 1.6 litres of water formed, warming the air, which rises further, gets colder, and so on.

The process results in the heat energy originally in the surface water on Earth, mostly in the oceans as explained, being shifted high in the atmosphere.

4: Eventually the air with the remaining water vapour reaches a height where it is so cold that all the water vapour turns to ice in the process called *deposition* discussed previously. This happens when the temperature has dropped to minus 50° C . This is about 18 kilometers high at the Equator, and 8 at the Poles. Above this point there can be no water vapour because the air is very thin, ice is too heavy to float in the very thin air. This point marks the top of the *troposphere,* and bottom of the *stratosphere,* as discussed in chapter 2. There is no water vapour in the stratosphere.

When the water vapour turns to ice it releases heat energy at the rate of one kilowatt hour per 1.4 litres of ice formed as discussed concerning the *latent heat of deposition* previously. This heat is then free to radiate into space, *because there is no water vapour to block it,* and is thus lost from the Earth.

So the end result of the process of *phase changes* from evaporation of liquid water at the surface through *condensation* and *deposition* high in the atmosphere is the *transfer* of large amounts of heat energy from the surface to space and away from Earth.

How much heat energy is lost from the Earth by this process? We can get some idea of this by considering what happens to the liquid water and ice formed high in the atmosphere by this process of warm moist air rising by convection, then cooling, the water vapour condensing to water and forming ice, giving up its heat in the process.

In the lower heights up to 2 kilometers high just water droplets form from water vapour when it condenses. If they are very tiny, around 10 microns size, they just float in the air and form *clouds.* They tend to initially form around *particles* of dust in the atmosphere, then grow with more condensation occurring on them. Droplets collide with others getting larger. When they reach about 1 millimeter in size they are then too heavy to float and fall as *rain.*

As they fall they take up *heat energy* from the lower warmer atmosphere cooling it, because the rain drops are colder due to their origin higher in the atmosphere. If they take up enough heat they may evaporate again into water vapour before reaching the surface. If they do not warm enough on the way down to evaporate they fall on the surface as rain, cooling it.

At heights between 2 and 6 kilometers a mixture of water droplets and ice crystals form from water vapour cooling. Clouds formed there contain a mixture of water droplets and ice crystals. The water droplets tend to be attracted and adhere to the ice crystals making them grow larger until they become too heavy to float in the air and then fall as rain or hail. Being colder from higher in the atmosphere they may not pick up enough heat as they pass through the lower warmer layers to evaporate to water vapour and so often reach the surface as rain or hail.

Above 6 kilometers up to the top of the troposphere at 18 kilometers at the Equator, 8 kilometers at the Poles, water vapour forms ice crystals and supercooled (as discussed before) water droplets which quickly are incorporated into ice crystals. The ice crystals grow in fine branches to form larger crystals. These make up the clouds at these higher levels. As the air temperature here is less than minus 10° C this process continues until the crystals become **snowflakes.** These are then too heavy to float in the air and gently fall. They may melt in warmer air on the way down to form rain, or in colder places especially in winter, or where the land is high as in mountain ranges, reach the ground as snow.

All of these processes – the formation of rain, hail or snow *which reaches the surface* – are called **precipitation.** In all of them cold water or ice, which has lost its heat energy, the original heat from the surface as explained, into space away from the Earth, reaches the surface and cools it.

Most precipitation occurs in warmer tropical parts of Earth because the higher surface temper results in more evaporation of surface water to water vapour. Less occurs in colder places near the Poles. 78% of precipitation occurs over the oceans because most evaporation, 86%, occurs from the oceans. Across the whole Earth the ***average precipitation per year*** is one meter. This is 1000 litres of liquid water per square meter of the Earth's surface. The surface area of the Earth is 5.1×10^{14} square meters (5.1 hundred trillion square meters). So every year 5.1×10^{17} litres of liquid water falls as precipitation. All of this is originally evaporated from surface water as discussed. One kilowatt hour of heat energy is required to evaporate 1.6 litres of liquid water. And that means that each year 3.2×10^{17} kilowatt hours of heat energy is removed from the surface and transferred to the high atmosphere, and then away from the Earth by these **water phase change** processes.

This is an **underestimation** of the heat removed by water phase changes from the Earth because some of the evaporated vapour does not return to the surface locally as rain, hail or snow. Some evaporates again on the way down as explained, receiving heat from the warmer lower atmosphere to do so. This warmer moist air then rises again to lose its heat higher when condensing or freezing to liquid water or ice. In this process also heat which was at the surface is shifted to high in the atmosphere to the top of the troposphere, bottom of the stratosphere, where it can radiate away into space.

The calculation also makes no provision for the greater amount of heat energy used and released in phase changes from water to and from ice, and water vapour to and from ice. Only liquid water to and from water vapor is considered.

Humans make and use about 1.2×10^{14} kilowatt hours of heat energy per year. This is a two thousandth of the minimal heat energy lost by water phase changes, evaporation from Earth in this natural cycle.

The process also can respond readily should the Earth's temperature increase, as has often occurred in the past as discussed in Chapter 3. Should the Earth get warmer evaporation and the other water phase changes will increase, resulting in more heat energy being shifted to the high troposphere where it can radiate away, avoiding further heating of the surface. In previous warmer periods of the Earth's history evaporation and precipitation were greater in a hotter moister Earth.

And in colder periods such as the ice ages the reverse happens. Evaporation decreases, less heat is lost from the surface, more heat is retained, the cold is less intense. In ice ages the climate is drier with less precipitation because there is less evaporation.

This is called a ***negative feedback mechanism.*** Negative feedback mechanisms ***regulate*** changes by tending to bring them back to a normal state. The existence of large amounts of liquid water on the Earth's surface means that the ***temperature*** of the surface is ***regulated*** by these water phase changes, which result in more heat energy being lost from the surface if it gets warmer, and less if it gets colder, evening out the temperature changes and easing extremes.

Clouds are very important for temperature control of the surface. On average 30% of the Earth is covered by clouds at any one time. They reflect incoming Sun heat energy back into space. Heavy thick clouds with a lot of water and ice can prevent up to 80% of the Sun's energy reaching the surface. More typically about 30% is so prevented.

As Earth receives about 10^{18} kilowatt hours heat energy each year from the Sun, it follows that 3×10^{17} kilowatt hours is reflected back into space due to clouds. This is more than 2000 times greater than all the energy humans make and use.

Clouds are another ***negative feedback mechanism*** regulating the surface temperature. If it gets hotter, more evaporation occurs and more clouds form, more Sun energy is reflected back into space, resulting in the surface cooling. Conversely if the surface gets colder, less evaporation occurs, fewer clouds form, more Sun energy gets to the surface warming it.

Ice crystal clouds reflect more Sun energy than water droplet clouds. As discussed above ice crystal clouds form at the top of the troposphere just below the stratosphere preventing the Sun's heat energy getting past. These clouds are usually wispy strands of ice crystals called ***Cirrus clouds.*** If they cover too much of the Earth's upper atmosphere they can reflect so much of the Sun's energy that so much cooling occurs that an ***ice age*** may occur.

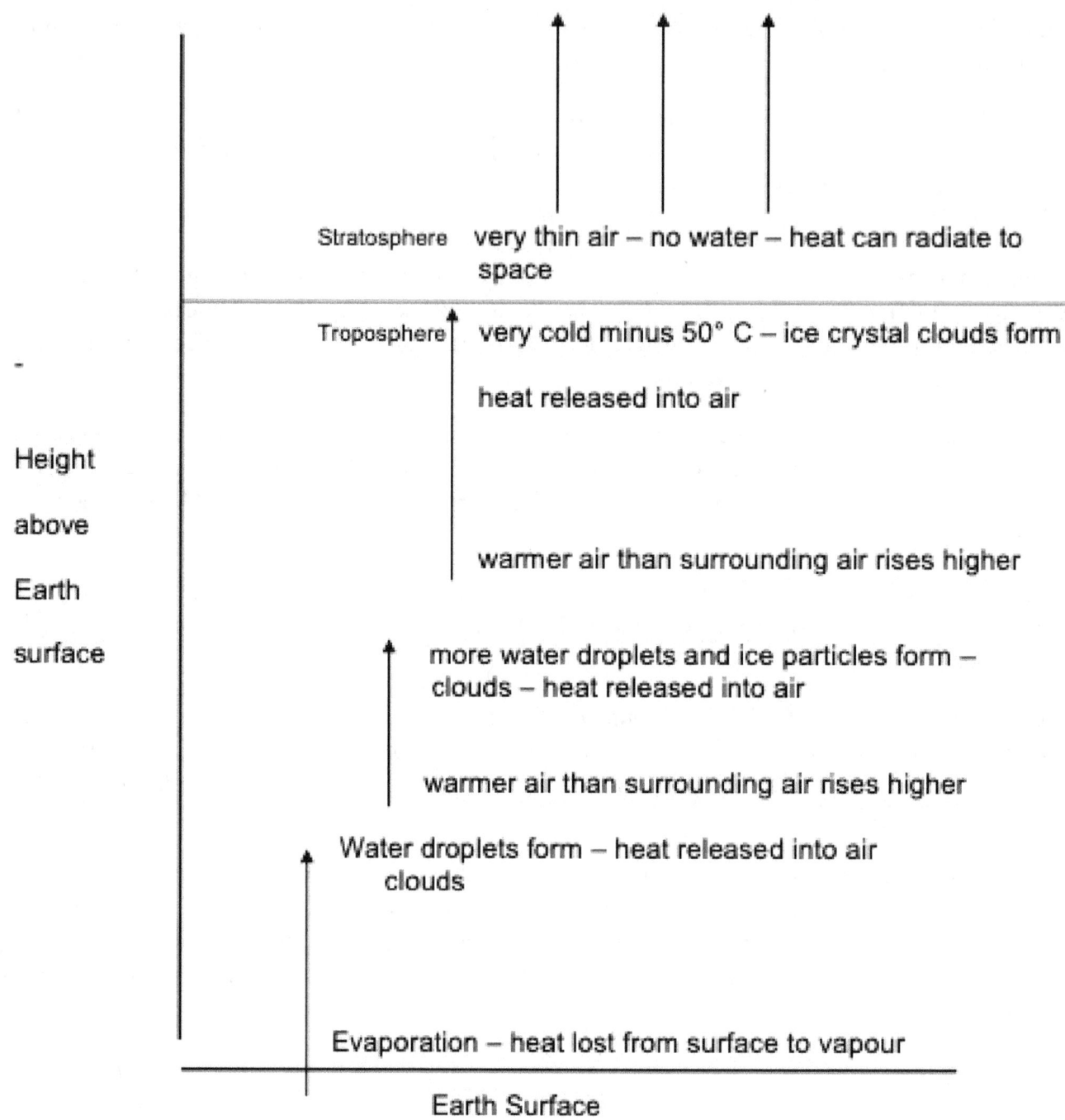

Figure 25: Showing how heat energy is moved from the surface – especially from the oceans – by **water phase changes** to the top of the troposphere, where the heat energy is free to radiate away into space because there is no water in the stratosphere to block its radiation. When the water vapour formed by evaporation from surface liquid water rises and then condenses back to water droplets, or freezes to form ice crystals, clouds form, and heat is released into the surrounding air keeping the process going.

To continue discussing *Ice Crystal Clouds* a vicious circle then occurs as ice and snow on the surface reflects up to 90% of the Sun's energy, making it even colder. Then more ice crystal clouds form in the atmosphere. This can become *positive feedback mechanism* where the cold causes the ice crystal clouds to form, they make it even colder, a vicious circle of change occurs leading to an ice age.

As Sir Fred Hoyle explains elegantly in his book "Ice" the formation of high ice crystal clouds high in the troposphere is related to the temperature there: if it very cold, less than minus 40°C, ice crystal clouds form; they enlarge and multiply forming a layer of cloud which reflects back at least 90% of the incoming Sun heat radiation, markedly cooling the planet. A layer just 10 microns thick would do this effectively. By comparison if the temperature where these very high clouds form is warmer than minus 40°C then ice crystal clouds are less common and the highly reflective layer does not form. The key to this is the amount of heat energy being shifted from the surface *especially the oceans* by water evaporating, then moving up the troposphere by convection, transferring heat energy away from the surface by this and phase changes. If the amount of heat energy from the surface is low, as over Antarctica, then high ice crystal clouds form. If it is high as over warm tropical seas then they cannot form. Ice crystal clouds are also called *diamond dust clouds.*

As discussed chapter 2 from 1945 to 1977 the Earth's surface cooled about 0.5° C, there were fears our warm period may cease and the current ice age recommence. Thankfully that did not occur (yet!) and these high altitude cirrus clouds continue to help regulate the Earth's temperature with the other clouds.

Heat Loss from Land:

While evaporation of liquid water to water vapour is the principal method of heat loss from the oceans that is not so for the 29% of the Earth's surface which is land. For land the principal method for losing the heat energy arriving from the Sun is by *radiating* it into the air above the surface.

Some heat loss by evaporation does occur from land areas, obviously from rivers, lake, very wet areas. And plants loose heat by evaporation of water called *transpiration.* And rain, hail and snow falling from the cold high atmosphere cool the surface, and subsequently may cool it more by evaporation.

Some heat loss from the land surface also occurs by *conduction* directly into the nearby air. But this is minor compared to radiation mainly because dry air needs only a very small amount of heat energy to raise its temperature. One kilowatt hour of heat energy can heat 3 million litres of dry air by one degree C, whereas it will only heat 8600 litres of water by one degree C. So cool dry air blowing over land will not cool it much compared to evaporation of water at one kilowatt hour every 1.6 litres of water evaporated.

Radiation from the land surface follows the pattern discussed before and shown in Figure 7 in chapter 1. That is copied below for convenience. The wavelength of the heat radiation is in the *infrared* part of electromagnetic radiation, it ranges from 4 to 90 microns wavelength as said before, longer than visible light which is from 0.4 to 3.2 microns.

The *radiation heat energy* from the land surface goes upwards into the air above. If the air is completely dry, has no water in it, then it is made of nitrogen, oxygen, argon gases mainly, with other trace gases like carbon dioxide, methane, as discussed on page 2.

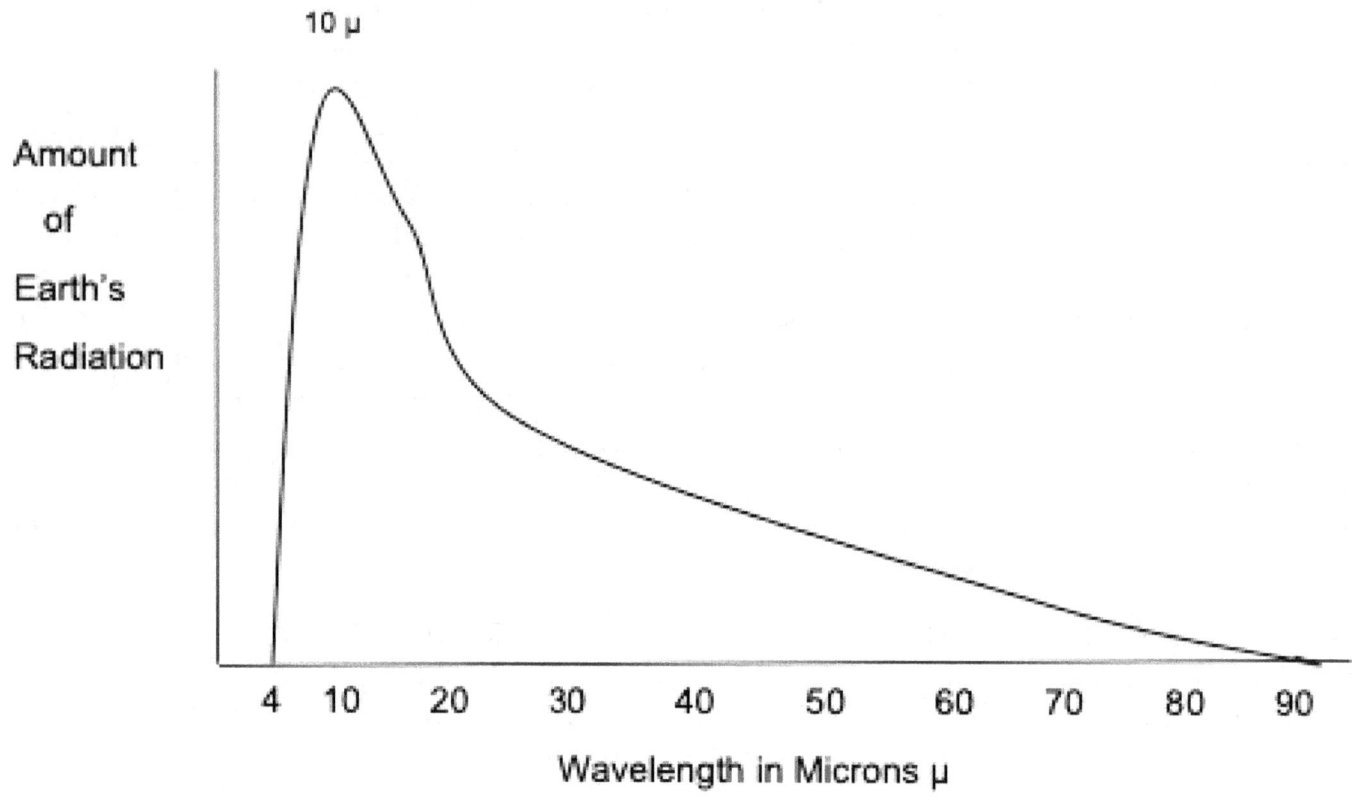

Figure 7: Earth's Energy (Heat) Radiation in the Infrared.

The *absorption* of radiation in these wavelengths from 4 to 90 microns by these gases has been discussed already in figures 8 to 11. To summarize the radiation passes straight through nitrogen and argon, oxygen absorbs a tiny amount at 6.2 microns wavelength, the trace gases of carbon dioxide and methane (which are discussed more later) absorb from 14 to 16 microns and from 7.2 to 9 microns respectively, all other wavelengths in the band 4 to 90 microns from the land radiation pass straight through.

So if the air is quite dry, no water vapour in it, the bulk of the land heat radiation can pass straight through the troposphere to the stratosphere, then escape into space cooling the land surface. There are some places on Earth where the air has very little water, as little as 0.01%, for example the Dry Valleys in Antarctica which has seen no rain for 2 million years, the Atacama Desert in Peru, parts of Namibia. In those places heat radiation can escape freely into space especially at night, when they get very cold. For example the Atacama Desert gets to minus 5° C at night despite reaching 30°C in the day.

But the air above the great majority of land contains significant water vapour usually ranging from 2% to 4%. And as discussed in chapter 1, Figure 9 copied below for convenience the water vapour makes a huge difference to what happens to heat radiation leaving the land surface:

Instead of escaping freely into space as occurs in dry air much of it is absorbed by the water molecules. These molecules heat up and pass some of the heat on to nearby molecules of the other gases in the nearby air by direct contact (**conduction**). As there are a large number of water molecules in the troposphere air, usually between 2 and 4 every hundred molecules of air, it is easy for conduction to pass some heat energy onto the other molecules – nitrogen, oxygen, argon and the trace gases – spreading the heat evenly through the air.

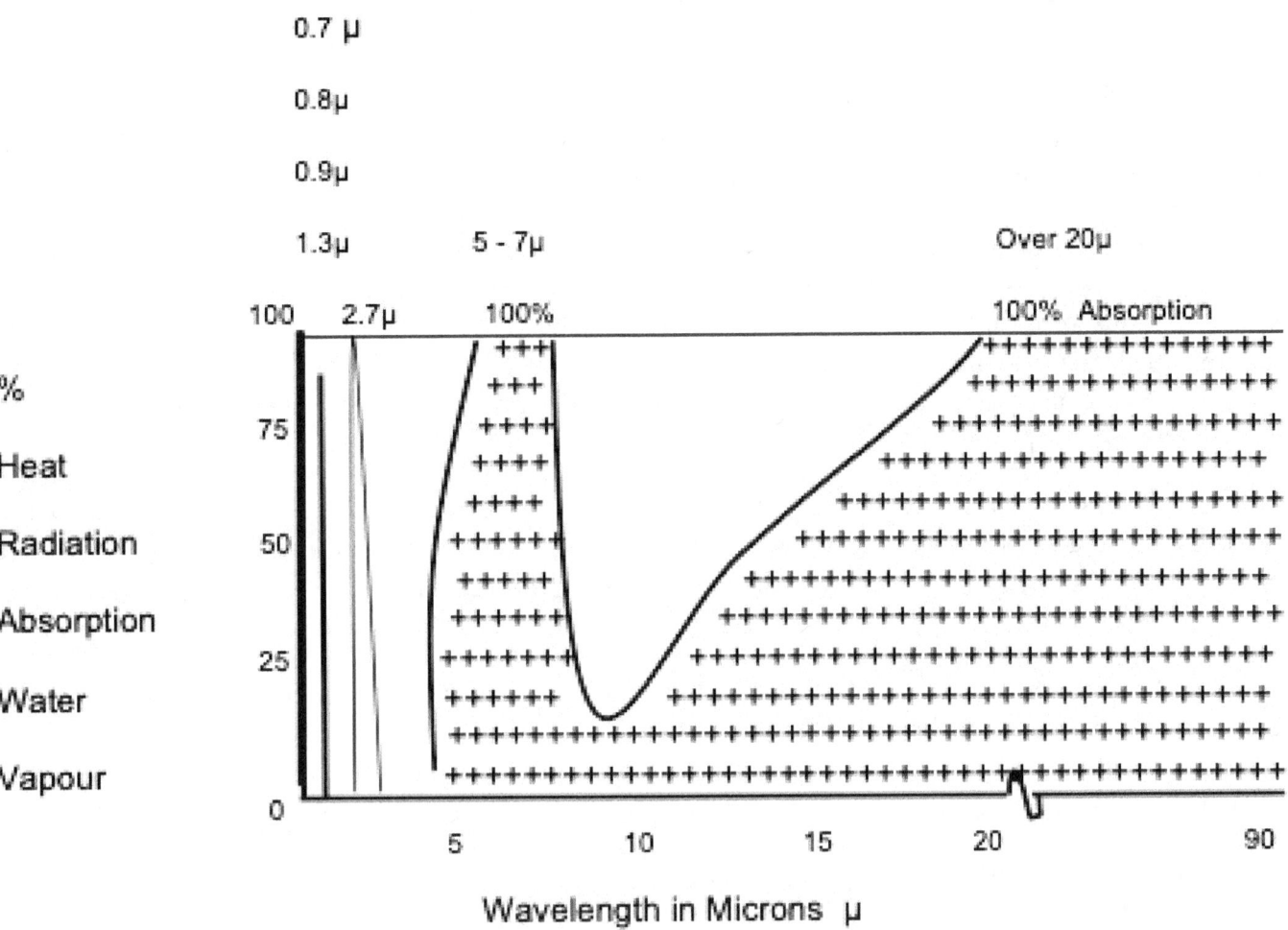

Figure 9: Radiation Absorption by Water Vapour Molecules H_2O

And some of the heat from a hot water molecule leaves it by blackbody radiation in all directions – up, down, sideways. The process from the water molecule absorbing heat energy, then re-emitting it as its own blackbody radiation is very fast – 10^{-14} seconds, or about one hundred million millionth of a second. The radiation only travels a tiny distance before it strikes another water molecule which absorbs it, thereby heating. So the heat energy is spread through the air.

The result is that the air heats up, it expands and becomes lighter than the surrounding air, so it rises higher in the atmosphere by **convection.** The heat is transferred higher in the atmosphere in the same way as explained already in chapter 3 including Figure 25 *if* the air has enough water vapour in it to go through the cycles of condensation, subsequent release of heat energy, cloud formation, eventual loss of the heat energy into the stratosphere and on to space as already described.

If the air has only a small amount of water vapour then it may rise by **convection** all the way to the stratosphere where any water vapour within it will turn into ice crystals – as the temperature there is too low – minus 50° C – for water vapour or liquid water to exist as already explained. The contained heat in this air from the surface originally is then radiated into space freely because the stratosphere has no water to block it.

The effect of all this is for water vapour in the atmosphere to block most of the direct radiation of heat energy from the land to space as would occur in completely dry air. Some direct radiation does occur, however, because of what is termed the **thermal window.**

Figure 9 shows that the absorption of heat radiation by water molecules decreases markedly from 9 microns wavelength to 10 microns, then increases gradually to become 100% by 20 microns. Heat radiation from about 9 microns to around 13 microns has a relatively free passage through water molecules. As said already the other gases in the atmosphere, including the trace gases of carbon dioxide and methane, have no effect from 9 to 13 microns either. So heat radiation from the land surface in this wavelength band is free to pass straight through to space and escape. This is called the *Thermal Window.*

Opinions vary as how much heat radiation from the land surface makes it through this window compared to the much greater proportion which is absorbed by water vapour and ultimately transferred to the stratosphere and space by *convection* with *water phase changes* as explained above. The range is from 2% to 17%. It varies depending on the amount of water vapour present obviously, maybe on average is around 7%.

It is clear that most of the heat radiation leaving the land surface must do so by initially heating the nearby low atmosphere because of the water vapour in this air, then the lighter heated air rising by convection, often involving water phase changes and cloud formation as discussed, eventually reaching the stratosphere where the heat energy can radiate away into space.

This is a much *slower* process than *direct radiation,* which occurs at the speed of light, 300,000 kilometers per second, to space. Because it is slower it results in heat energy build up and higher temperatures on the surface.

It is considered that this slowing effect increases the surface temperature by 33°C compared to if it did not exist. Instead of the average temperature on Earth being 14°C it would be minus 19°C. At this temperature Earth would be a frozen ice ball, arguably.

But this argument is fatuous because of all the factors already discussed concerning the Earth – its distance from the Sun, length of daily rotation, allowing the existence of large amount of liquid water, which in turn is the great regulator dictating temperature on our planet.

The effect is usually termed *the greenhouse effect.* We are all aware that greenhouses usually made of glass roofs and walls are much warmer than their surroundings, allowing crops, plants needing warmer environments to flourish.

But there is an error in naming the slowing of heat moving away from the Earth's surface *the greenhouse effect.* A greenhouse works because the glass roof and walls prevent warmer air rising out of the building, they trap it inside. **Convection,** the major method for transfer of heat energy in liquids and gases as explained in chapter 1, cannot occur from a greenhouse to outside. But convection is very able to occur in our atmosphere. It is much faster *radiation* which is blocked by this effect.

Transfer of Heat Energy from Warmer Equatorial Regions to Colder Polar Regions:

There is a further major effect of the Earth being a rotating ball with a great amount of water. That is the inevitable transfer of heat energy from the warmer Equatorial regions to the colder Polar ones. This evens out the surface temperature of both areas considerably.

Figure 18 above shows the difference between the Sun's heat energy arriving at the regions near the Equator compared with those near the Poles. Figure 18 is copied below for convenience.

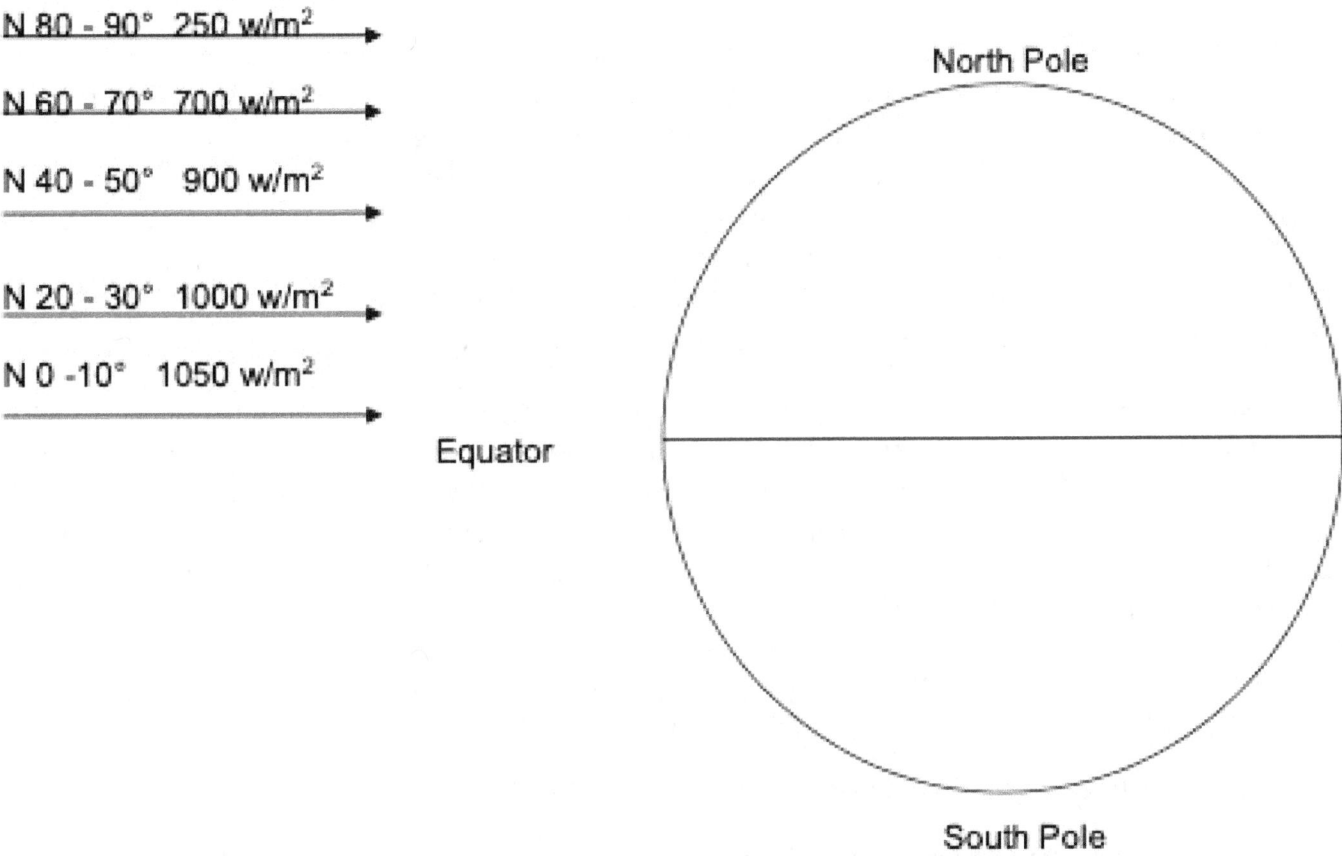

Figure 18: showing the Sun's radiation power reaching the Earth's surface at various latitudes at midday on a clear day at the Equinox when the Sun is directly above the Equator. As one goes away from the Equator and closer to the Pole the Sun appears lower in the sky at midday, shadows are longer, and the Sun's radiation is spread over a bigger area and is less powerful.

The differences between Equatorial regions and Polar ones are much greater than shown in Figure 18 because of the *albedo,* or reflection of Sun energy, once it has arrived at the surface. The Polar regions have a great deal of ice and snow with albedo of up to 90%, meaning that instead of 250 watts per square meter arriving finally to heat the surface, as little as 25 watts per square meter arrives and is absorbed by the surface. And in the Equatorial regions much of the surface is ocean, which has a very low *albedo,* around 6%, meaning that over 900 watts per square meter finally is absorbed by the surface there.

The greater amount of heat energy arriving close to the Equator is transferred to colder Polar regions by the following process:

1: The heat energy from the Sun arriving at Equatorial regions results in considerable convection of warmer air vigorously upwards in the troposphere with a lot of water vapour from evaporation of the ocean. The sea surface temperature here is about 27° C.

2: As the warmer moist air rises in expands into a much greater volume (Boyles Law above – the atmospheric pressure decreases with height) and so it pushes north and south of the Equator, towards the Poles.

3: This air then loses its heat energy to the stratosphere and space, cools, as a result decreases its volume, becomes heavier and sinks towards the surface. It reaches the surface at about 30° North and South of the Equator. It causes the air pressure at that area to increase, becomes a "High". This air then at the surface is drier and warmer than other air there, and is associated with fine weather. In fact the hottest temperatures on the planet occurs at about 30° north and south of the Equator due to this hot dry air descending with little water vapour to assist cooling by evaporation. Places such as central Australia, the Sahara Desert can reach 50° C because of the dry hot air originally from the Equator. Whereas the Equator area is cooled by the evaporation process and air temperatures there are milder, around 25 to 30° C as a result.

The descending air is made to *rotate* by the Earth being a rotating ball in what is called the **Coriolis Effect.** This results in this air rotating clockwise when viewed from above north of the Equator, and anticlockwise south of the Equator. In Australia and New Zealand these are called **anticyclones** and are associated with warm sunny dry weather, and winds which go anticlockwise around the centre of the anticyclone. The sea level atmospheric pressure is higher in anticyclones, sometimes up 1030 hectopascals. For this reason they are called **high pressure systems.**

In the northern hemisphere these high pressure systems rotate *clockwise,* that is surface winds go around the high pressure area clockwise. Again they feature fine sunny dry weather.

The usual sea surface temperature at these latitudes is about 16° C.

The air rising at the Equator has to be replaced by air moving in from north and south at the surface. This movement of surface air towards the Equator is altered from being directly from the North or South because of the Earth being a rotating ball, and so the air flows from the northeast or southeast and is called the **trade winds.**

The whole process of hot moist Equatorial air rising, then shifting away from the Equator, then cooling and falling to Earth at about 30° latitude north and south, then the trade winds completing the circuit back to the Equator – is called the **Hadley Cell.**

4: The process then repeats itself. Warmer moist air above the oceans at latitude 30° then rises, and is forced to head towards the Poles by the Hadley Cell air moving away from the Equator. This air in turn loses heat energy when it gets higher by condensation, cloud formation as described before. It cools, shrinks and becomes denser, so sinks towards the surface reaching it at about 60° north and south latitude. This cycle of air rising at 30° latitude, then falling to 60° latitude is called the *Ferrel Cell.*

5: The final step of transfer of heat energy from the hotter Equator to the cold Polar regions is completed by another cell – the *Polar Cell.* The sea surface temperature at 60° latitude is about 5° C. But *comparatively* warmer moister air here rises, then moves towards the Poles, loses its heat energy high in the troposphere, then falls towards the Poles completing another cell cycle. The Artic Ocean at the North Pole has a sea surface temperature of about minus 2° C.

This process where vast amounts of heat energy are transferred from hotter regions near the Equator to colder ones close to the Poles is shown in Figure 26.

This movement of heat energy away from equatorial regions considerably evens out the temperatures across the planet, making areas far from the Equator much warmer than they would be.

Once the heat energy gets to Polar regions it is more easily radiated away to space, cooling the planet. Factors helping this include:

- The *troposphere* is much thinner here, only 8 kilometers thick compared to 18 at the Equator. This means that the *stratosphere* is much closer to the surface, and this has no water in it to block free radiation of heat energy into space.

- The great *albedo,* reflection, of the ice and snow here means more heat energy is lost into space.

- It is so cold here that the troposphere air has much lower amounts of water vapour, liquid water as droplets in clouds, or ice crystals in clouds – the atmosphere is *freeze dried* by comparison with warmer regions. This means that heat energy can more freely radiate away into space without the blocking effect of water.

Antarctica also has a different very cold severe wind called a *katabatic wind.* The name is derived from a Greek word *katabaino* which means to go down. Figure 28 shows cold dry air descending onto Antarctica close to the South Pole which is on the high Antarctic Plateau at 2800 meters above sea level. The cold air is heavy, therefore sinks and flows downhill across the continent, into valleys, down slopes, eventually reaching the sea as a powerful freezing cold wind at temperature down to minus 80° C, and wind speeds up to 320 kilometers per hour.

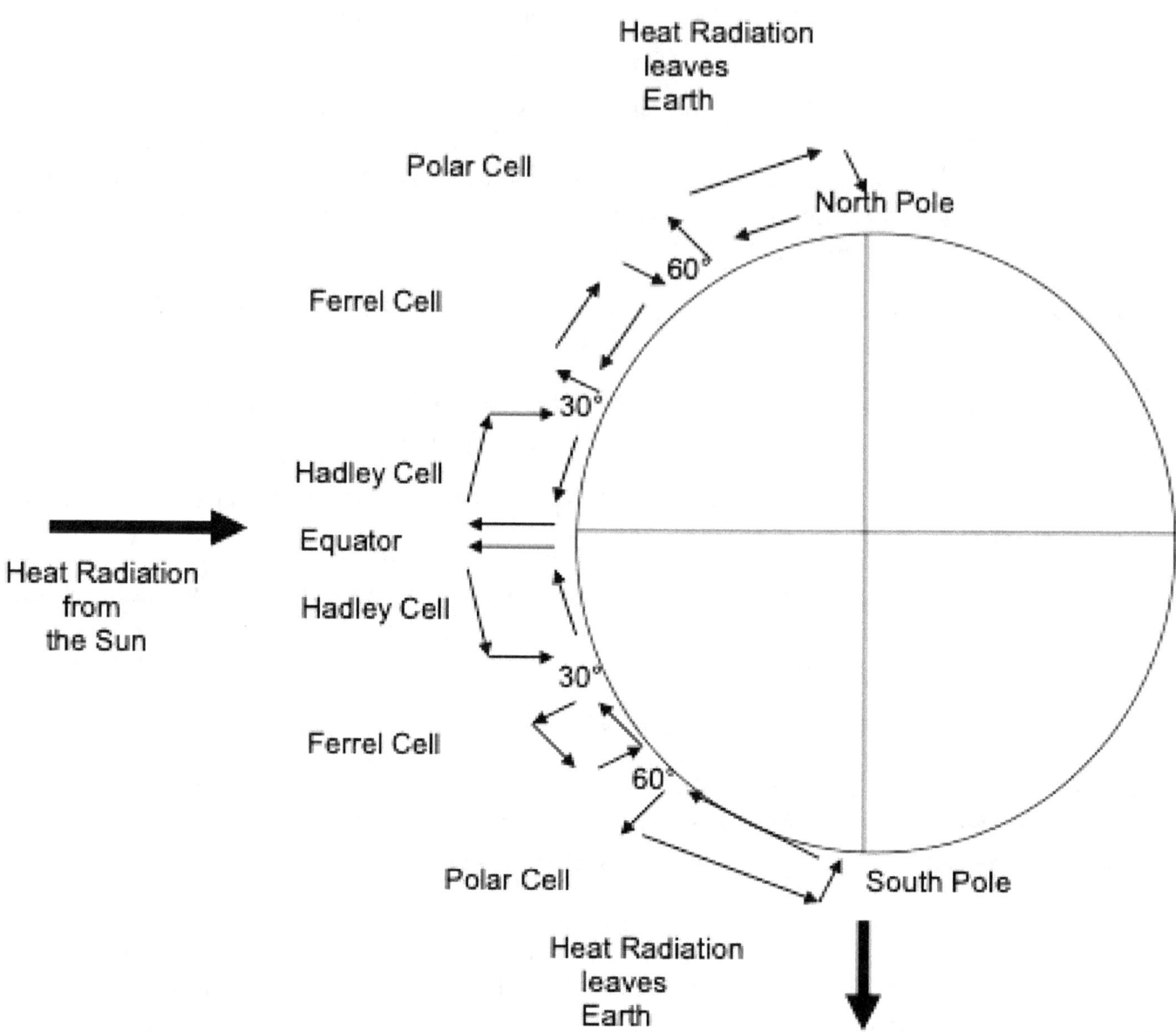

Figure 26: showing that much more heat energy from the Sun arrives at the Equator and this is transferred to the cold Polar regions by passing through the Hadley, Ferrel and Polar cells. At night time the same cells operate but less intensely especially at the Equator.

As the Antarctic Katabatic air flows downhill it compresses due to higher atmospheric pressure lower down and this makes it a little warmer, like pumping air in a bicycle pump, and following **Boyles Law** discussed previously. This is the opposite process to air cooling as it rises, expands due to lower pressure with height, discussed often before.

Similar katabatic winds occur in many places where there is high land with cold air above it, which then flows down powerfully. These include the Bora winds in the Adriatic Sea, the Oroshi winds in Japan, and the Barber winds on the west coast of the South Island of New Zealand.

The formation of the easterly trade winds in the tropics as part of the Hadley Cell has already been discussed. As has the rotation of the high pressure air systems at about 30° north and south of the Equator due to the *Coriolis Effect.*

The Coriolis Effect is due to the ball shaped Earth rotating and causing forces which are a combination of centrifugal forces due to the rotation. Gaspard-Gustave Coriolis, a French mathematician, in 1835 worked out that a cannonball fired towards the north pole in the northern hemisphere would deviate to the right, or eastwards, because of the Earth rotating ball effect. The same deviation occurs for air moving *north* away from the Equator in the Hadley cell causing the high pressure to be spinning *clockwise* when it descends to the surface as a high pressure fine area at 30° North.

In the southern hemisphere the cannonball would also be deviated to the left, also to the east, as is the Hadley cell area moving south from the Equator, causing the high pressure system when it falls to the surface at 30° South to rotate *anticlockwise.*

If the cannonball was fired from the South Pole towards the Equator the same force would cause it to deviate to the right, to the east. This means that air moving away from Antarctica as the surface part of the Polar cell northwards curves to the east and spins *clockwise.* These clockwise rotating very cold masses of air form *depressions* containing cold moist air, the moisture picked up as they pass over the Southern Ocean which has a sea surface temperature close to 0° C. The surface atmospheric pressure in these depressions can drop as low as 970 hectopascals. The clockwise rotating cold air forms strong *westerly* winds along the northern edge of the depressions, the centres of which are usually about 60° South. They are continually forming and marching from west to east around Earth in the Southern Ocean.

In the Southern Hemisphere, as mentioned above, the high pressure *anticyclones* with their winds rotating anticlockwise around their centres, which are at about 30° South, have *westerly* winds also along their southern borders. These, combined with the westerly winds caused by the depressions further south, form the belt of strong westerly winds which circle the Earth at about 40° South – the *roaring forties.*

In the Northern Hemisphere the equivalent weather high pressure areas at 30° North, low pressure cold depressions at 60° North, and strong westerly winds at 40° North, all form too but the direction of rotations is reversed – because the *Coriolis effect* works the opposite way as explained before – so the high pressure areas have clockwise winds, the depressions have anticlockwise. The strong westerly winds are weaker because so much of the Northern Hemisphere is land, not ocean.

The description above is the situation when the Sun is directly above the Equator at the *equinoxes* of March and September shown in Figure 21 on page 41. In the Southern summer everything moves about 20° to the south, The belt of fine warm weather anticyclones moves to about 45° South, giving us in Australia and New Zealand easterly winds along the northern edge of the anticyclones. And the cold depressions move further south too, with their cold wet weather.

In the winter the warm fine anticyclones move to about 15° North, and the cold depressions to about 45° South where they have a greater influence on our weather.

The reverse processes occur in the Northern Hemisphere.

Other Ways Water Vapour from Evaporation at the Surface Rises to Lose Heat Energy Higher into Space:

So far the rising of moist air laden with water vapour from evaporation of the oceans and land surface water by ***convection*** driven by surface temperature has been discussed, ultimately resulting in massive heat energy loss from Earth as explained.

There are other situations where surface air is forced upwards, resulting in similar loss of heat energy. They are:

1: Mountain ranges, high hills, rising land results in air flowing across them rising causing the same loss of heat energy by condensation, cloud formation, rain, radiation of heat energy to space as explained for convection. This is commonly seen on the windward side of coastal land, for example the western side of New Zealand facing the Tasman Sea, a very wet cold place. The western side of British Columbia and the western British coasts are others.

2: Masses of cold heavy air from depressions closer to the Poles as discussed above push under warmer lighter air to the north in the Southern Hemisphere and South in the Northern Hemisphere, pushing the warmer air higher. This results in it cooling, and the same processes of condensation, cloud formation, rain, heat loss into space as already explained for convection.

3: Opposing winds may collide forcing the air to move upwards resulting in the same processes. This commonly happens in weather situations.

4: The troposphere is naturally turbulent and disturbed, mostly because of its high content of water as explained, and also because its temperature decreases rapidly with height from an average 14° C at the surface to minus 50° C at its top. This makes convection much more powerful and increases turbulence. In turn this means that air is often shooting upwards as part of the turbulence, resulting in condensation, cloud formation, rain, heat loss also.

Violent Ways Upward Movement of Lower Air Occurs:

Thunderstorms:

Thunderstorms are caused by violent upward movement of moist air. The upward movement is due to convection, but made severe by a lot of heat at the surface, more commonly the ocean, or converging winds resulting in upward air movement, or land forms such as hills, mountains giving it a big push upwards.

The result is a big updraft of moist air sometimes climbing at 100 kilometers per hour or more. This in turn results in sudden cooling, condensation, freezing of water to ice, cloud formation. The clouds are towering very high ones called ***Cumulo-Nimbus.*** The violent updraft often pushes the tops of the clouds into the lower stratosphere where they can be dangerous for aircraft flying there because of the risk of very gusty winds.

The rapid updraft and resultant cooling frequently cause hail, including large dangerous hailstones up to 9 centimeters across crashing to Earth damaging roofs, cars, people. Other effects including lightning due to static electricity caused by the violent air movement, the formation of water spouts and tornadoes, violent surface winds up to 130 km per hour caused by air rushing in to replace that moving upwards, and acid rain caused by nitric acid formation from Nitrogen gas in the air.

The average thunderstorm, about 15 kilometers across, results in about 500 million kilograms, or 500,000 metric tons, of water vapour shifting upwards violently, condensing, losing its heat as explained ultimately to space. That is about 300 million kilowatt hours of heat energy transferred from the surface to space. They are a spectacular method of cooling the surface.

Worldwide there are about 16 million thunderstorms each year. An estimate of the heat energy lost from Earth as a result is therefore 3.8×10^{15} kilowatt hours, or 38 thousand million million kilowatt hours, about 30 times the total energy creation and use by humans, and about one hundredth of the annual heat loss from Earth by evaporation and other water phase changes as discussed before.

Thunderstorms have not changed in their frequency over many decades of scientific observation.

Cyclones, Typhoons, Hurricanes:

These great circular storms form due to high temperature of surface ocean water which must usually be over 26° C. This results in the familiar story of high evaporation rates to form water vapour, then vigorous convection upwards combined with the Coriolis effect explained above to produce a violent rotating storm featuring very high winds up to 280 kilometers per hour, intense condensation, cloud and rain formation, all of which can be very damaging especially to small islands in tropical oceans, or tropical coastal regions.

In the southern hemisphere these storms rotate clockwise and in the northern hemisphere anticlockwise. In the ocean off Australia, New Zealand and the nearby Pacific Islands they are called *cyclones;* in the Caribbean Sea and off the Eastern seaboard of North America *Hurricanes;* and in Asian oceans *Typhoons.* But they are all essentially the same. Especially they all need a big supply of heat energy from the warm ocean to keep going. As soon as they move over waters cooler than 26° C, or over land where their supply of heat energy decreases, they lose the power which drives them and subside to lower energy *tropical depressions* with gentler winds.

Like thunderstorms these storms rise high in the troposphere, as high as 15 kilometers. They do not usually get into the stratosphere.

They are another massive violent way for the Earth to lose heat energy from the surface to space. Cyclones may use up to 6×10^{11} kilowatts, or six hundred billion kilowatts of energy. A powerful cyclone may result in 100 billion kilowatt hours of heat energy are transferred from the ocean to space. This understandably cools the ocean considerably.

The number of cyclones occurring affecting Australia seems to have decreased over the last 40 years. There is no evidence that this type of violent storm is increasing in frequency worldwide.

References and Further Reading for Chapter 4: Water.

www.usgs.gov/special-topic/water-science-school/science/how-much-water-there-earth

www.en.wikipedia.org/wiki/Extrterrestial_liquid_water

https://moon.nasa.gov/news/155/theres-water-on-the-moon/

www.waterencyclopedia.com/Re-St/Sea-Water-Freezing-of.html '

https://marsed.asu..edu/mep/atmosphere

University Physics. Sears and Zemansky, MIT and City College New York. 2nd Edition pages 329 – 333. 1955.

www.doe.virginia.gov/testing/sol/standards_docs/science/2010/lesson_plans/grade6/matter/sess_6-6abc.pdf

https://keisan.casio.com/exec/system/1224562962

https://en.wikipedia.org/wiki/atmospheric_temperature

Calculation latent heat of evaporation: Latent heat is 2256 kJ/kg water, equals 626 watt hours per kg, equals 1.6 liters water per one kWh.

Calculation latent heat of fusion: latent heat is 333 kJ/kg ice, specific gravity ice 0.92, equals 92.5 kWh / kg ice, equals 0.085 kWh/liter ice, equals one kWh melts 12 liters ice.

Calculation latent heat of sublimation: latent heat is 2838 kJ/kg ice, equals 0.788 kWh/kg ice, equals 0.72 kWh/liter ice, equals 1.4 liters per kWh.

www.usgs.gov/special-topic/water-science-school/science/how-much-water-there-on-earth?

www.pubs.acs.org/doi/10.1021/acs.jpca.8b07926

www.britannica.com>science>climate>meterology.

https://agupubs.onlinelibrary.wiley.com/doi/full/10.1002/2014RG000449

https://en.wikipedia.org/wiki/Water_vapor .

Ice: a chilling scientific forecast of a new ice age. Sir Fred Hoyle FRS. Pp 132-133.

www.Nov79.com/gbwm/ntyg.html

https://en.wikipedia.org/wiki/Horse_latitudes

https://oceanservice.noaa.gov/facts/tradewinds.html

www.bom.gov.au/weather-services/severe-weather-knowledge-centre/severthunder.shtml

https://www.nssi.noaa.gov>svrwx101>thunderstorms

www.en.wikipedia.org>wiki>thunderstorm

https://www.reseachgate.net/publication/285731520_thunderstorm_distribution_and_frequency_in_australia Y Kueshov et al 2002.

Global Precipitation and thunderstorm frequencies. Aiguo Dai; Journal of Climate; vol 14 issie 6 pages 1092-1111; 15 March 2001.

https://rmets.onlinelibrary.wiley.com/doi/full/10.1002/gdj3.75

Thunderstorm occurrence at ten sites across Great Britain over 1884 – 1993. Geoscience Data Journal; Maria Valdivieso et al; 2 August 2019

www.bom.gov.au>.....>understanding Cyclones

https://en.wikipedia.org/wiki/Effects_of_tropical_cyclones

www.bom.gov.au/cyclone/climatology/trends.shtml

www.climatlas.com/tropical/

https://en.wikipedia.org/wiki/Tropical_cyclones_by_year

https://www.csiro.au/en/research/natural-disasters/floods/faqs-on-tropical-cyclones

www.smithsonianmag.com/innovation/can-we-capture-energy-hurricane-180960750/

www.books.google.com.au (cyclone energy).

Chapter 5

The Oceans

71% of the Earth's surface is covered by the oceans. The total volume of water on the planet is 1.38×10^{18} cubic meters, or 1.38 million trillion cubic meters. Of this 96.5 % is in the oceans as salt water. Where water is located is shown below:

- Oceans 96.5%
- Ice caps, snow, glaciers 1.74%
- Groundwater 1.69%
- Ground ice and permafrost 0.02%
- Lakes 0.013%
- Rivers 0.0002%
- Soil moisture 0.001%
- Swamps 0.0008%
- Biological – plants, animals 0.0001%
- The atmosphere 0.001%

Clearly almost all the water is in the oceans.

The *average depth* of the oceans is 3.7 kilometers. The deepest is in the Pacific Ocean where the Challenger Trench is 10.9 kilometers deep. By contrast if all the water in the atmosphere was to fall to the surface it would be only 2.5 centimeters (one inch) deep.

Ocean water is very salty. About 3.5% of sea water is salt. While there are over 50 different salts present, by the most common are sodium chloride, also called common salt, and magnesium chloride making up 99% of the salt present.

The presence of the salt changes the behaviour of the ocean water considerably. One of the most profound changes is altering the *density* of water as it gets colder. Ordinary fresh water is most dense, or heavy, at 3.5°

C. As it gets colder than that it expands, gets lighter, and when it freezes it is much lighter so fresh water ice floats to the surface. We commonly see water expanding when it turns to ice and breaks containers in our freezers! And as fresh water gets warmer than 3.5° C it gets progressively lighter also.

This means that if fresh water gets very cold the water will sink to the bottom at 3.5° C, then rise towards the surface again as it gets colder still. As the water temperature in the great depths of the ocean is between 0.3 and 2° C if the oceans had just fresh water they would be quite turbulent as the colder lighter water floated up through heavier warmer layers of 3.5° C.

The salt in sea water changes this property. Salty sea water just gets denser and heavier as it gets colder right down to when it freezes to ice at minus 2.3° C, a little colder than fresh water which usually freezes at 0° C, as discussed above. This steady increase in sea water's density as it gets colder means that the coldest sea water sinks to the bottom, does not rise again through lighter warmer layers and makes the oceans less turbulent, more placid, than if they were fresh water.

When sea water gets so cold it freezes the ice is much lighter and floats as ice bergs. This is vital for our planet as explained in chapter 4.

Temperature of the Oceans:

The *average* temperature of the oceans is 3.5° C. There is a big variation from minus 2° C to plus 40°C.

The highest temperatures are near the surface. Here the water is heated by the Sun's heat radiation during daylight. As already discussed the sea surface reflects only about 6% of incoming Sun heat radiation, it absorbs 94%, so the oceans, covering 71% of the Earth's surface, absorb a massive amount of Sun heat radiation each day. As discussed already on average across the Earth's surface 6 kilowatt hours of Sun energy reach every square meter. That means that each day the oceans absorb 2.1×10^{15} kilowatt hours, or over two thousand trillion kilowatt hours of Sun heat energy. This is more than 6000 times human energy creation and usage.

The Sun heat radiation only penetrates a few centimeters into the sea before it is absorbed completely. The warm surface layer is then mixed deeper by surface turbulence, waves, surface currents. The amount and depth of mixing varies from considerable in warmer tropical waters to not at all in cold polar seas. In warmer seas the mixed layer may be 100 meters deep, a similar temperature to the surface, around 25° C in the tropics, 15° C mid-latitudes.

Below the mixed surface layer the temperature gradually falls over the next 1000 to 1500 meters depth to about 2° C. This drop is called the **Thermocline,** which means temperature slope.

Below the Thermocline the temperature is relatively constant all the way to the bottom, which, as said already, averages 3.5 kilometers down, up to 10.9 kilometers down. The temperature in these deep cold waters is between minus 2° C and 2° C.

Figure 27 shows the temperature change with depth, how this varies from summer to winter in mid latitudes, and is constant in the tropics throughout the year. The solid line on the left at 2° C shows that for Polar seas it is a constant cold temperature all the way to the bottom with no thermocline there.

What is clear is that the great bulk of the ocean is very cold, between minus 2° C and 2° C. This is so for about 90% of the planet's sea water. Only the top 100 meters, about 2%, is warmer than 10° C. The worldwide average of this thin top layer is 17° C. This is warmer than the average surface air temperature of 14° C.

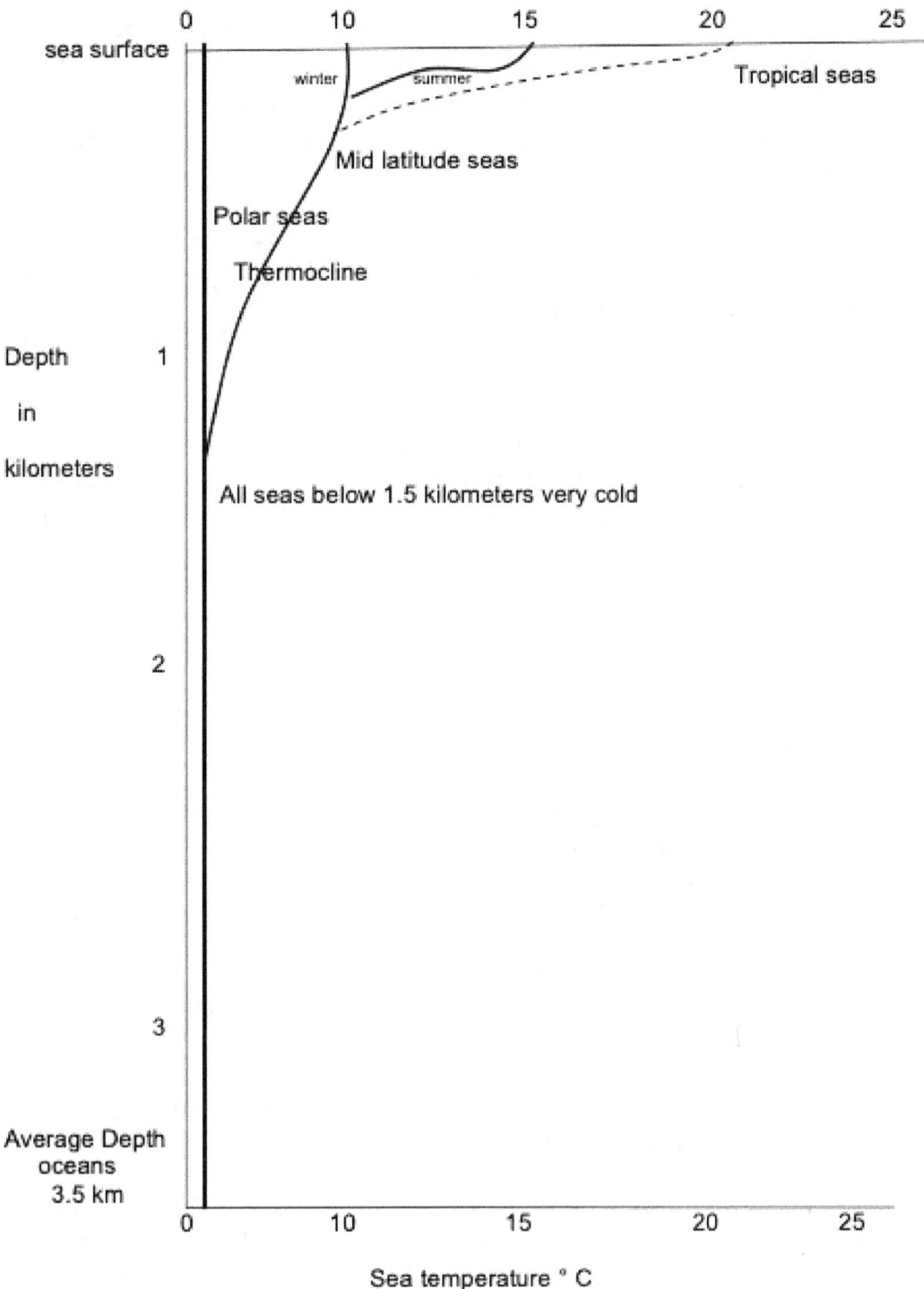

Figure 27: The Temperature of the oceans at various depths.

The source of all the very cold deep water is related, as explained before, to the presence of a large land mass at a Polar region. Before Antarctica shifted to the South Pole by plate tectonics 40 million years ago the deep ocean water was at about 12° C. Since then huge quantities of ice averaging two kilometers thick ranging up to 16 kilometers have built up on Antarctica.

At the edges of Antarctica the ice falls into the sea and melts forming very cold water which being heavier than warmer water sinks towards the bottom. This has steadily dropped the deep water temperature to 2° C. Also evaporation of water at the surface results in colder water, due to the heat energy used to form water vapour as discussed already, which sinks also. All that keeps the top 100 meters of the seas warmer than 2° C is the continuous massive flow of heat energy from the Sun.

Because the effect of the Sun's massive heat radiation is greatest at the Equator and nearby seas warm surface water formed there tends to float on the surface, build up as more forms, and push away from the Equator towards colder regions displacing colder surface waters. Because of the Earth's daily rotation and being a sphere *Coriolis* effects on these warm surface currents flowing away from the tropics cause them to mostly flow on the *western* sides of the great oceans.

An example is the Gulf Stream, which flows from the warm Caribbean Sea to Britain and Norway, making a huge difference to their climates causing them to be much warmer and moister. The Gulf Stream transfers about 76 trillion (a trillion is a million million) kilowatt hours of heat energy from the tropical Caribbean to these colder areas every day.

Other examples of warm surface currents shifting huge amounts of heat energy from tropical to colder seas are:

- The Eastern Australian Current from the tropical western Pacific Islands to New Zealand and the Eastern Australian coasts.
- The Mozambique Current and the Aquinas Current going from the Indian Ocean along the Eastern seaboard of Africa to the Southern Ocean.
- The Brazil Current from the Equatorial Atlantic Ocean to the South Atlantic along the eastern coast of South America.
- The Kuroshio Current flowing from the Equatorial Western Pacific along the coast of Japan to the North Pacific.

By these means huge quantities of heat energy are transferred from warmer tropical areas to colder Polar regions, in a similar way to the major atmospheric flows already discussed.

The reverse happens also concerning cold waters formed in Polar regions. Although many of these cold waters forming from ice dropping into the sea around Antarctica, or from heat energy loss from evaporation, cooling, and lack of Sun heat energy arriving there, will sink to great depths as already explained, some will form cold currents closer to the surface. These tend to move towards the Equator along the *Eastern* side of the Great Oceans, again because of the *Coriolis* effect. The arrival of these cold waters at warmer places has a huge influence on the climate, life forms there.

Examples of these colder currents near the surface are:

- The Western Australian Current from the South Indian Ocean along the Western coast of Australia.

- The Peru or Humboldt Current from the South Pacific/Southern Ocean along the coasts of Chile, Peru, western side of South America.

- The East Greenland and Labrador Currents from the Artic to the North Atlantic.

- The Benguela Current from the South Atlantic along the western coast of Africa to equatorial Atlantic Ocean.

By these means massive amounts of heat energy are shifted tending to even up the Earth's surface temperature.

There is also some slow movement of deep cold water driven by colder water sinking, and Coriolis forces. This very deep cold water current flows in a complex way around the oceans completing a circuit about every 300 years. While much of it involves deep cold water, parts involve warmer surface waters. This occurs because in some places cold deep water wells up towards the surface and warms from the Sun. It later cools and sinks deeper. It is called the **Great Ocean Conveyor Belt** because it is a complex circuit like a conveyor belt. It flows:

- Gulf Stream surface water cools in the cold North Atlantic, then sinks deeply and flows south to join the deep Polar current which flows around Antarctica.

- The deep cold water then goes north to the Indian and Pacific Oceans, then flows to the South West Pacific, then north along the Tonga Trench, then south again to the South Atlantic, it rises towards the surface, then goes north to the Gulf Stream again to complete the circuit.

Other names for this include the *Abyssal Circulation* and the *Thermohaline Circulation.* It moves about a million cubic meters per second of sea water, whereas the Gulf Stream shifts 30 times that per second.

There is a good map of this on Wikipedia.

In addition to these massive movements of sea water with their contained heat energy there are regular large variations in ocean climate, currents and wind. These are called *Climate Oscillations.* There are nine of these:

El Nino / Southern Oscillation: this has a major effect on the climates of Australia, New Zealand, and also America. There are three phases:

Neutral Phase: normally cold water in the *Humboldt Current* from Antarctica flows along the South American coast in the Eastern Pacific, then near the Equator turns westward driven by the easterly trade winds and warms from the Sun. As a result there is a build-up of warm water and moist conditions in the western Pacific including Australia and New Zealand. The surface air pressure is higher on the eastern side and lower on the western side of the Pacific Ocean.

El Nino, also called *the Man, or The Christ Child:* During this phase there is warming of the eastern Pacific Ocean off the coast of South America due to weakening of the usually cold **Humboldt Current.** This results in a decrease in the sea water nutrients along this coast and poorer fishing there. In addition the surface air pressure on the eastern side lowers and is higher than usual on the western side of the Pacific Ocean. The waters off Australia and New Zealand are colder and the climate much drier there, often there are severe droughts.

La Nina, also called ***the Woman,*** or ***the Girl Child,*** is the opposite phase. The cold Humboldt Current increases, the waters of the Eastern Pacific get colder by between 5 and 9° C, the fishing improves off South America. The surface air pressure is higher there and the climate cold and dry. On the western side of the Pacific including Australia and New Zealand the sea is warmer, the air pressure lower, the climate much wetter, floods often occur, thunderstorms and cyclones are more common. The low air pressure can result in the sea level being 60 centimeters higher in the western Pacific than the neutral phase.

The big variation in surface atmospheric pressure with these cycles is called the ***Southern Oscillation.*** During El Nino the pressure is higher over Australia and New Zealand, and during La Nina it is lower.

This climate oscillation causes very big changes across the Pacific and elsewhere. Just the sea surface temperature changes of up to 9° C, and sea level changes of up to 60 centimeters, illustrate this.

The other major Climate Oscillations are:

- The Pacific Decadal Oscillation.
- North Pacific Oscillation.
- North Pacific Gyre Oscillation.
- Atlantic Multidecadal Oscillation.
- North Atlantic Oscillation.
- Arctic Oscillation.
- Antarctic Oscillation.
- Indian Ocean Dipole.

They all cause big variations in climate, sea surface temperature and sea level for their regions.

The huge amounts of very cold water beneath the thin surface layer of the oceans form a massive cold reserve, with the potential to use up vast amounts of heat energy.

To raise the temperature of one cubic meter of water by 1° C it takes 1.16 kilowatt hours of heat energy (see calculation below).

This means that to raise the temperature of all of the ocean water, about 1.32×10^{18} cubic meters, by just 1° C, would take about 1.5×10^{18} kilowatt hours of heat energy. This is more than all the heat energy arriving at the Earth's surface each year from the Sun, as discussed in Chapter 3. Clearly this is not going to happen in a hurry given the huge normal losses to space.

And even if one looks at raising the temperature of just the mixed layer referred to above, about 100 meters thick, for just the mid latitudes and the tropical regions, about 1.5% only of all the ocean water, by 1° C only, some 2.2×10^{16} kilowatt hours of heat energy would be required. This is about 2% of all the Sun's heat energy arriving at the Earth's surface each year. This is 180 times the amount of energy used and created by humans each year.

The relevant question is: is it possible to raise the temperature *only of the surface of the ocean,* not the whole of it, by heat energy moving from the atmosphere? This is discussed next:

Clearly to raise the temperature of sea water, or anything else, heat energy has to be added to it in the amounts discussed above. And to **lower** the temperature heat energy is lost from the water, or anything else, to the surroundings warming them. The amount of heat energy making a change in temperature of 1° C, either warmer or colder, is called the **Specific Heat** of the substance. It varies a great deal for different substances. For liquid water it is, as said above, 1.16 kilowatt hours per cubic meter of water.

A cubic meter of water has a mass (weighs) 1000 kilograms. The same mass of dry air, 1000 kilograms, needs only 0.28 kilowatt hours of heat energy to increase its temperature by 1° C, less than a quarter the amount liquid water does. And if 1000 kilograms of dry air cools by 1° C it puts out the same amount of heat energy, 0.28 kilowatt hours, to its surrounding.

As explained already dry air is uncommon, most air near the surface in the troposphere has between 2% and 4% water vapor. Always over the oceans there is water vapour. Water vapour has a specific heat of 0.56 kilowatt hours to raise the temperature of 1000 kilograms of it by 1°C. And the same amount of heat energy is lost from it to the surrounding if it cools by 1°C.

So for normal moist air above the ocean with, say 4% water vapor, 1000 kilograms of it would contain at most 40 kilograms of water vapour. The specific heat of 1000 kilograms of this moist air is 0.3 kilowatt hours for a temperature change of 1° C.

The total mass of the Earth's atmosphere is 5×10^{18} kilograms. Of this 80% is troposphere, where all the water vapour is. This means that if the temperature of the whole troposphere dropped by 1° C it would release 1.2×10^{15} kilowatt hours of heat energy which we are trying to argue would be available to heat the surface 100 meters of the oceans as discussed above.

But to raise this 100 meter surface layer of the ocean by 1° C would take 2.2×10^{16} kilowatt hours of heat energy. This is 18 times as great as that released by the whole troposphere cooling by 1° C. So in order to increase the temperature of the surface 100 meters of the ocean by 1° C all the heat released by cooling the atmosphere by 18° C would have to be transferred to the ocean. That would mean a drop in the average air temperature of 14°C world-wide to minus 4° C, well below freezing. And as heat was lost from the troposphere in this hypothetical case its temperature would plummet, heat energy does not shift from colder parts to warmer ones, the process is not possible. Figures 28 and 29 explain this further.

Figure 28: Showing the relative amounts of heat energy needed to change temperature by 1°C. The larger sphere represents the amount needed to change the whole ocean. This is 36% greater than all the energy arriving from the Sun in a year. The middle sized sphere that to change just the top 100 meters of ocean in tropical and mid latitudes. And the small sphere shows the amount to change the whole of the troposphere, which is 80% of the whole atmosphere. To change this 100 meters by 1°C would require 18°C change in the atmosphere. It follows it is impossible to change the ocean temperature by heat from the atmosphere.

Or it could be argued that only the surface 5 centimeters of the ocean, a twentieth of the 100 meters discussed above, would change by 1° C if all the troposphere cooled by 1° C. But the slightly warmer surface 5 centimeters would quickly mix with the rest of the top 100 meters due to wave action, turbulence and currents as discussed already, the change would be meaningless.

Another possible argument is that a ***water phase change*** could occur in the troposphere to release more heat to warm the ocean. We have already discussed that huge amounts of heat energy are shifted from the surface to the high atmosphere by evaporation because it takes one kilowatt hour of heat energy to evaporate 1.6 litres of liquid water to vapour. So more heat could be released from the atmosphere if water vapour cooled back to liquid water there. But that process only occurs at very cold temperatures high in the atmosphere, and so the movement of that heat released down to the warmer ocean is not possible. The heat energy rises to the stratosphere and space as already discussed.

In addition to these considerations of the vast amounts of energy needed to raise the temperature of the ocean, and that the amount of heat energy in the atmosphere is far less than required, there is the practical matter of the ***direction*** heat energy from the surface and atmosphere moves: it virtually always goes up, not down. All the processes that have been discussed – evaporation from the sea surface, convection of warmer lighter air with contained water vapour, phase changes higher resulting in more heat energy there, and finally radiation of the heat energy away into space form the stratosphere, occur upwards. They cannot be reversed to occur downwards to the ocean. Figure 25 in chapter 4 outlines this.

The only source of enough heat energy to warm the ocean surface is the Sun. as discussed earlier in this chapter; each day 2.1×10^{15} kilowatt hours arrives from the Sun at the ocean, enough to warm the surface 5 meters by over 2° C. It is not possible to conclude significant ocean surface warming occurs from the atmosphere.

And so a key argument of climate alarmists, that a hotter atmosphere from human carbon dioxide emissions is heating the oceans, is impossible.

Therefore the oceans are the "Big Boss" controlling the Earth's surface temperature being more than 2500 times the mass of the atmosphere, shifting vast quantities of heat energy around the planet, and preventing the surface overheating by their huge reservoir of coldness, and causing massive amounts of heat energy to be lost to space by the resulting water cycles in the atmosphere explained above.

All finally due to the distance of the Earth from the Sun, its mass allowing the retention of liquid water on the surface, and its daily rotation evening its temperature as discussed above.

References and Further Reading for Chapter 5: The Oceans.

If a site does not open try a search of the subject – often it will turn up then.

www.worldatlas.com/geography/deepest-oceans-and-seas.html

www.usgs.gov/special-topic/water-science-school/science/how-much-water-there-earth?

www.web.stanford.edu/group/Urchin/mineral.html

www.ocean.stanford.edu/courses/home/chem/lecture_03.pdf

www.nature.com/suitable/knowledge/library/key-physical-variable-in-the-ocean-temperature-102805293/

Calculation ocean Sun heat energy absorption: $6 \text{ kWh/m}^2 \times (3.6 \times 10^{14})$ m² ocean area $\times 0.96$ (for albedo) $\times 24$ (hours per day) $= 4.8 \times 10^{16}$ kWh/day.

www.en.wikipedia.org/wiki/Thermocline

Ice. Sir Fred Hoyle. Hutchinson & Co 1981. Page 101.

Emiliani, C. Decline in Pacific Ocean bottom water temperature over the past 35 million years. Scientific American, Vol 198, 1958.

www.en.wikipedia.org/wiki/Deep_ocean_water

https://www.antarctica.gov.au>ice-sheet

www.theguardian.com/enviroment/2014/nov/24/antarctic-ice-thicker-survey-finds

www.earthhow.com/ocean-currents/

www.en.wkipedia.org>wiki>Gulf_Stream

Calculation heat energy for one m^3 water 1° C: specific heat water 4184 joules/kg/°C = 4.184×10^6 joules/m^3/° C divided by 3.6×10^6 joules/kWh = 1.16 kWh/m^3/° C.

Calculation heat energy from Sun versus heat needed to raise temperature oceans 1° C: 1.5×10^{18} (kWh for oceans) divided by 8×10^{17} (kWh from Sun in one year).

www.storymaps.arcgis.com/stories/d62f89509b504633a2c7516ab12d7767 (has good maps Climate Oscillations).

Calculation specific heat air: specific heat dry air 1.0035 j/g/°C = 1.0035×10^6 divided by 3.6×10^6 (joules per kilowatt hour) = 0.28 kWh/1000 kg/°C.

Calculation specific heat water vapor: specific heat water vapor 2.009 j/g/° C = 2.009×10^6 j/1000 kg/° C divided by 3.6×10^6 (joules per kWh) = 0.56 kWh/1000 kg/° C.

Calculation specific heat moist air: 40 kg water vapor needs 0.04 X 0.56 kWh = 0.0224 kWh, plus 0.28 kWh for 1000 kg dry air = 0.3 kWh/1000 kg/° C. this overestimates slightly because of the molecular weight of H2O compared to the other air gases.

https://www.britannica.com>demystified>science

Calculation heat loss by troposphere: mass troposphere 4×10^{18} kg (80% total atmosphere mass) = 4×10^{15} 1000 kg's X 0.3 kWh/1000 kg = 1.2×10^{15} kWh

Chapter 6

Ice and Snow

Ice:

There are two places on Earth with significant amounts of ice:

- Antarctica contains by far the greatest amount of ice, 3×10^{16} cubic meters, thirty thousand trillion cubic meters. The ice averages 2 kilometers thick and ranges to 16 kilometers thick.

- Greenland contains 3.7×10^{15} cubic meters of ice, about 12% that of Antarctica. The average thickness of it is 1.5 kilometers, ranging to 3 kilometers.

About 1.5% of the water on Earth is in these ice sheets. The ice has formed from ocean water by the following processes already explained in Chapter 4 on water.

- Surface ocean water evaporates, mostly in the warmer tropical and mid latitude regions, to form water vapour which rises by convection, cools and undergoes water phase changes condensing to liquid water droplets or fusing to form ice crystals in clouds, thereby shifting heat energy to the stratosphere. The air moves in the Hadley Cell, then the Ferrel Cell, then lastly the Polar Cell as explained to reach the Polar regions.

- The cold quite dry air of the Polar Cell descends – because it is cold from the edge of the troposphere here which is 8 kilometers up at minus 50° C as explained before – to the Polar regions. It still has some water which has formed snow due to the cold. The snow falls to the surface.

- In the northern polar region this is the Arctic Ocean, where the snowflakes melt on contact with the sea water to become liquid water because the ocean water is at a higher temperature than freezing which for sea water occurs at minus 2° C, colder if the sea water has more salt. This requires the snowflakes made of ice crystals to take up heat from the sea water at a rate of 1 kilowatt hour per 11 litres of water formed, the latent heat of melting as explained in Chapter 4. This cools the Arctic surface sea, which averages close to minus 2° C.

- In the southern polar region is the large continent of Antarctica, which has been there for 40 million years. The snow falls onto the continent which is quite high. Its height above sea level averages 2500 meters, the South Pole is at 2800 meters, the highest point is 4000 meters. By contrast the average height of Australia is 330 meters.

- Because Antarctica is so cold, with an average temperature inland of minus 57°C, and because snow has a very high **albedo** meaning it reflects 90% of the weak Sun heat energy arriving on it, as explained in Chapter 3, which is nothing in the long polar nights, the snow cannot melt and remains on the surface.

- Although Antarctica is comparatively dry with low precipitation of water, all as snow, on it the snow gradually builds up, gets covered with more snow which then compresses the deeper fluffy snow crystals into solid ice by squashing the air out of the snowflakes. This has resulted over the 40 million years that Antarctica has been at the South Pole in thick ice sheets averaging 2 kilometers, ranging up to 16 kilometers, thick across the continent. This ice is made of fresh water with no salt. While most of the ice is on dry land above the sea, parts bulge over the coastline and rest on the sea floor, in places 2.5 kilometers below sea level. The water here is very cold hence the ice melts very slowly, most remaining as huge ice sheets beneath the sea resting on the sea floor.

- The build-up of massive ice sheets results in the formation of glaciers, slow moving rivers of ice which move towards the lower coastal regions by gravity. When the ice reaches the sea it comes in contact with liquid sea water there at close to zero° C, but above the freezing temperature of sea water of minus 2°C obviously. So the ice melts and falls into the sea, a process called *ice calving,* a process accelerated by surface waves and surface warmer water. It takes 1 kilowatt hour of heat energy from the sea water for every 12 litres of ice which forms 11 litres of liquid water, again the latent heat of melting as explained in Chapter 4. This cools the sea water further and results in very cold water which sinks in the ocean as also explained before in Chapter 5 adding to the massive amounts of very cold deep water in the oceans.

Usually about 10^{15} litres of ice (a thousand trillion litres) falls into the sea around Antarctica each year and melts using up about 10^{14} kilowatt hours of heat energy from the nearby Southern Ocean to do so. This cooling effect is similar in quantity to all human heat energy creation and usage, and about one ten thousandth of all the heat energy arriving from the Sun each year.

There is no evidence this rate of Antarctic *ice calving* has changed significantly over the 14 thousand years since the ice age paused enabling our current warm Holocene period. There are variations.

The process is similar but smaller for Greenland. Greenland's average temperature is about minus 25° C, a little warmer than Antarctica, and some melting of surface ice does occur as well as calving. World-wide similar calving processes occur in many places where glaciers reach the sea and melt.

The Relevant Question: is the planets amount of ice, which is mostly in Antarctica and Greenland, changing significantly? The answer is that there is no conclusive evidence that it is. For example the Intergovernmental Panel on Climate Change Third Scientific Assessment (2001) reported an estimate of 2233 gigatonnes (a gigatonne is a billion tonnes, one tonne is 1000 kilograms) of ice formed each year world-wide, and about 2612 gigatonnes melted each year. But the range of error of the studies was 608 gigatonnes, meaning that no conclusion could be made.

Other studies are also uncertain. A recent one by Bamber and others published in 2018 using new satellite technology and computer modelling suggested that Antarctica might have lost an average of 109 gigatonnes of ice each year from 1995 to 2016. In assessing the significance of this Antarctica has a total 27 million trillion tonnes of ice (2.7×10^{19} tonnes), the change is less than a ten millionth.

The same authors described an annual change in the Greenland ice sheet of 247 billion tonnes. This compares to the total mass of this ice sheet of 3.4 million trillion (3.4 X 10^{18}) tonnes. The change is about a millionth.

These reported studies also have ranges of error from 8 to 50%, understandable for this type of difficult scientific work. That they are using new technology, which is clearly sensible, means it is difficult to compare their results with earlier work, which in turn failed to show a significant change. So the conclusion that there is no evidence of significant change in the Earth's ice is the only reasonable one to make.

Clearly the reported changes are very tiny compared to the huge size of the objects being studied making it even more reasonable to conclude that no significant change has been found.

Geothermal Heating is important for parts of the Antarctic and Greenland Ice sheets promoting deep melting where the sheets rest of ground. Recent studies using aircraft surveys with special sensing equipment have shown this is greater than suggested by earlier work using satellite data or ground based techniques.

In Antarctica geothermal heating affects especially the West Antarctic Ice Sheet, a huge area. There are many volcanoes under the ice and the flow of heat to the overlying ice is huge, ranging up to 1.7 million kilowatt hours per day per square meter. This clearly produces significant melting of the deep ice.

In Greenland even greater geothermal heat has been found in deep fiords beneath the northeastern coast transferring up to 2.4 million kilowatt hours per square meter per day. Greenland has many areas with hot springs, include a 28 thousand square meter area of hydrothermal vents releasing huge amounts of hot water up to 60° C. These features have only recently been assessed by surface based studies.

Geothermal heating continues day night, summer and winter. It clearly is important and will be more accurately appreciated as methods to study it improve.

Sea Ice:

Sea ice is different to these massive fresh water ice sheets resting on land. It is formed from freezing of sea water at minus 2° C. It varies from summer to winter greatly. Sea ice is located in the Arctic Ocean and on the edge of Antarctica. Sea ice contains salt, the amount varies depending on how quickly the sea water has frozen to ice. If quickly it has more salt; if slowly the salt tends to be extruded as more salty water and the resulting ice is less salty.

Modern studies of sea ice have been greatly improved by satellite imaging and computer analysis from 1978.

The Antarctic sea ice covers about 20 million square kilometers in winter, about the same size as the USA and Canada combined. The area shrinks to 4 million square kilometers in summer, about the size of India. The amount of Antarctic sea ice is increasing at about 1% per decade since these studies started in 1978.

The Arctic sea ice covers about 16 million square kilometers in winter, about the same size as Russia. In summer it shrinks to about 4 million square kilometers also. Since modern satellite studies started in 1978 Arctic sea ice has decreased by about 10% compared to earlier historical records dating from 1953, obviously pre-satellite studies. This trend has occurred for both summer and winter sea ice. The *volume* of Arctic sea ice in winter, measured by satellite imaging and computer analysis, has decreased from 18 billion cubic meters in 2005 to 15 billion in 2013.

The reasons for these changes in Arctic sea ice remain uncertain. An important factor may be the **Arctic Oscillation,** a weather system variation mentioned in the previous chapter, similar to the Southern Oscillation / El Nino system which has a big influence on the weather in Australia, New Zealand and America. In the 1950s and 1960s the Arctic Oscillation resulted in colder weather, and since 1970 warmer. The **Beauford Gyre** is a rotating ocean current system in the Arctic Ocean which can shift warmer water into it. More windy conditions, which have been present in some recent years, discourage sea ice accumulation. Like the Earth's climate overall the situation is very complex, and making quick simple conclusions regarding causes of trends is not possible sensibly.

A major limiting factor when drawing conclusions from modern studies is that they are very limited in the time of study, going back only to 1953, and for satellite based studies, to 1978. That is tiny in geological terms, as is all of human history. We know that the Arctic region was very much warmer from 950 AD to 1250 AD during the Medieval Warm Period mentioned in Chapter 3. Then the Vikings lived comfortably in Greenland and Iceland, and sailed the seas there unimpeded by large amounts of sea ice.

It is interesting to consider the amount of heat energy which would be required to melt these huge amounts of ice:

For the Arctic sea ice at the end of summer, about 10 trillion litres, about a trillion kilowatt hours of heat energy would be needed to melt it to sea water.

For the Greenland ice sheet, about 3.7×10^{18} (3.7 million trillion) litres, about 3.5×10^{17} kilowatt hours would be needed, all the heat energy arriving on Earth from the Sun in 4 months.

For the Antarctic ice sheet, about 3×10^{19} litres (30 million trillion) about 3×10^{18} kilowatt hours (three million trillion) would be needed. This is all the Sun's energy arriving in almost 3 years.

This underlines the conclusion that these ice regions are far from fragile and not going to disappear anytime soon. If they were the cooling effect on the oceans would be immense illustrating they are part of the negative feedback temperature control system for the planet, a huge reserve cooling mechanism. Concerning the idea they can be significantly melted from atmospheric heating the same arguments discussed in the previous chapter for the oceans in principle apply here making it an impossible hypothesis. Speaking practically only a tectonic plate shift of Antarctica away from the South Pole could melt the ice and result in the ocean temperature rising, perhaps to 12°C average as it was 40 million years ago before Antarctica moved to the Pole.

Snow:

Snow covers about 11% of the planet's surface in the northern hemisphere winter and this drops to 6% in the northern summer. Virtually all the snow in the southern hemisphere is located on Antarctica, as so much of that hemisphere is ocean. Far more of the northern hemisphere is land allowing for the snow cover to vary there from winter to summer.

In winter in the northern hemisphere about 46 million square kilometers, 18% of the area of the whole hemisphere, are covered. This drops to about 18 million square kilometers in summer, 7%.

Snow covers virtually all of Antarctica all year round because the continent is so cold there is no melting with summer.

The amount of the Earth covered with snow has varied but not changed overall in modern times. Some recent years have had more snow than the average since modern studies began.

Snow has a high *albedo,* fresh clean snow reflects 90% of the Sun's heat energy occurring in the wavelengths 0.4 to 3.2 microns. It is not possible for clean snow to be melted by that. But snow, and ice too, do absorb the longer infrared heat radiation from water and gases in the atmosphere well, wavelengths from 4 to 90 microns. And if the air above snow is significantly warmer than zero °C there will be heat transfer to the snow by conduction, as well as by radiation.

So warm air is the most powerful factor melting snow, except where the snow falls into the sea which immediately melts it.

In practice summer melting of snow occurs in places where the air temperature gets warm enough to shift enough infrared heat energy into the snow surface. This is usually in mid and higher latitude places and mountains with warm summer air temperatures.

Some special examples of this include places where warm winds are caused by water vapour condensing into liquid water droplets releasing the *latent heat of condensation* referred to already, a kilowatt hour for every 1.6 litres formed. This considerably warms the air which then melts snow quickly. Examples are winds in the European Alps where they are called *Fohn Winds*. Another is winds warmed by Gulf Stream waters flowing over Britain affecting snow melt.

Snow melt in Antarctica is tiny because the air is so cold, average temperature in summer on the coast is around minus 10° C, inland minus 40°C. There is therefore little infrared radiation and heat energy to transfer to the snow and melt it. So it packs down as discussed in the description about ice to form the massive ice sheet.

For Greenland, where the average summer air temperature is around 5° C, more melting occurs in summer, not in winter. The warmest summer in recent times was 2003 when it averaged 6° C.

Snow is an example of a *positive feedback mechanism* concerning surface temperature. If it gets colder, more snow forms over a larger area, that reflects more Sun heat energy from the surface, it tends to get colder still. And if it gets warmer less snow forms, more Sun heat energy is absorbed, the surface gets warmer. In practice this is more than counteracted by the more powerful *negative feedback mechanism* of liquid water evaporating to vapour as discussed in the chapter on water. But it is considered to be a factor in causing ice ages, as well as the formation of high altitude ice crystal clouds, another positive feedback mechanism discussed in Chapter 4.

Glaciers:

There are more than 400,000 glaciers on Earth, many of which are in Antarctica or Greenland associated with the massive ice sheets there. 215,000 are elsewhere and the total volume of these is 160,000 cubic kilometers, which is 160 trillion cubic meters (1.6×10^{14} cubic meters). The total mass of all the ice in these glaciers not in Antarctica or Greenland is 140,000 trillion tonnes (1.4×10^{17} tonnes - a tonne is 1000 kilograms). So each glacier averages about 700 billion tonnes of ice. But all the ice in these glaciers is a small fraction of the ice in the polar ice sheets, about one two hundredth.

Glacial thickness ranges from 50 meters to 500 meters for most glaciers, many are about 300 meters. That means there is immense weight under the glacier where it rests on ground. That in turn often results in the ice melting to water there due to a lowering of the usual freezing point of water of 0° C due to great pressure. This phenomenon is called **Regelation.**

Regelation was discovered by Michael Faraday, the famous English physicist, in 1850. At, for example, at 1300 times the usual sea level air pressure 1300 hectopascals, that is 650 thousand hectopascals, the freezing point drops to minus 4° C. It can be thought of as the ice adapting to pressure by shrinking, as water expands when it turns to ice, 11 litres of water making 12 litres of ice as already discussed. And it shrinks turning back to water.

It is important for glaciers because the water formed under the glacier lubricates its flow downhill over the ground, usually rock, below. In lower warmer altitudes melting of the glacier starts underneath it, and rivers, streams, flow from beneath it. The same process occurs under ice skates making them very slippery.

Geothermal Heat is important also for many glaciers, especially in geologically thermally active regions. Iceland is such a place and geological heat flow beneath the seven large glaciers there is up to 7.2 kilowatt hours per square meter per day, a significant melting heat for the glaciers. There are many other similar areas around the world. And it may be that lower level background geothermal heat is important. It is a field far from exhaustively studied, and there is much uncertainty about the accuracy of results. Studies of Canadian glaciers, for example by traditional hot water borehole techniques have shown a warmer zone at the bottom of the glaciers due to geothermal heat from the underlying rock. It is possible, even likely, that this is an important factor in glacial change.

A feature of ice is that heat energy takes a long time to travel through it, it is a good *insulator*. Heat applied at the top of the Antarctic ice sheet may take 10,000 years to reach the bottom 2 kilometers down by *conduction* through the ice. This is the principle of Eskimo ice houses, igloos, which can be warm inside with a fire in the dark long cold winter despite being made of ice. Snow caves used by mountaineers in emergencies are similar. This means that warm air applied to the surface of ice takes time to melt the surface first, then the next surface layer emerging after melt. The warmth does not quickly penetrate the ice and melt it all.

It also means that *geothermal* heat applied to the bottom of a glacier tends to be concentrated there and warm the nearby ice encouraging melting; if ice were not a good insulator this heat would spread quickly through the ice and have less effect.

Clean ice has a high *albedo,* like clean snow, and reflects 90% of the Sun's heat energy reaching its surface. Further, any Sun heat energy, wavelength from 0.4 to 3.2 microns, penetrates deeply into the ice which is transparent to it. It goes in about 10 meters before being absorbed. This heat energy is not absorbed at the surface. That means that the heat energy is spread into a large volume of ice and is not enough to melt it, or even change its temperature significantly. This property if ice is important for the stability of large ice areas, including glaciers. If the Sun's energy was absorbed in a few surface millimeters it would heat it considerably and melt it. But it does not.

These properties mean that glaciers need a very large amount of heat energy to melt, and they only find this if they flow downhill, aided by their huge weight and the formation of lubricating water underneath them as explained by *regelation,* to much warmer lower altitude places where the air temperature and the ground temperature on which they rest are much warmer than freezing; or they fall into the sea.

The melting of ice at lower warmer altitudes is aided by it, like snow, absorbing infrared heat blackbody heat radiation with wavelengths from 4 to 90 microns, which it, like snow, readily does.

There have been many studies of trends in glaciers over the last 70 years especially, since 1978 using satellite data. But only a very small number, about 1%, have had studies of their thickness, volume and mass. Most of these are located in Europe especially Scandinavia where the interested scientists and universities are; there are very few studies from the wilder parts of the world.

Some available studies have suggested a trend of shrinkage of studied glaciers from about 1970, especially in the last 20 years. It is suggested that about 300 billion tonnes of ice is being lost from all the glaciers outside of the polar ice sheets each year. That is a tiny fraction of the mass of these glaciers, about one 400,000th.

Forty one have been studied carefully since 1970 by the World Glacier Monitoring Service. The overall trend is that the glaciers are shrinking averaging about 25 meters decrease in thickness over the 50 years. They report the changes in meters of water equivalent, m.w.e, per square meter of glacial ice surface. This is the same as a tonne (1000 kilograms) of ice per square meter. Commonly the change is about a meter m.w.e. per year for these glaciers. In some regions, however, notably the Asian area including the Himalayas, some are increasing. To put this in perspective many glaciers are very thick, 300 to 500 meters thick.

In addition to the uncertainty created by only a very small proportion, 1%, of the glaciers having been studied, there is more uncertainty concerning errors in the *methods* of study.

The most accurate methods are based on surface assessment of the shape and thickness of the glacier. The thickness is most difficult, requiring hot water bore hole drilling, or radio-echography, or radar gravimetry, or seismic methods, all of which require work on the surface itself. This has only been done on a very small number of glaciers, and then only from about 1969 in most cases.

To get around these practical difficulties methods using satellite imaging to detail the surface shape, then computer modelling to estimate thickness, have been applied. The modelling is very complex, requiring estimates of ice flow patterns in the glacier from its surface shape. Farinotti and Huss are amongst those developing this, and initially tried it on four Swiss glaciers. They then went on to apply it to all 215, 000 glaciers in the Randolph Glacier Inventory outside of the great ice sheets, which did include some at the edge of the Antarctic sheet. The study used data from satellites concerning the surface shape and height of the glaciers, especially the Advanced Spaceborne Thermal Emission Reflection Radiometer (ASTER DEM). This does not work well in the Antarctic where the Radarsat Antarctic Mapping Project (Ramp) was used, another satellite. The computer modelling was complex, about 5 different models were used.

They concluded that, compared to previous estimates, glaciers in Asia including the Himalayas had shrunk by 27%; those in the Arctic region by 17%; and those in the Andes South America by 20%. The Antarctic periphery glaciers they studied seemed to have increased by 24%. These estimates compare to those in the late 1970s using earlier methods obviously. The authors point to many uncertainties and potential errors in the estimates.

An illustration of that error is found in a study by Robatel et al of the Argentiere Glacier in France. They used traditional surface based survey methods of ice thickness and compared that to the satellite/computer modelling methods of Farinotti and Huss. They found their traditional method resulted in thicknesses up to 500 meters, whereas the satellite based method resulted in only 300 meters. Their ice volume for the glacier was double that for the method of Farinotti, 1.2 billion cubic meters versus 0.6 billion cubic meters.

Another study finding large error inherent in estimating glacier volume and change was that of all Austrian glaciers by Helfricht et al. They used aircraft based equipment surveys, and compared results of 2016 to 1969, which suggested 30% shrinkage. But they thought the error in the old and modern estimates was at least 30%, so firm conclusions that shrinkage is occurring are difficult.

The World Glacier Service advocates caution in interpreting the data so far, stating that climate change and glacier mass is complex, there are local variations in weather and glacial structure. The same reservations about the modern study being of geologically short duration expressed above also apply here.

Certainly a great deal remains unknown about the 215,000 glaciers world-wide outside of the polar ice sheets. Sensible conclusions about very long term trends are therefore not yet possible. The geologically minute time they have been scientifically studied, and small number studied in detail, 1%, underline this.

If all these glaciers were to suddenly melt into the sea the sea level would rise by 0.32 meters. That compares to 56 meters if the entire Antarctic ice sheet was to melt, illustrating the small proportion of the World's ice in these glaciers.

References and Further Reading for Chapter 6 : Ice and Snow:

If a website cannot be found from the reference try a search on the subject, it will often appear.

www.en.wikipedia.org/wiki/ice_sheet (Ice sheet sizes)

https://www.britannica.com>place>Greenland-Ice-Sheet

www.cambridge.org/core/journals/journal-of-glaciology/article/thickening-of-the-western-part-of-the-greenland-ice-sheet/423E1011DB6ED099991F7F57E83AB9E

Weidick,A. 1991. Present-day expansion of the southern part of the inland ice. Gronlands Geologies Undersogelse. Rapport 152,73 – 79. GoogleScholar

Krabill, W., Thomas, R.K., et al. 1995. Greenland ice sheet thickness changes measured by laser altimetry. Geo-phys. Res. Lett., 22(17), 2341 – 2344.

Fleming, K. Lambeck K. Constraints on the Greenland ice sheet since the last Glacial Maximun from sea level observations and glacial-rebound models. ANU Canberra 2003. Quaternary Science Reviews 23 (2004) 1053 – 1077

www.en.wikipedia.org/wiki/Climate_of_Antactica

Ice Shelf Stability. Doake, C.S.M. www.curry.eas.gatech.edu/Courses/5225/ency/Chapter10/Ency_Oceans/Ice-shelf_Stability.pdf 2001

Calculation ice calving energy Antarctica: 1000 gigatonnes normal ice calving Antarctica per year = 10^{15} kilograms = 10^{15} liters = 10^{14} kWh/year

www.britannica.com/science/glacier/Net-mass-balance

The Land Ice contribution to seas level rise during the satellite era. Bamber, J.L. et al. Environmental Research Letters, Vol 13, no. 6.

www.iopscience.iop.org/article/10.1088/1748-9326/aac2fo/meta

www.psc.apl.uw.edu/research/projects/arctic-sea-ice-volume-anomaly/

www.en.wikipedia.org/wiki/Cryosphere

www.nsidc.org/cryosphere/sotc/sea_ice.html National Snow and Ice Data Center

www.epa.gov/climate-indicators/climate-change-indicators-glaciers

www.antarcticglaciers.org/glaciers-and-climate/estimating-galcier-contribution-to-sea-level-rise/

Farinotti, D et al. A consensus estimate for the ice thickness distribution of all glaciers on Earth. Nature Geoscience https://doi.org/10.1038/s41561-019-0300-3

www.livescience.com/24168-glacier-volume-sea-level-rise.html

Huss, M et al. J geophys res. 11 Oct 2021. (Glacial changes)

www.sciencedaily.com>releases>2019/02

www.climate.nasa.gov/interactives/global-ice-viewer/#/1

Zemp, M et al. global glacier mass changes and their contributions to sea level rise from 1961 to 2016.

Ice. Sir Fred Hoyle. Page 70 -72. (ice and snow melting, Sun and infrared effects).

www.ncdc.noaa.gov/sotc/global-snow/202102

www.worlddata.info/america/greenland/climate.php (Greenland summer air temperature)

www.nature.com/articles/s43247-021-00242-3 Dziadeck, R. et al. Communication Earth and Environment, 2, 162, (2021). (Heat flow beneath West Antarctica)

www.downtoearth.org.in/news/climate-change/heat-from-the-earth-s-interior-is-melting-greenlan-s-glaciers-study-59554 Rusgaard, S. et al. High Geothermal heat flux in close proximity to the Northeast Greenland Ice Stream. Scientific Reports 8, 1344 (2018).

www.cambridge.org/core/journal-of-glaciology/article/method-to-estimate Farinotti, D. Huss, M. et al. A method to estimate the ice volume and ice thickness distribution of alpine glaciers. 8 Sept 2007. Cambridge Uni Press.

www.frontiersin.org/articles/10.3389/fearnt.2018.00112/full Robatel, A. et al. Estimation of Glacial thickness from surface mass balanced ice flow velocities: a case study in Argentiere Glacier France.

https://doi.org/10.3389/feart.2019.00068 Helfricht, K. et al. Calibrated Ice Thickness Estimate for all Glaciers in Austria. (2019).

www.agupubs.onlinelibrary.wiley.com/doi/full/10.1029/2012JK002371 Larour, E. et al. Ice flow sensitivity to geothermal heat flux of Pine Island Glacier, Antarctica.

www.glaciers.gi.alaska.edu/sites/default/files/Notes_thermodynamics_Ascwardenpdf (geothermal heating of glacier bases)

Classen, D. and Clarke, G. Basal Hot Spot on a surge type glacier. Nature, 229,481-483 (1971)

Chapter 7

Sea Level

The International Panel on Climate Change Working Group II Assessment Report 5 in 2014 (IPCC WGII AR5) pointed to sea level rise from 1901 to 1990 being 1.4 millimeters per year, increasing to more than 2 mm per year after 1970, and being greater again at 3.6 mm per year from 2006 to 2014. The total increase in sea level from 1901 being 200 mm.

The IPCC predicted a likely increase of between 430 mm to 840 mm (almost a meter) by 2100, with a steady increase in the rate of rise. They stated the level would continue to rise for centuries.

The predictions were based on satellite pulsed radar altimetry sea level measurements processed by computer modelling to produce Global Mean Sea Level (GMSL) information. Data from many world-wide land based tide measuring stations was also processed by computer modelling for the same outcome.

The satellites involved include TOPEX/Poseidon (1992 – 2006); Jason-1 (2001 – 2013); Jason-2 (2008 – 2019); Jason-3 (2016 – present); and Sentinel-6 (2020 – present).

As a result of the prediction many authorities for coastal settlements started preparing. The measures included restricting coastal development. On the Eastern seaboard of Australian, for example, Shoalhaven and Eurobodalla Shire Councils restricted new coastal development and the rights of property owners to develop their properties, resulting in billions of dollars decline in value, and considerable anxiety for residents. There are many similar examples world-wide.

Fear of the consequences of sea level rise resulting from the predictions has affected many societies, especially low lying islands and areas.

By contrast some long established very secure land based measuring stations with data examined by independent scientists have not shown similar sea level changes.

170 world-wide tide gauges with at least 60 years of data collection in a special database called the Permanent Service for Mean Sea Level (PSMSL) showed an average change of just 0.43 mm per year. PSMSL is based in Liverpool, U.K. and is part of the British Oceanography Centre. And it is considered many of these gauges could be affected by subsidence – the tide station sinking lower due to the ground they are on sinking – as this was not often measured accurately. Subsidence results in a false increase in apparent sea level measurement, like putting a ruler further into the water.

As an example the Kronstadt sea level measuring station in Finland, part of the PSML database, found that sea level rise from 1841 to 1993 was about 12 mm (from the graph given in the paper referenced below). This is a very tiny rise, less than a millimeter per year. The authors point to difficulties in the series due to the land where the instruments are sinking due to increased loads on the pier it is on, intensive draining of underground water in the area, and nearby dam construction. These all tend to increase the apparent sea level rise.

Another study by Ekman concerning suggests a ***drop*** in sea level overall averaging 3.9 mm per year from 1885 to 1984 concerning observations for Stockholm in Sweden. This may be due to the land rising due to loss of the huge weight of ice which was there before our Holocene warmer period. A similar sea level fall due to ice loss is found in Hudson Bay in Canada, Alaska, and the U.S. Northwest.

Another example of land sinking resulting in sea level rise is the eastern seaboard of the United States. This is detailed in a study by P. Huybers and others from Woods Hole Oceanographic Institution reported in the Harvard Gazette referenced below and published in Nature. The eastern seaboard is sinking significantly resulting in an apparent rise in sea level there. The sinking is considered to be due to very long term changes due to the massive Laurentide ice sheet which covered Northern North America from 95,000 to 20,000 years ago disappearing during our warm Holocene period. The weight of the ice sheet levered the adjacent eastern seaboard and Atlantic sea bed up, so when it disappeared the leverage went and sinking has resulted. It is expected to continue for thousands of years with consequent sea level rise there.

There have been several studies of land sinking along the very populated U.S eastern seaboard affecting sea level. These include one published in Nature by Harvey and others in 2021 referenced below. This finds that the land is often sinking by 2 mm per year, and some places by several millimeters where the sinking is due to underground water being used by humans, and effects of river systems. These effects explain most of the apparent sea level rise there. This region is the heartland of many climate alarmists who are not always forthright in explaining that apparent rising sea levels there are largely due to the land sinking.

Another area with subsiding land is New Zealand, where some parts, including the eastern coast of the North Island and the northern end of the South Island are subsiding at up to 6 mm per year producing a similar apparent sea level rise. This is due to the country being on the boundary of the Pacific and Australian tectonic plates with instability, earthquakes and volcanic activity.

Another study by Woodworth concerning the record at Liverpool from 1768 to 1996, the longest record in the U.K., shows an increase of just 35 centimetres in the 228 years, 1.5 mm per year. But a lot of ***oscillation*** is shown in the results, there is not a steady upward trend but variations. For example the highest reading was 514 centimeters in 1981, other highs of 508 centimeters were found in 1906, 1957, 1958, 1959. The last readings in the series after the high in 1981 were all lower including 497 in 1992 illustrating the absence of a steady upward trend.

There is more data on the PSMSL website including some good maps showing those parts of the world where the sea level is rising, and many other locations where it is falling. And good descriptions of complicating factors.

There are some very long standing secure land based tide measuring devices in Australia:

- **The *Isle of the Dead*** off Tasmania, which is very stable geologically, was visited by J.C. Ross in 1841 and he inscribed into the rock cliff face at the high tide level a deep mark. He was careful to make the mark exactly at the high tide level averaged over many calm days, he said.

In 1889 the mark was visited and assessed by a scientist, J. Shortt. He carefully studied the mark and the sea level concluding it was then 340 mm above the mean high tide sea level, implying the sea had dropped in level, or the land risen, or the mark position been incorrect, perhaps because of tide estimation difficulties in 1841, the most likely explanation. This became the new benchmark for future observation.

In 1985 B. Hamon, another scientist, studied the mark again concluding it was now 360 mm above the sea implying a drop in sea level since 1889 of 20 mm, given there was no identified rise in the rock cliff.

The mark is currently 300 mm above the sea, suggesting a rise in sea level of just 40 mm in 130 years since 1889, 0.3 mm per year, similar to the PSMSL 170 world-wide tide gauges change of 0.43 mm per year, and a marked contrast to the IPCC claim of more than 2 mm per year.

- **Fort Denison** in Sydney Harbour has had a tide measuring gauge since 1886 and the record is continuous since. This has shown an average increase in sea level of 0.65 mm per year over the 130 years, a total 84 mm apparently. A feature of the record is that it varies up and down, there are periods of a few years when it increases, then years when it decreases, it ***oscillates***. It is the overall gradual change that shows the long term apparent trend of 0.65 mm per year.

But the land the gauge is on is subsiding, sinking at a rate of 0.89 mm per year as shown by detailed geological work including Global Position System (GPS) data as well as land based survey methods. Which means there is no rise in sea level detectable in Sydney since 1886.

The key to this disagreement between the long standing reliable land based measurements and the dire IPCC prediction lies in the error inherent in satellite radar altimetry. These are considerable. Factors include accurately knowing where the satellite is, calibrating the radar equipment including for water vapour in the atmosphere which alters the readings, many other complexities. Then the data is computed to give a world-wide sea level using complex computer models, which make many assumptions. Even NASA admits the error is at least 25 mm, and says it is greater than the increases claimed. This very understandable in the circumstances.

The paper referenced below from the Ocean University of China by Xiangying Miao suggests error from 120 to 64 mm, reducible perhaps to 28 mm by special methods.

Another paper referenced below by Fu and others discusses achieving accuracies within 70 mm by special techniques.

Even the more robust GPS system which uses multiple satellite for global navigation purposes has trouble being highly accurate for land heights. It is much more difficult to accurately measure the moving surface of the sea.

The only reasonable conclusion to make is that the land based long established tide gauges, such as the PSMSL list, Fort Denison and the Isle of the Dead where subsidence is known, are currently the only believable sources of reliable sea level changes; and effectively they show little change.

Many familiar with their own coastline will intuitively agree with this.

This is perhaps remarkable when various factors affecting sea level are considered:

- The great depth of the oceans averaging 3700 meters puts into perspective a minor change of 0.4 mm per year found in the PSMSL data.

- The sea surface is very changeable due to waves, weather, storm surges, tides, major cyclones, typhoons and hurricanes causing low pressure and rises in sea level. Waves are commonly 6 meters, sometimes up to 20 meters. Storm surges can be as high as 8 meters. Tides range up to 12 meters.

- The major climate oscillations discussed before have big impacts on sea level – for example the Southern Oscillation/El Nino change may alter the level up to 600 mm.

- Tectonic plates move around the ocean at up 150 mm per year displacing the sea. The ocean floor is sinking in many places including the Arctic.

- Many coastal areas have large rivers discharging into the sea. Deposition of silts can raise the land level and sea bed.

- Undersea volcanoes are common and can pump huge quantities of lava into the ocean changing sea levels.

- Removal of massive weight on land areas results in it rebounding upwards causing sea level nearby to drop. Our Holocene warm period resulting in loss of ice mentioned above is an example, as is seasonal glacial melting.

Scientifically there are two types of sea levels:

- **Absolute Sea Level:** this is measured from the centre of the Earth.

- **Relative Sea Level:** whereas this is measured relative to a local reference, usually a stable land point, or at least one with known rates of movement up and down.

There are plenty of reasons for the level to change, so it is remarkable it is so stable.

A special case is low lying coral atolls in the Pacific where satellite/computer modelling data has caused great alarm. But recent studies have shown that these islands are mostly stable or increasing in size, only 20% are smaller. The study periods are generally from the 1940s to the present, but in the case of Tuvalu go back to 1897, making conclusions convincing and reassuring. It has been found that human developments such as removing ground water, sea wall construction, have a greater impact on island size changes than sea level change.

This reflects the knowledge that coral is very adept at adjusting to sea level change, an observation made by many scientists including Charles Darwin. If, say, old volcanoes on which coral islands form sink, the coral quickly grows to reach the surface. If they rise the coral is pushed up above the sea, dies, becomes coral sand and tends to erode away.

The recent report by the Australian government AIMS (Australian Marine Institute of Marine Science) that the northern two thirds of the Great Barrier Reef has the largest amount of coral for the whole 36 year period since observations started is another illustration of the health of coral. It can be threatened of course by many of man's activities including sea pollution, fishing methods and building activities.

14,700 years ago the current ice age paused and soon the sea had risen 130 meters to about its present level. Coral islands kept up with this big rise by growing appropriately.

All this more reliable information is clearly very gratifying to Pacific Island people, those living in low lying areas world-wide, as well as coastal land owners in wealthy countries.

An interesting feature is the *oscillation* in sea level found in reliable land based measuring systems as mentioned for Fort Denison. The sea level may rise for a few years, then fall, oscillating around a mean level. This means that short term observations over a few years may detect an apparently big change, but when evened out of many years, as with the long standing measurements in the PSMSL data base, can be put in perspective.

This marked oscillation from one year to the next in the short term is also shown in the PSMSL databases referenced below. The long term record evens this out and allows interpretation of long term changes.

Sea level, like virtually all other aspects of our life and world, turns out to be very complex with many factors influencing it in different places and globally. In some parts it is rising, others falling. It is arguably impossible to measure within a few millimetres because of all the variations each day, amongst which are tides, weather, land and sea bed movements. Those who profess ability to do so by promoting their methods, especially satellite ones with their errors as discussed, then announce dogmatically that the global level is rising by so many millimeters, then make alarmist predictions and argue for massive changes in human behaviour, which would be certain to be futile concerning sea level given the complexities, need to be approached with great caution.

References and Further Reading for Chapter 7: Sea Level:

www.ipcc.ch/working-group/wg2/

https://earthobservatory.nasa.gov/images/150192/tracking-30-years-of-sea-level-rise

www.cmar.csiro.au/sealevel/sl_data_cmar.html

https://www.epa.gov/climate-indicators/climate-change-indicators-sea-level

https://en.wikipedia.org/wiki/Sea_level_rise

Xiangying Miao et al. estimating the baseline error of wide swath altimeters using nadir altimeters via numerical simulation. Journal of Ocean University of China. 21, 681 – 693 (2022)

Fu, L.L. et al. The Surface Water and Ocean Topography Mission. Academic Press. New York 2012.

Xiaoyun Wan et al. Effects of Interferometric Radar Altimeter Errors on Marine Gravity Inversion. Sensors 2020, 20(9). 2465; https://doi.org/10.3390/s20092465

https://www.mdpi.com/1424-8220/20/9/2465/htm

https://tidesandcurrents.noaa.gov/sltrends.shtml

Bogdanov, V. I. Technical Report on levelling and sea level rise at the Russian GPS-points of BSL GPS 1993 Campaign. Helsinki. Reports of the Finnish Geodetic Institute. No. 95:2, pp. 9 – 20.

Bogdanov, V.I. et al. Mean Monthly Series of Sea Level Observations (1777 – 1993) at the Kronstadt Gauge. Reports of the Finnish Geodetic Institute. ISBN 951-711-237-8. ISSN 0355-1962. 2000.

Ekman, M. The World's longest continued series of sea level observations. Pure and Applied Geophysics. Vol. 127, pp. 73 – 77.

Woodworth, P.L. et al. High waters at Liverpool since 1768: the U.K.'s longest sea level record. Geophysical Research Letters. 26 (11), 1589 – 1592.

www.psmsl.org

www.psml.org/data/longrecords/ReportsFGI_2000_1.pdf

https://news.harvard.edu/gazette/story/2019/02/study-of-sea-level-rise-finds-land-sinking-along-east-coast/

Harvey, T.C. et al. Ocean Mass, sterodynamic effects, and vertical land motion largely explain US coast relative sea level rise. Nature, Communications Earth and Environment, 2 article 233(2021) https://www.nature.com/articles/s43247-021-00300-w

www.sciendo.com/pdf/10.1515/quageo-2015-0003 Parke, A. The Isle of the Dead Benchmark, the Sydney Fort Denison Tide Gauge and the IPCC AR5 Chapter 13 Sea Levels Revisited. Quaestiones Geophricae 34(1) 2015.

www.researchgate.net/publication/280914977_Destruction_or_persistence_of_coral_islands_in_the_face_of_20th_and_21st_century_sea_level_rise McLean, R. and Kench, P. Wiley Periodcals. Inc. Vol. 6, 6 September/October 2015.

Darwin, C.R. On certain areas of elevation and subsidence in the Pacific and Indian oceans, as deduced from the study of coral formations. Proceedings of the Geological Society of London. No. 2, pp. 552 – 554 (1842).

www.en.wikipedia.org>wiki>The_Structure_and_Distribution_of_Coral_Reefs

https://www.aims.gov.au/monitoring-great-barrier-reef/gbr-condition-summary-2021-22

Chapter 8

Carbon Dioxide

Properties and Characteristics:

Carbon Dioxide molecules have a single Carbon atom centrally and two oxygen atoms attached to it on opposite sides by strong bonds. The three atoms are in a straight line, the molecule is symmetrical. Because it is made of three big atoms CO2 gas is heavier than air, about 1.5 times the weight of air. By contrast water vapour molecules have one heavy oxygen molecule and two very light hydrogen ones.

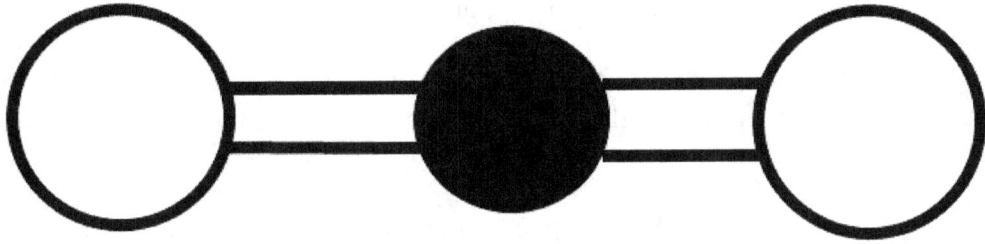

Figure 29: a Carbon Dioxide molecule.

On Earth CO2 is always present just as a gas. The gas is clear, colourless and has no smell. At sea level atmospheric pressure, 1013 hectopascals as discussed in Chapter 4, CO2 turns straight into solid CO2 when cooled to minus 78° C. this is called **dry ice.**

CO2 exists in liquid form only at very high pressures above 5179 hectopascals, about five times the atmospheric pressure at sea level here. It exists as a liquid at pressures above that for temperature from minus 56° C to 31° C. Above 31° C it can only exist as a gas no matter how high the pressure is.

Of the other planets Venus has an atmosphere which is 96% CO2 . It is thought that if Earth has been 5 million kilometers closer to the Sun it would have been similar as liquid water could not have existed here.

Mars has also about 96% CO2 in its atmosphere but it is very thin and very cold, down to minus 140°C in winter. **Dry ice,** solid CO2 , forms at the poles then like snow.

No other Solar System planets have any CO2 but some moons of other planets do. These are Titan, the largest moon of Saturn, and Enceladus, the sixth largest. The CO2 has been detected in trace quantities in their atmospheres.

Early Earth History:

CO2 is formed from Carbon atoms and Oxygen atoms combining in a chemical reaction which releases large amounts of heat energy. It usually needs a temperature of around 500° C to start. When the Earth formed 4.5 billion years ago these temperatures were greatly exceeded. The very early Earth was essentially a very hot rock with many volcanoes. The volcanoes put out huge quantities of CO2 made by Carbon and Oxygen combining readily in this environment. Hydrogen and Oxygen also combined to form large amounts of water, and Nitrogen and Hydrogen combined to form Ammonia, NH3 . These reactions, like the formation of CO2 , release large amounts of heat energy.

It is inevitable that when Carbon, Oxygen, Hydrogen, Nitrogen, all present during Earth's early formation, are together in a hot environment they will form CO2 , water and ammonia. In the process free oxygen is used up completely.

These gases were spewed out to form the atmosphere, then the water vapour condensed to form the oceans. Ammonia is very soluble in water and dissolved in the ocean, as did most of the CO2 which is also very soluble. The atmosphere at this time, from about 4 billion years to 2.5 billion years ago, a period called the **Archaen,** was composed of nitrogen mostly, with large amounts of CO2 ranging from 5% to 70% in various studies, smaller amounts of methane (CH4), about 5%, even cyanide gas, about 0.01%.

There was virtually no oxygen, and only tiny amounts of hydrogen, which had been the most common gas when Earth first formed 4.5 billion years ago. All the oxygen had been used up as said and the hydrogen left over from water formation disappeared into space being the lightest gas and therefore not held back by gravity. There is uncertainty about the temperature of the surface and atmosphere at this time but it was certainly above freezing as the oceans were liquid, it may have been as high as 50° C. It is thought there was a great deal of evaporation from the oceans and rain, as well as ongoing volcanic eruptions. In addition there was ongoing bombardment from huge asteroids striking the planet probably up to about 3.5 billion years ago as the Solar System tidied up into the current eight planets with their moons. One of these strikes formed the Moon.

Early life started about 4 billion years ago it is now thought. The ocean soup of water, ammonia, dissolved CO2 and methane promoted micro-organisms which did not need oxygen, **anaerobic** organisms. When these died they sank to the ocean floor forming muddy soils. Much of the CO2 formed carbonate solid salts which also sank forming rocks such as limestone on the seafloor.

As said oxygen, essential for life today, was practically absent from the Earth's atmosphere then. Clearly there were huge quantities locked up in water, CO2 , and rocks.

Photosynthesis:

2.5 billion years ago there was a massive revolution on Earth. Quite suddenly huge numbers of **Blue-Green Algae,** also called **Cyanobacteria,** proliferated in the oceans. These contained cells containing **Chlorophyll.** This is a complex molecule made up of 55 carbon atoms, about 70 hydrogen, 4 nitrogen, 6 oxygen and a magnesium atom. It is a **hydrocarbon,** and has a ring structure and a tail. Similar hydrocarbons are found in **Haemoglobin,** the protein which transports oxygen in our red blood cells.

By a complex process called **Photosynthesis** chlorophyll in these cells takes in CO2 and water, H2O, and turns them into oxygen (O2) and **Carbohydrates.** These are strings of carbon atoms with hydrogen atoms and oxygen atoms attached to them. Usually there are two hydrogen atoms and one oxygen atom for each carbon atom. An example is **glucose,** a molecule of which is C6H12O6 . In order to make one molecule of glucose 6 molecules of CO2 and 6 of water H2O are combined using Sun energy in the plant cell containing chlorophyll to make the glucose, and 6 molecules of oxygen O2 are released.

6 X CO2 and 6 X H2O goes to 1 X C6 H12 O6 (glucose) and 6 X O2 (oxygen)

It is as though all the oxygen in the carbon dioxide was released into the air and the carbon added to water to make glucose, and many other carbohydrates.

Carbohydrates are the building blocks for all plants including their wood which is made of **cellulose,** a long complicated carbohydrate. And as all animals either eat plants for food, or eat other animals which have in turn fed on plants, **photosynthesis** is the foundation of almost all life on Earth. In addition because oil, petroleum gas and coal are derived from old buried plants it is the source of all fuels of this type used by humans.

The energy for this process comes from the Sun. The chlorophyll absorbs Sun radiation in the visible part of the electromagnetic spectrum which as said in Chapter 1 is from 0.4 to 0.7 microns. The absorption is shown in Figure 30:

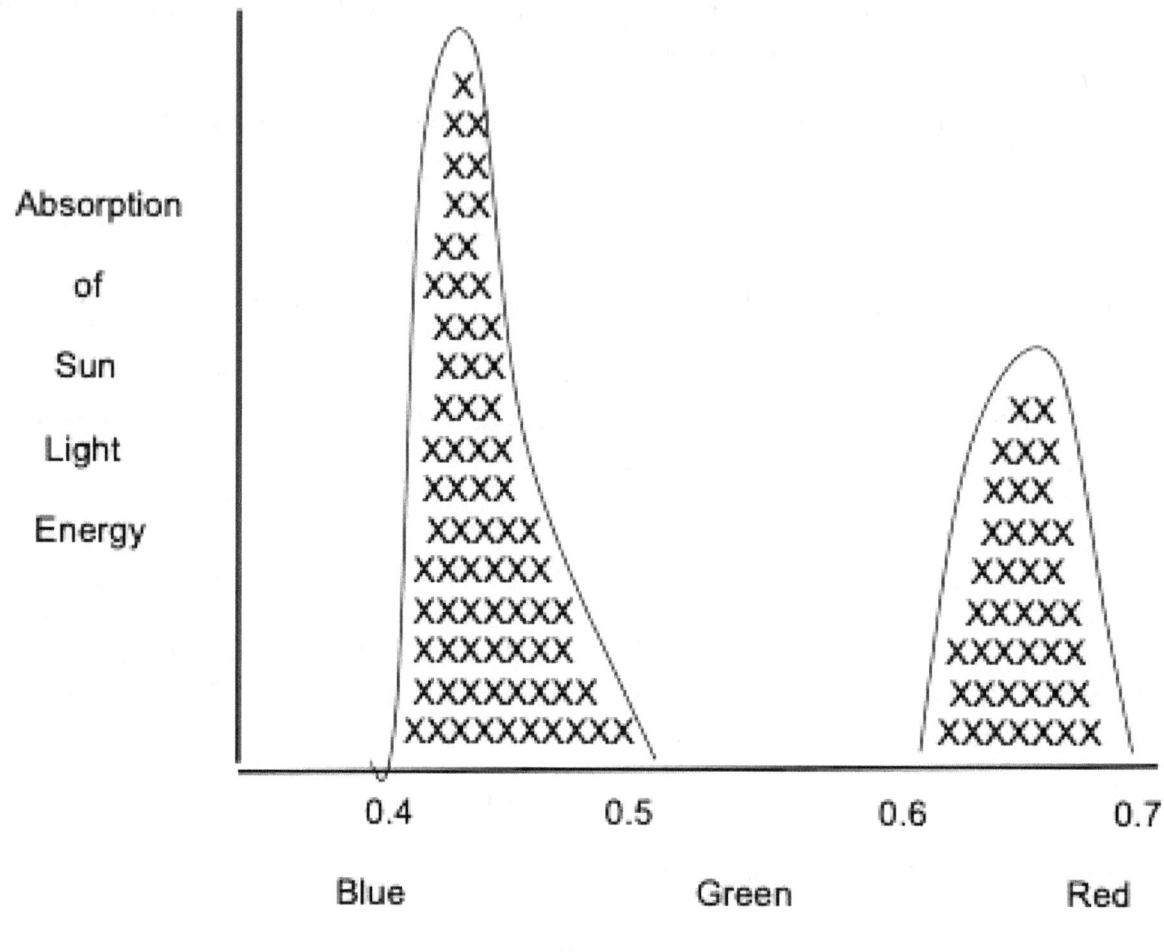

Figure 30: the wavelengths of visible sunlight absorbed by Chlorophyll plant cells.

As is shown the chlorophyll absorbs Sun energy in two bands, one from 0.4 to 0.5 microns, blue visible light, and the other from 0.6 to 0.7 microns, red light. It does not absorb in the green part of the light wavelengths, so this is reflected back and plants appear green to us.

While chlorophyll in primitive microbe cells first developed about 3.5 billion years ago the process took off massively 2.5 billion years ago with the rapid proliferation of Blue-green algae. The very high amounts of CO_2 in the warm oceans provided them with huge food supplies. They ate this voraciously and put out oxygen as a waste product. This resulted in the atmosphere and ocean rapidly gaining huge quantities of oxygen. This is called the **Great Oxygen Event.**

The Blue-green algae multiplied so much they ate virtually all the CO2 in the ocean. More flowed in from the atmosphere and they ate all that too until there was virtually none left when most of them starved and died. They gradually recovered and became the dominant life form for another 1.5 billion years, in the process creating an atmosphere similar to ours: about 20% oxygen, 79% nitrogen, and less than 1% carbon dioxide.

The new oxygen rich atmosphere allowed more complex organisms to evolve as it gave them far more energy to use than the previous **anaerobic** (non-oxygen using) microbes had. This led to multicellular organisms, plants and animals, commencing about 700 million years ago.

Free oxygen is virtually unique to Earth in the Solar System. Tiny amounts have been found in the high atmospheres of Europa, a moon of Jupiter, and Rhea, a moon of Saturn. This oxygen was made by high energy radiation striking water molecules in the high atmosphere and breaking them up into hydrogen and oxygen.

It is clear that **Photosynthesis** is an immensely powerful process using the Sun's energy to transform Earth's environment and resulting in the layer of life today, the **biosphere.**

It is only present on Earth in the Solar System, and has not so far been found elsewhere in the studied universe.

Each year global photosynthesis uses about 8×10^{11} (800 billion) tonnes of CO2 and produces about 5×10^{11} (500 billion) tonnes of oxygen. In the process it uses about 2×10^{16} kilowatt hours of Sun energy, about a 50th of all the Sun's energy arriving on Earth and 200 times the annual human usage. About two thirds of all photosynthesis occurs in the surface layers of the oceans in tiny algae called **phytoplankton** and one third in land plants. Most of the oxygen produced in the oceans is used by ocean animals and plants for their respiration, but it can escape to the atmosphere freely. And some of the CO2 used by phytoplankton in the oceans comes from ocean sources and some from the atmosphere, as it did freely in ancient times. There is uncertainty about these proportions.

Over the billions of years since photosynthesis first started large amounts of dead plant material rich in **hydrocarbons**, which are derived from carbohydrates minus most of the oxygen, have been deposited initially on the surface, then going deeper with more soils deposited on them. These form the world's fossil fuels, coal, oil and petroleum gas. About 1.5×10^{16} kilograms of CO2 (15 million billion kilograms) would have been required to make this using photosynthesis. As such it is a huge store of past Sun energy.

If atmospheric CO2 fall below 0.015% (150 parts per million) photosynthesis stops. If that happened plants, including ocean algae, and animals would suffer massive starvation. Conversely if CO2 increases photosynthesis does also, processing more CO2 to make more oxygen, and resulting in plant life proliferating markedly followed by increasing animal life. CO2 levels of 0.06% (600 parts per million) increase photosynthesis by 40% compared to CO2 levels of 0.03% (300 parts per million). Photosynthesis is therefore another **negative feedback mechanism.** But it is weak compared to ocean effects to be discussed next which dominate CO2 levels. This is shown by atmospheric CO2 levels dropping markedly with ice ages despite photosynthesis slowing, and rising in warm periods despite plant activity increasing.

Carbon Dioxide and the Oceans:

Carbon Dioxide is quite soluble in water. We know this from our everyday use of carbonated drinks, sparkling wines, beer, soda syphons all of which use CO_2 gas dissolved in the drink. The amount dissolved in water including the sea depends on three properties of the environment:

1: **Temperature of the water, sea:**

The colder the water is the more CO_2 dissolves in it, and the warmer it is less dissolves. Seawater at close to freezing, 0° C, dissolves 1.05 milligrams of CO_2 per litre; at 10° C, 0.96 mg per litre; at 20° C, 0.7 mg/L; at 30° C, 0.58 mg/L; and at 40° C, 0.38 mg/L. The same occurs with warm champagne when the CO_2 tries to escape.

This means that cold polar seas at close to freezing hold double the amount of CO_2 that warm tropical waters at close to 30°C do. And it means that the vast amounts of deep cold water in the oceans just above freezing hold huge amounts of CO_2.

When the colder deep waters move in the great ocean currents to the surface in warmer areas they lose the carbon dioxide to the atmosphere. And when warmer waters move to colder areas they take up CO_2 from the air.

Similarly if the Earth generally becomes colder, such as during an ice age, the oceans dissolve much more CO_2 causing the atmosphere to lose it into the sea, and dropping the atmospheric level down. It has been thought to drop to as low as 0.017%, 170 parts per million, during ice ages.

And during warm periods massive amounts of dissolved CO_2 are vented from the sea to the atmosphere raising the level there. In such periods the Earth's atmosphere has been up to 9 - 14° C warmer on average than 14° C average temperature now. And the oceans have been much warmer, indeed were 10° C warmer in the deep ocean 40 million years ago, 12° C compared to 2° C now.

2: **Sea Pressure:**

The *pressure* in the sea increases markedly with depth, as anyone who swims knows. For every 10 meters depth the pressure goes up by an amount equal to sea level pressure, say 1000 hectopascals. So for deep cold ocean water at 3000 meters the pressure is 300 times sea level air pressure, 300,000 hectopascals. The amount of CO_2 able to be dissolved increases due to pressure increase quite rapidly down to about 1000 meters, for example it has doubled at 10 meters. Below that it increases only slightly. This effect has been present ever since the Earth's oceans and carbon dioxide formed more than 4 billion years ago, therefore is very stable, unchanging. It clearly increases the capacity of the oceans to dissolve CO_2.

3: **Sea Saltiness:**

The salt content of sea water does decrease the amount of CO_2 dissolved by about 5 to 8%, more if the water is colder than if it is warmer, compared to the amount that can dissolve in fresh water. At the usual saltiness of the oceans, 3.5%, this is a small factor only.

Currently about 1.4×10^{17} kilograms of carbon dioxide is dissolved in the oceans, compared to 2.7×10^{15} kilograms in the atmosphere. The oceans hold more than fifty times as much CO2 as the air.

It was discussed above in Chapter 4 that the deep sea temperature has dropped by 10° C in the last 40 million years, largely because of the shifting of Antarctica to the South Pole then. This change would result in the oceans taking up about 10% more CO_2, or five times all the CO2 in the atmosphere at any one time.

The huge amount of carbon dioxide dissolved in the oceans and the temperature of the ocean determines the amount in the atmosphere. To say the amount dissolved in the oceans is determined by the amount in the atmosphere is to put the cart before the horse because of the huge amount in the oceans. For a given sea temperature the ocean will take up a certain amount of CO_2. At the surface some of the dissolved CO2 has enough heat energy – temperature – vibration of molecules - to jump out of the liquid sea into the air above. As more CO2 builds up in the air above some molecules loose heat energy and drop back into the sea again. For each different sea surface temperature a balance point is reached where the amount leaving the sea equals the amount going back in. This is the ***partial pressure of CO2*** for that ocean surface temperature. It is similar to the balance reached for water and water vapour discussed in Chapter 4.

The partial pressure of a gas above a liquid in which the gas is dissolved is measured by ***Henry's Law.*** This simply states that the partial pressure is proportional to the amount dissolved for a particular temperature. To enable the partial pressure to be calculated if one knows the amount dissolved, or the reverse, that is the amount dissolved to be calculated if one knows the partial pressure, there is number called ***Henry's Constant*** for that temperature applied. It equals the partial pressure divided by the amount dissolved. The number is higher for higher liquid temperatures and lower for lower temperatures, reflecting the increased amount dissolved at lower temperatures.

Henry's Constant (called K_H) = partial pressure gas above liquid (called P_i) divided by Amount gas dissolved per amount liquid

So to work out the partial pressure of a gas above a liquid in which it is dissolved:

Partial pressure = amount dissolved X Henry's constant for that temperature

Concerning the oceans which have been around for about 4 billion years, as has CO_2, the total amount of CO2 dissolved has had all that time to stabilise and so is very constant, and, as discussed, is a huge amount, 50 times the amount in the atmosphere. This means that practically speaking for the Earth the partial pressure of CO2 depends on the sea temperature. If the sea temperature goes up, less CO2 can be dissolved and the partial pressure in the atmosphere rises; if the sea temperature goes down, more CO2 is dissolved and the partial pressure decreases. And the partial pressure for that sea temperature is measured by Henry's Law.

This means that the concentration of carbon dioxide in the air is dominated by the sea temperature. The horse is before the cart. If atmospheric CO2 rises above the partial pressure for that temperature the sea quickly takes in more to bring it back to the balance point; and if CO2 falls below the partial pressure for that sea temperature CO2 vents quickly from the sea to increase it back to balance. Not only is the sea the ***big boss*** of atmospheric carbon dioxide concentrations, but it is the massive ***negative feedback mechanism*** to control changes.

In addition to this reasoning based on the huge amount of CO2 contained in the oceans there is now further strong evidence that the temperature of the oceans determines atmospheric CO2 levels; *not* the other way around, changes in atmospheric carbon dioxide determining ocean and surface temperatures. For example:

- Lui and other scientists studied the timing of ice volume changes and CO2 atmospheric changes when the present ice age paused 14,000 years ago: they found that the ice volume decreased up to 500 years before the CO2 increased, concluding that the CO2 change *followed* the ice melting, and did not cause it.

- Petit and other scientists studied cores from the deep ice of Antarctica at the Vostok Russian Base at Princess Elizabeth Land, inland Antarctica, which has the lowest recorded temperature on the planet so far at minus 89.2°C. They particularly studied the last 1000 years concluding that CO2 atmospheric levels clearly followed temperature changes not the other way around. Other studies of the ice cores have shown the same thing.

- Roe and other scientists studied various records of the present ice age, which started 34 million years ago, especially its latest phase, the Quaternary glaciation which started 2.5 million years ago. They also concluded there was clear evidence that ice volume changes were *followed* by CO2 changes, not the other way around.

- In a very interesting modern study Demetrios Koutsoyiannis and Zbigniew Kundzewicz from the University of Athens using data from NOAA and NASA satellites, the Mauna Loa Observatory high on a volcano in Hawaii, the Barrow Observatory in Alaska, the South Pole, many land CO2 and temperature measuring stations in the world-wide CRUTEM and Global Air Sampling Network, as well as the Hadley Centre and East Anglia Climatic Research Centre in the UK, did a detailed mathematical analysis on the question of: which came first? The temperature change or the CO2 change? Or, as they put it, the Hen or the Egg? They only looked at modern data from 1980 to 2019. They concluded that temperature change was first and CO2 change followed about 6 months to a year later. Their study was in part stimulated by observations that the Covid pandemic lockdowns, resulting a drop in human CO2 production of up to 17% world-wide, did not result in any expected (by some) drop in CO2 atmospheric levels. This is in keeping with CO2 as a follower, not a leader, for global temperature.

Another interesting study was performed by Ernst-Georg Beck, a Biology and Biotechnology Professor at Freiburg, Germany, published in 2008. He studied historical records of CO2 atmospheric measurements from 1812 to 1970, finding 134 he considered reasonably accurate. Amongst his conclusions was that it was more likely CO2 levels followed temperature change that the reverse. He also found many records of much higher atmospheric CO2 levels than claimed for the period by others – up to 470 parts per million in 1830 for example. This made his study contentious for others especially as it negated the idea that CO2 levels were low then and only rose during the 20th century. This work is further discussed later.

The Nature of Dissolved Carbon Dioxide in the Oceans:

More than 99% of CO2 dissolved in the oceans is simply as molecules of CO2 in amongst the water molecules. But a very small amount reacts with the water molecules to form hydrogen ions (H^+, same as protons) and either bicarbonate ions (HCO_3^-) or carbonate ions (CO_3^-).

0.12%, one in every thousand molecules of CO2 dissolved in seawater, does this. This mixture of hydrogen and bicarbonate, carbonate ions is called **Carbonic Acid.** It is a very weak acid compared to say, sulphuric acid or hydrochloric acid which can burn skin on contact severely.

In seawater the carbonate ions formed, CO_3^-, combine with calcium ions and magnesium ions also dissolved in the sea to form salts, calcium carbonate and magnesium carbonate, if the concentrations of the ions are high enough to make this happen. At lower concentrations they remain as separate ions. The upper limit they can remain separate is called *saturation concentration* for those ions.

This is the same process as for many other ions dissolved in sea water, for example sodium (Na^+) ions and chloride (Cl^-) ions combine when they reach saturation concentrations to form common salt.

Magnesium (Mg^{++}) and calcium (Ca^{++}) ions are amongst the most common ions in seawater, about 1300 and 400 ions per million molecules in the water (0.13% and 0.04%). Seawater is **saturated** with these ions, it cannot take anymore without magnesium carbonate and calcium carbonate salt forming. The salt formed drops to the sea floor and helps form *limestone* rocks. In the process carbonate, originally from dissolved CO2, is removed from the sea and changed into limestone rock. This decreases the amount of carbonate ions back to less than saturated levels, thereby decreasing the carbonic acid level and reducing the weak acid effect.

Another way carbonate ions are decreased in the ocean is by living creatures, especially molluscs and coral which are profuse. They convert calcium and carbonate ions into their hard shells and coral formation, the hard limestone like materials called **Aragonite** and **Calcite.** On their death they fall to the sea floor helping to form more limestone. The process also removes carbonate ions, and therefore dissolved CO2, from the ocean. These creatures have been around for 500 million years and have converted a huge amount of carbon dioxide into limestone and other carbonate rocks over that time.

And going back further the **Blue-green algae** which proliferated 2.5 billion years ago also caused calcium carbonate to form solid platforms called Stromatolites acting as bases for further proliferation. These ultimately formed rocks and used up more CO2. .

These processes help regulate the amount of CO2 dissolved in sea water, and hence the carbonic acid level controlling any acidity due to this weak acid. They are another powerful *negative feedback mechanism* controlling ocean, and ultimately atmospheric CO2 . And as said above the great bulk, over 99%, of CO2 in the oceans is dissolved as intact molecules and has no effect on acidity. Only one in every thousand CO2 molecules reacts with water, H2O, to form H^+ and HCO_3^-, Carbonic Acid, by ionization. As a result the oceans are slightly alkaline, not acid.

Over the 2.5 billion years since Blue-Green algae and later shell and coral forming animals have been converting CO2 to **sedimentary** carbonate rocks about 10^{20} kilograms of carbon dioxide has been so processed (100 billion billion kilograms). This is 130 thousand times as much as is currently in the atmosphere, and 2500 times as much as is currently dissolved in the ocean.

This is not completely a one way street in that some of the carbonate rocks get recycled back to the oceans and atmosphere by volcanic eruptions, and also by venting of molten rock from deep ocean rifts at the boundaries of tectonic plates. Each year about 2×10^{11} kilograms of CO2 is returned this way. But this is equally counterbalanced by some tectonic plates pushing under their neighbour, called *subduction*, which buries 200 billion (2×10^{11}) kilograms of CO2 in the form of carbonate rocks each year.

To make us seem even smaller these huge quantities of carbon containing substances, carbon dioxide, carbonate rocks and hydrocarbon fossil fuels in the Earth's atmosphere, oceans and crust are dwarfed by all the carbon substances deeper in; the mantle and core contain about 3×10^{22} kilograms of carbon in substances including carbides and diamonds. About 1000 times the amount of carbon in substances in the Earth's crust, oceans and atmosphere.

It is considered that a very long time into the future all the CO2 will have gone through the stages in the atmosphere and oceans and finished up in limestone and other rocks. Life will then be impossible.

Transport of Carbon Dioxide from the Sea to the Air, and Air to Sea:

There is now a substantial amount of information about the mechanism by which CO2 moves from the sea to the air, and air to sea. Many research groups have studied this in recent years because of great interest in the influence of oceans on atmospheric CO2.

One of the most interesting studies is by Mariana Ribas-Ribas and colleagues at the Oldenburg University, Wilhelmshaven, Germany. The used a large wind tunnel filled with natural sea water called the "Aeolotron" to study the speed CO2 moves across the sea-air boundary. The speed was measured in centimeters per hour, cm/h, and is called the "gas transfer velocity" in the study report published in 2018.

It has been shown in several previous studies that wind over the sea, and consequent wave formation, have a big effect markedly increasing the speed CO2 moves across this boundary. Because the study was in a sea water wind tunnel with powerful fans they could change the strength of the wind and resulting waves. The fans could generate winds up to 40 knots (70 kilometers per hour).The tunnel was sealed and so they could alter the concentrations of CO2 in the water and the air and then measure very accurately changes in CO2 using infrared equipment in both the water and air. From these they could compute the speed CO2 had moved across.

They were also interested in the influence of naturally occurring substances called surfactants on this speed. They introduced these into the tunnel and studied changes.

Previous studies have mostly combined observations of wind, waves, CO2 levels, temperatures done at sea and then calculated by computer modelling the CO2 transfer velocity.

The meaningful part of the transfer of CO2 across the boundary takes place in a very thin layer at the surface called the *sea surface microlayer*. This is less than a millimeter thick. It contains minute bubbles of gas, including CO2, being transferred across. It also contains micro-organisms including tiny plankton and algae. It has been shown that these actively shift CO2 across by their cell metabolism.

They found that for gentle wind speeds the transfer speed was often around 5 to 20 centimeters per hour. But for high wind speeds up to 40 knots it was often 150 to 200 centimeters per hour – 2 meters per hour.

The natural surfactants on the surface about halved the speed, but only in gentle winds; in heavy winds with waves they had little effect, presumably because they were broken up by the waves and wind.

Another very interesting study was by Takahashi and others from a number of American scientific institutions including the Scripps Institution of Oceanography, University of California in San Diego published in 1997. They studied 250,000 measurements at sea of CO_2 levels in the sea and air, temperatures which all involved sea and air sampling by 250 expeditions from 1970 to 1995. They excluded equatorial measurements and ones taken during 6 years of El Nino weather cycles because those warm waters emit big quantities of CO_2 upsetting measurements. They finally had 16,500 measurements to study from world-wide oceans. From computer analysis of all this using modelling they worked out the transfer of CO_2 across the sea air boundary globally.

They concluded that most oceans were *in equilibrium* concerning ocean and air levels of CO_2, that is there was no or little net transfer across the boundary because the partial pressure of CO_2 in the air was at a level in balance with the CO_2 in the sea. Strong wind areas were more likely so balanced than calm areas.

In warmer seas CO_2 tended to be emitted from the oceans into the atmosphere for the reasons discussed already, warmer water dissolving less CO_2 and having a higher atmospheric partial pressure. Colder Polar seas tended to absorb CO_2, again as expected from the previous discussion, colder water holding more CO_2. They calculated from their computer models that warmer seas in the Pacific, Atlantic and Indian oceans perhaps emitted 4 billion tonnes of CO_2, colder seas in the Atlantic absorbed about 5 billion tonnes, and the Pacific generally was about equal concerning amounts emitted and absorbed. They estimated that overall world-wide the oceans absorbed perhaps 4.5 billion tonnes more than they emitted of CO_2. They considered the error in this estimation could be as high as 75%. Their findings are in line with other studies. There are good diagrams in their paper online.

What these and other studies show is that huge quantities of CO_2 pass across the sea – air boundary globally quickly. While *net* transfers have been estimated it is very difficult to estimate *gross* transfers, that is the actual amounts of CO_2 shifted across both air to sea and sea to air globally. But it must be much larger than the net transfer of 4.5 billion tonnes mentioned above. One estimate even suggested 100 billion tonnes. Other estimates suggest about 90 billion tonnes annually.

A hint at how big this gross amount is can be guessed from Mariana Ribas-Ribas's paper about the Aeolotron discussed above. A maximum CO_2 transfer speed of 2 meters per hour was found at 40 knots windspeed as said. If one extrapolates this across all the world's oceans a possible estimation of 39,000 billion tonnes of CO_2 transfer emerges as a guessed maximum. Even if one takes only 1% of this, implying not 40 knots but gentle breezes world-wide, it is still 390 billion tonnes, almost 5 times greater than the estimated net balance from other studies.

Therefore modern studies point to rapid transfer of massive amounts of carbon dioxide between the oceans and atmosphere. And this fits in with our everyday experience of the speed of CO_2 fizzing out of champagne and carbonated drinks when the cork is popped decreasing the pressure in the bottle suddenly. That this is much greater if the wine or beer is warm, resulting in eruptions of froth, than if it is properly chilled is a clear example of the greater solubility of CO_2 in water at colder temperatures than warmer.

Carbon Dioxide Mixing and Distribution in the Atmosphere and Oceans:

It is widely accepted that carbon dioxide is effectively and rapidly spread globally in the atmosphere especially in the turbulent troposphere with its powerful storms, wind and rain as already discussed. There are many stations measuring CO2 levels around the world and they are all in close agreement concerning the level. For example stations Mauna Loa (Hawaii), Barrow (Alaska), the South Pole, and American Samoa are all in virtual lockstep concerning their CO2 levels. An explanation for this is that there is thorough mixing of the atmosphere world-wide. If that were not so there would be big regional differences in levels.

Similarly there is considerable mixing of the oceans as discussed Chapter 5 with massive currents both in surface and deep waters, and turbulence of the top 100 meters especially. In addition dissolved CO2 readily diffuses from higher to lower concentration areas in the sea aiding the evening out of levels. There are some very deep still regions where there may, however, be little mixing and very deep pools of concentrated CO2 may be present there. But generally there is mixing and evening out of sea CO2 within the limits set by sea temperature as discussed, higher levels for cold water, lower for warm water especially in the top 1000 meters.

And there is free and rapid transit of CO2 across the sea – air boundary as discussed.

All this means that if an intense localised increase in CO2 occurs, say from a massive volcanic eruption or huge bushfires, the increase quickly spreads elsewhere in the atmosphere and oceans, and evens out again.

The point is important as it has been stated, including by the International Panel on Climate Change (IPCC) that CO2 released into the atmosphere remains there for decades and evening out is very slow; but that cannot be so in view of this evidence for thorough mixing.

Carbon Dioxide transfers from Atmosphere to Land:

Each year *photosynthesis* takes up about 800 billion tonnes of CO2 world-wide as explained earlier in this Chapter. There is variation in estimates of how much of that occurs in land based plants ranging from 260 to 440 billion tonnes. The plants grow using carbon dioxide to make carbohydrates including sugars, wood, leaves.

Rain:

The only other natural process transferring appreciable quantities to land is *rain.* As raindrops fall through the troposphere they absorb CO2. The amount they contain as they drop follows the rules discussed before concerning **Henrys Law**, it depends on the partial pressure of CO2 in the air around the raindrop and the temperature. As the raindrop falls the partial pressure increases as the lower atmosphere is denser encouraging more CO2 to dissolve in the drop. But the temperature gets warmer, discouraging that. The net effect is a small decline in the amount of CO2 the drop has as it falls.

Rain therefore transfers CO2 gas from the atmosphere to the surface dissolved in water drops. Some of it will end up back in the atmosphere again through evaporation but a lot will remain in liquid water in rivers, lakes, swamps, ponds, and some will run back into the sea. Rain that falls on the sea clearly goes into it.

An estimate of the amount of CO2 leaving the atmosphere each year in rain was carried out by Liu and others. They estimate that about 10^{12} kilograms, or a billion tonnes, is involved. Other estimates based on the Earth's average precipitation of one meter every square meter of the Earth's surface suggest a much larger amount is possible. For example a crude estimate based on the average world precipitation of one meter and average world surface temperature of 14° C gives an estimate of 5×10^{15} kilograms, or 5000 billion tonnes.

As more knowledge is gained about this there will clearly be more certainty. It seems likely that very large amounts of CO2 are leached from the troposphere each year by rain.

Carbon Dioxide Transfers to the Atmosphere:

Volcanic Activity:

Since 1978 in instruments on satellites have been able to more accurately measure emission of gases by volcanoes worldwide and this has complemented ground and aircraft based measurements, which are collated by the Network for Observation of Volcanic and Atmospheric Change, NOVAC, a multinational scientific body.

CO2 from volcanoes can be measured by direct sampling of volcanic gases, but the satellite measurements use an indirect method. By ***spectroscopy*** the instruments measure absorption of light passing through the instrument especially which wavelengths are absorbed. The satellite instruments were designed to measure ozone atmospheric levels. The first of these, launched in 1978, was the Total Ozone Mapping Spectrometer (TOMS) superseded by the Ozone Monitoring Instrument (OMI) in 2004. They measure ozone and a gas called Sulphur Dioxide, SO2, which only gets into the atmosphere from volcanic activity and gas emissions from beneath the Earth's crust, from the ***Mantle.***

The proportion of CO2 compared to the amount of Sulphur Dioxide is known for volcanoes, and so the amount of CO2 is calculated. Many assumptions and computer models are used in working out world-wide gas emissions, and the researchers involved are the first to recognize potential errors in the results.

The US Geological Survey estimates global CO2 emissions from volcanic activity is between 0.13 and 0.44 billion tonnes each year since 1978.

Another group, Shinohara and others, in 2008 estimated 0.5 billion tonnes annually. And another, Fischer and others, in 2019 estimated up to 0.09 billion tonnes annually (93 million tonnes).

An interesting feature especially of Fischer's 2019 study is that gentle gas emission from volcanoes and their surroundings puts out over 95%, and eruptions put out less than 5% of the total gases emitted including CO2.

Gases including CO2 are also emitted from the ***Mid Ocean Ridges*** between the tectonic plates on the ocean floor. Huari and others estimated the total CO2 from these at 58 million tonnes annually, or 0.058 billion tonnes.

These are small amounts of CO2 compared to other sources. The principal effect of volcanic gas emissions on the Earth's surface temperature is due to the large amounts of sulphur dioxide emitted. This turns into collections of particles in the atmosphere which reflect and scatter sunlight, preventing it reaching the surface. This results in marked cooling after big eruptions. For example the eruption of Mount Pinatubo in the Philippines in 1991 put 20 million tonnes of sulphur dioxide into the atmosphere and resulted in world-wide surface cooling of almost 1°C, the effect lasting 3 years. Krakatoa and Tambora eruptions in 1883 and 1815 had similar cooling effects which lasted for years. And in 1783 the Laki Fissure of Iceland erupted severely putting an estimated 120 million tonnes of sulphur dioxide, six times more than Pinatubo, and exacerbated already cold times due to the Little Ice Age from 1300 to 1850 discussed in Chapter 3.

Bushfires, Wildfires:

Information on wildfires, called bushfires in Australia, is collected by the Copernicus Atmosphere Monitoring Service (CAMS) by MODIS satellites run by NASA as part of the Global Fire Assimilation System.

The MODIS satellites use a heat detecting device to scan the Earth's surface from 700 kilometers up and they cover the whole planet every 2 days. They can detect fires larger than 30 meters across reliably. From this data the amount of CO_2 produced by fires is calculated using computer modelling with its inherent assumptions.

The system has been operating since 2003. That year 8.4 billion tonnes of CO_2 was calculated as produced world-wide by wildfires. In the most recent year, 2021, 6.4 billion tonnes was estimated. A slight decline has occurred since 2003 with some years with more, 2012 and 2015, but 2003 seems to be the largest in this record.

Australia had significant bushfires in 2020. 120 million hectares burnt then compared to an average year when 30 million hectares would be burnt. Many quote estimates of 700 million tonnes (0.7 billion tonnes) of CO_2 produced by the fires that year.

Not everyone agrees with this estimation of CO_2 produced globally by wildfires of about 8 billion tonnes annually. For example, research by Christopher Johns and others of the Northern and Regional Australia Research Programme, part of Future Directions International, an independent Australian analysis group, has made the following findings from observations on Australian bushfires, then applied them world-wide:

- In Australia in an *average* year 30 million hectares would burn.

- Between 18 and 109 tonnes of CO_2 are produced for each hectare, meaning that in an average year between 0.6 and 3.6 ***billion*** tonnes of CO_2 would be produced from bushfires.

- In an *exceptional* such as 2020 120 million hectares burn. That would mean that between 2.1 and 13 ***billion*** tonnes of CO_2 would result.

Applying these rates of CO_2 production world-wide, there are on average 450 million hectares burnt. This results in estimates from 8 billion to 50 billion tonnes of CO_2 produced from global wildfires into the atmosphere.

The principal reason for these big differences between the CAMS estimates and this Australian group appears to be different opinions about how much CO_2 is produced by wildfires per hectare burnt. It seems that atmospheric CO_2 is calculated from oxygen spectroscopy in these MODIS satellite measurements and not measured directly, unlike other satellites, which creates errors. Clearly more research is needed. But the Australian group is likely to be accurate, at least for Australia, as they are well placed to do quality practical research on site, perhaps less dependent on computer modelling.

It has long been known that wildfires result in subsequent increased plant growth from ash fertilization of burnt areas. It used to be considered that virtually all the increased CO2 is taken up by photosynthesis for this over the next year. And it is known that ash falling into the ocean downwind of fires has the same effect stimulating photosynthesis in phytoplankton creating huge blooms in the sea which eat up the extra CO2 . Many consider that fires are effectively neutral concerning their ability to raise CO2 levels for this reason. Even the IPCC has promoted that view.

In addition to effects on plant regrowth fires produce a large amount of **charcoal** which is pure carbon, an excellent fuel producing large amounts of heat, as our ancestors knew for generations. The charcoal gets incorporated in soil where it fertilizes new plants, and may stay stored there for many years. It is estimated that 12% of the carbon in burnt forests ends up as charcoal stored in the soil.

What is beyond doubt is that the natural world easily copes with huge increases in CO2 production from wildfires in a short time. As commented by W. Tang, R Matear and others from the Australian Commonwealth Scientific and Industrial Research Organisation (CSIRO):

"—greater appreciation of wildfires, pyrogenic aerosols, nutrient cycling and marine photosynthesis could improve our understanding of the atmospheric CO2 and global climate systems."

Historically bushfires are part of the natural world in places, most famously Australia where the plant life often needs fire including for seed germination. For example the huge Tasmanian Mountain Ash trees only germinate seeds during rare intense fires every 5000 years. Fire has been part of the natural cycle in Australia for 15 million years.

Aboriginal First Australians used fire intelligently for thousands of years to grow crops and prevent huge conflagrations. In very recent times since European settlement there are good records of regular huge bushfires.

One of the largest fires in the geological record in Australia lasted for 30 years in the 11th century.

Land Plants Dying, Decaying and Respiration of Plants and Animals:

When plants and animals die they invaded by bacteria, fungi, earthworms, insects, lice, slugs, snails and millipedes which feed on them and turn them into *organics*, substances with a lot of complex carbon matter including humus, lignins, making up rich soil nutritious for more plants to grow on as every gardener knows.

About 80% of the carbon, originally from atmospheric CO2 , in the dead plants escapes back into the atmosphere again as CO2 but about 20% is retained long term in soil as organic matter. This has gradually resulted in large amounts of carbon being stored in soil – about 2.5×10^{15} kilograms, or 2500 billion tonnes. To make this would have required 9×10^{15} kilograms of carbon dioxide, about three times all the CO2 in the atmosphere at any one time, and about one fifteenth of the CO2 present in the oceans as explained above.

Living plants and animals use oxygen and put out CO2 in by ***respiration* to live.** There is great uncertainty about how much CO2 is produced by respiration globally. There have been many ground, sea and aircraft based studies, and satellites have provided more information in the last 20 years. But, as explained before, satellites use oxygen uptake information from spectroscopy to enable calculation of CO2 production on the land surface by plants and animals, and have very limited use for ocean work. There is extensive use of computer modelling with its assumptions. So there is much uncertainty about the results.

To quote a professor from a prestigious American university who has participated in many studies about this:

"--- the (carbon dioxide) emission scenario is beset with many difficulties."

He and others point to measurements concerning ocean CO2 being especially difficult due to huge variability from one place to another; difficulties sampling enough to make results more accurate; that many of the methods have only been in use for 20 years, a very short time geologically; and that ocean CO2 can only be measured by satellites studying atmospheric changes, very indirect. All this leads to huge errors which the researchers are the first to admit.

In addition respiration output of CO2 by animals is barely studied to date. There is one recent study on human and livestock respiration published by Cai and others. But virtually nothing on wild animal respiration, which clearly is difficult to study. A general assumption seems to be made that all animal respiration including humans, livestock and wild animals is about 1% of all respiration. But that is questioned by Cai and others who suggest it could be much higher.

There also are great difficulties measuring how much CO2 comes from respiration of living things and how much results from decay and decomposition. Complicating that is that much of decomposition is done by microorganisms, bacteria, fungi, insects and they all put out CO2 in respiration to add to the CO2 put out by the breakdown of carbohydrates (sugars, wood), proteins and fats in dead plants and animals. The total quantity is easier to estimate. Hence they are grouped together here.

Examples of global outputs of carbon dioxide from plant and animal respiration **on land** are:

- By C. Green and K. Byrne: 220 billion tonnes CO2 annually.
- By Q. Cai and others: 219 billion tonnes CO2 annually.

Concerning **Ocean respiration:**

- By Prof. Yadvindermahli, Oxford: 330 billion tonnes CO2 annually (derived from oxygen photosynthesis – see calculation in references below).

There is great uncertainty about how much of the ocean respiration CO2 is transferred to the atmosphere. Probably most stays in the ocean.

Examples of global outputs of carbon dioxide from **land decay and decomposition** including soil animal respiration studies are:

- C. Green and K. Byrne: 200 billion tonnes CO2 annually.

- By Q. Cai and others: 183 billion tonnes CO2 annually.

Concerning Ocean decay and decomposition there are very few studies. One recent one by K.S. Grace and others from Rutgers University suggested that about 37 billion tonnes of CO2 are produced each year by marine life decay, of which 16% is waste product excretion. This sinks deeper and much of it ends up as carbonate rocks as previously discussed.

Human Production of Carbon Dioxide:

This is also called *Anthropogenic Carbon Dioxide Emissions.*

Human respiration is estimated at about 0.01% of all land respiration – about 2 million tonnes of CO2 perhaps are produced by humans breathing each year. Human breath, like animals generally, contains up 40,000 parts per million CO2, 4%. There is truth in the believe that plants like being talked to!

Human use of *fossil fuels,* oil, coal and petroleum gas is more accurately measured. The consumption of these can be collated, theoretically at least, for all countries, and therefore the resulting CO2 production when they are used calculated.

Carbon Isotope Studies:

Also the CO2 resulting from fossil fuels can be estimated by sampling air and water, and measuring the amount of an *isotope* of carbon in it. Carbon atoms usually have 6 protons and 6 neutrons, but the isotope atoms have one or two extra neutrons, so instead of being C^{12} are C^{13} and C^{14}.

Carbon 13 and 14 atoms are made in the high thin atmosphere as a result of atoms of gases there, oxygen, nitrogen, argon mostly as previously explained, being shattered by high energy radiation from the Sun. This results in loose protons and neutrons flying around, and when a neutron crashes into a carbon atom in CO2 it may end up jammed in the nucleus. This makes a carbon 13 atom. About 1% of Carbon atoms are carbon 13 from this process in the high atmosphere.

If a neutron crashed into a carbon 13 atom, it may get jammed also in the nucleus which becomes a carbon 14 atom, with 6 protons and 8 neutrons. This is a very rare process and only about one in every trillion (a million million, 10^{12}) carbon dioxide molecules in the high atmosphere are changed from having carbon 12, the usual atom, to having carbon 14.

Another way carbon 14 atoms may form in the high atmosphere is by one of the 7 protons in a nitrogen atom, which usually has 7 protons and 7 neutrons, changing into a neutron. This results in the atom having 6 protons and 8 neutrons, becoming carbon 14.

Carbon 14 atoms are unstable and *decay* very slowly. During this one of the neutrons becomes a proton so the atom changes from carbon 14 to nitrogen 14, the usual form of nitrogen, which as said has 7 protons and 7 neutrons.

The *rate* of this decay is very slow. After 5,730 years half of it has decayed. That means it takes about 50,000 years for it all, practically speaking, to have gone.

The extensive shifting of CO2 by transfers to land and sea plants by photosynthesis, then into animals who eat plants and other animals and eventually soil, and also by other processes, has been discussed already. This includes massive transfers to the oceans, then to rocks. All of these then contain a very small amount of carbon 14, one in every trillion carbon atoms. If it is more than 50,000 years since the carbon in them came from the high atmosphere all the carbon 14 would have changed to nitrogen, there will be none left. This is the case for fossil fuels which were made millions of years ago.

Efforts can be made to measure the minute amount of carbon 14 using Accelerator Mass Spectrometry. This is an exquisitely demanding delicate process. The aim of measuring it is to try to determine how much CO2 in the air or ocean has come from fossil fuels which have a lower amount of carbon 14. But measuring such minute differences when only one in every trillion CO2 is a carbon 14 one is extremely arduous and difficult. There is a very good description of this at www.gml.noaa.goc/ccgg/isotopes/decay.html . To briefly summarise:

- Weekly air samples are taken from 60 sites world-wide and passed to the Global Monitoring Laboratory (GML) of the American National Oceanic and Atmospheric Administration (NOAA) at Boulder, Colorado. Exquisite care has to be taken with this.

- Some of these are used for carbon 14 measurement. Initially the other gases are removed by intense freezing and heating, then the CO2 is changed to pure carbon, graphite, by more intense heating. A tiny amount of graphite results.

- The graphite is then bombarded with Caesium atoms which negatively electrically charge the carbon atoms. These are then passed through a particle accelerator machine to separate nitrogen 14 atoms which if not removed would be counted as carbon 14 being the same mass. Another machine is then used to positively charge the carbon atoms. Of these, as said, one in every trillion is carbon 14.

- They are than passed through an ***Accelerated Mass Spectrometer*** which has an extremely strong magnetic field. The carbon 14 atoms behave slightly differently and are able to be individually counted by amazing equipment.

- These results are compared with standard air sample results and a difference in carbon 14 content calculated. This difference is then expressed as a figure called "delta carbon 14". Arriving at this involves a great deal of complicated mathematics.

A similar but much less demanding process is involved in measuring the more common, 1%, carbon 13 atom amount. Again a difference compared to a standard air sample is found, this is called "delta carbon 13".

Delta carbon 14 in ocean plant and land plant CO2 is usually about plus 45 relative to the standard; but fossil fuel CO2 gives a result of minus 1000, reflecting the very low amount of carbon 14 in these fuels millions of years old.

Delta carbon 13 is about the same for land plants and fossil fuel CO2 , results about minus 26. It is slightly less, minus 10, for ocean plant CO2 . This small difference is due to different uptake of carbon 12 and 13 by land and ocean plants.

A later step in the calculations to arrive at these values concerning tiny differences is to multiply them by 1000; this means that minus one becomes minus 1000, as for delta carbon 14 for fossil fuels discussed above. And plus 0.045 becomes plus 45 for its value for the standard. And the difference between these two is just 1.045, not 1045. This clearly makes these minute differences look greater.

Delta carbon 14 is being used to estimate the amount of CO2 from fossil fuel use going into the atmosphere each year, which information is also available from collating fuel use for each country world-wide as explained. In addition it is being used to try to estimate the length of time CO2 put into the atmosphere stays there, and how much is transported to land plants and animals by photosynthesis, or the oceans. Because of good mixing in the air evening out CO2 levels quickly these calculations are very complicated and require considerable use of computer models.

On average each year 27 kilograms of carbon 14 containing CO2 is made in the upper atmosphere. But the amount made varies considerably from one year to the next. This is then mixed with 2.7×10^{15} kilograms of CO2 in the atmosphere. At every step in the process to arrive at a Delta Carbon 14 result there are built in errors, very understandable concerning the problems detecting changes in concentration of something present one in every trillion molecules of CO2. .

Any *decrease* in Delta 14 carbon, that is it becomes more minus, can be interpreted as meaning that the CO2 is being mixed with CO2 from fossil fuels which contain less or no carbon 14 as said. But there are other causes for such a decrease:

- Less may be made in the high atmosphere.

- More may be taken up faster than computer modelled by photosynthesis and the oceans.

- Carbon 14 from nuclear bomb tests, a large amount, may decrease faster than modelled.

- The nuclear industry also produces large amounts of carbon 14 at many steps, this adds to interpretation difficulties; varying pollution from large cities changes carbon 14 levels.

Carbon 14 **Radioactive Dating** has been used since it was invented by Walter F. Libby in 1949 with immense benefit, especially in archaeology. But it is much more accurate when larger amounts of carbon are available for analysis, around 50 grams. For atmospheric CO2 the available sample is much smaller, about 0.4 milligrams, less than one hundredth of 50 grams. And there are many more processes for the air to go through to enable analysis, all with errors, compared with larger samples in, say, archaeology. And even this work inevitably involves problems; for example:

- Equipment calibration.

- Variations in carbon 14 production rates from one year to the next.

- Contamination during collection, transport and processing of samples.

- Man made nuclear industry and nuclear bomb carbon 14 contamination. Large amounts are made this way.

- Local natural variations in carbon 14 concentration which could alter interpretation. For example deep sea CO2 has less carbon 14 because it is older, atmospheric distribution is not always completely even.

- Pollution of air near cities, industries resulting in lower carbon 14 results can lead to misinterpretation.

These are just some of many problems. There is good discussion at www.pnas.org/content/117/24/13300 .

Therefore a tendency to be blinded by this wonderful science needs, as always, to be balanced by an understanding of its problems. It does not negate all other considerations including the ones discussed already.

Human Production - Summary:

The latest estimates for carbon dioxide production by human use of fossil fuels is 31 billion tonnes for the 2020 year. It has been about this level for the last few years. It was 31 billion tonnes in 2010.

The following are examples of individual country CO2 production:

- China   34%,   8.7 billion tonnes.
- USA   15%,   4.7 billion tonnes.
- India   7%,   2.2 billion tonnes.
- Russia   5%,   1.6 billion tonnes.
- Japan   3%,   0.9 billion tonnes.
- Germany   2%,   0.6 billion tonnes.
- Australia   1.5%,   0.47 billion tonnes.

Australia produced 0.47 billion tonnes in 2018. This compares to China with 10.6 billion tonnes.

The world-wide human use of fossil fuels includes:

- Electricity generation   20%
- Transportation   32%
- Manufacturing and construction 37% (some say 54%)

Some interesting ones are:

- Aviation   3%
- Cement   3%
- Shipping   3%
- Trucks, Buses, heavy vehicles   10%

- Home use for lighting, heating etc   11%
- Fuel used for agriculture and fishing   1.7%
- Rail   1%

These human production estimates are produced by multiple sources of information including individual nations measuring their use and production of fossil fuels. But, as one would expect, this process is far from perfect. These estimates are produced from what is termed a "bottom up" process. This is in contrast to a "top down" estimation which involves adding data about atmospheric CO2 measurements from ground, aircraft or satellite information.

For the "bottom up" estimate not only fossil fuel production and use is recorded but also other information including population density, nighttime electricity use, power plant operation, road vehicle use and others. Individual nations produce data on these and it is collated globally by the Carbon Dioxide Information Centre (CDIAC) in Oak Ridge Natural Laboratory, USA.

But, as can be imagined, there are many inaccuracies in this. Not least a great variation in the ability of countries to accurately collate the required information. It is estimated that the error range is about 5% for advanced (OECD) countries, 15 – 20% for China, and 50% for third world countries, maybe more. A conservative estimate is that the global estimate of 31 billion tonnes CO2 human fossil fuel production globally for the last few years could be out about 8%. This has been ignored in the above figures.

Additional Carbon Dioxide Production by Humans:

An unstudied but likely large CO2 producer by humans is **wood burning.** Wood is used as a fuel for cooking and heating by the majority of people, especially in the less technologically advanced countries, but also in advanced colder ones. The CAMS satellite system discussed above concerning wildfires cannot pick up the small fires involved, nor can other satellite techniques distinguish between CO2 produced from this source and other natural sources. Because wood burning is a comparatively poor source of heat compared to fossil fuels, and it is used by so many, it is probable a large quantity of CO2 is produced in relation to human production. But it has not been scientifically studied to give global information.

A guess at how CO2 is passed into the atmosphere by this can possibly be made by considering the estimates for wildfires discussed before, ranging from 8 to 50 billion tonnes annually. Perhaps the amount for wood burning for cooking and heating world-wide could be from 4 to 10 billion tonnes annually?

A saving grace is that these fires leave behind charcoal and ash some of which is deposited in soil boosting fertility and later growth of plants as does the smoke and ash passed up in to the atmosphere from them, like wildfires.

A Summary of Carbon Dioxide Movements:

This summary considers all the various producers and consumers of CO2 discussed in this Chapter. Estimations are all in Billions of Tonnes per year.

Land to Atmosphere:

Respiration	219	-	220
Decay	183	-	200
Wildfires	8	-	50
Human Fossil Fuels:	31		31
Human Fires	4	-	10
Volcanic	0.13	-	0.44
Atmosphere to Land:			
Land Photosynthesis	260	-	440
Rain	1	-	10 + ?
Ocean to Atmosphere:			
CO2 Transfer	90	-	390 ?
Atmosphere to Ocean:			
Ocean Photosynthesis	0	-	100 ? (unknown)
Rain	1	-	15 + ?
CO2 Transfer	90	-	390 ?
Totals:			
--- *To the Atmosphere:*	535.13	-	901.44
--- *From the Atmosphere:*	352	-	955

These results illustrate the great uncertainties in measuring the **natural world** carbon dioxide movements on our planet. The reasons for concluding there are large uncertainties are explained for each item in this Chapter.

The only items for which there is reasonable certainty are human fossil fuel use, which can be measured as explained, and land respiration. But that apparent certainty may be only because the estimates agree with each other rather than true certainty.

Concerning human production of CO2 into the atmosphere from fossil fuel use these figures result in estimates of from 3.4% to 5.7% of the total CO2 entering the atmosphere including natural sources. And total human CO2 *if* wood burning is included ranges from 3.8% to 4.5% of the total including natural sources.

Concerning Australian human production of CO2 from fossil fuel use it is estimated at 1.5% of total human production as said above. Which means it is about 0.05% , one two thousandth, of total CO2 entering the atmosphere. It is not likely Australian carbon dioxide production from domestic wood burning is significant because it is a technologically advanced country with common use of electricity.

To put this in some perspective the 31 billion tonnes human fossil fuel CO_2 production annually into the atmosphere mixes with 2700 billion tonnes already there; about 1%. And this compares with an annual natural world turnover of between 350 and 950 billion tonnes. And 140,000 billion tonnes in the oceans.

Not everyone agrees there is this level of uncertainty, or at least they *say* there is much more certainty allowing them to make detailed statements about the ***net balance*** of CO_2 entering and leaving the atmosphere. To be fair they may be making a best guess at CO_2 movements trying to estimate a net balance. Examples are:

Green and Byrne: (Billions of Tonnes of CO_2).

Land to Atmosphere:

Respiration	219.6
Decay	197.6
Fire	14.6
Human Fossil Fuels	23 (for 1990s)
Human Land Use change	5.8
Human wood use	3.6

Atmosphere to Land:

Photosynthesis	440.6
More vegetation	10.9

Ocean to Atmosphere:

Respiration	330.5

Atmosphere to Ocean:

Photosynthesis and Absorption	336.7

Their figures are in billions of tonnes of ***carbon,*** they have been converted to CO_2 here.

They consider they can make a ***net balance of CO2 movements*** statement which is:

To the Atmosphere:

Human Fossil Fuels	23 billion tonnes of CO_2

Human Land Use change	5.8

From the Atmosphere:

More vegetation	10.9
To the Ocean	6.2

Net Balance — **11.7 To the Atmosphere.**

A casual observer unfamiliar with all the uncertainties would form the impression that human fossil fuel use was clearly dominant in creating this balance from this presentation.

Another example:

Carlson and Steinberg:

To the Atmosphere:

Land Respiration	183
Human Fossil Fuels	20.1 (for 1990s)
Human Land Use changes	5.8
Ocean Emission	329.4

From the Atmosphere:

Primary Production (? Photosynthesis)	186.6
Human Land Use changes	1.8
Ocean absorption	336.7

These figures give a net balance of 13.2 billion tonnes of CO_2 added to the atmosphere annually. The authors do state: "—but scientific understanding of the causative processes is hindered by the complexity of <u>terrestrial ecosystems.</u> ".

It is suggested that a reasonable person considering all the factors discussed in this Chapter would conclude that a great deal is unknown. And that the natural world's use, creation and transfer of carbon dioxide is huge and dominant, human contributions are tiny by comparison.

Methods of Measuring Carbon Dioxide:

Non Dispersive Infra-Red Technology NDIR

This is by far the most common method for measuring CO2 now. It is used extensively by scientists and also by others interested in CO2 air concentration for air quality reasons, for example in enclosed spaces, aircraft, greenhouses.

There are many cheap NDIR meters available. These are usually accurate to within 100 parts per million which is useful for air quality work. Interestingly CO2 in human and animal expired air reaches 40,000 parts per million, 4%. Work place limits are set at 5000 parts per million, 0.5%. And humans begin to feel tired with inspired air containing 1000 parts per million, 0.1%.

For scientific work where great accuracy is sought the equipment is more much more expensive. Accuracies of less than 5 parts per million require more than $10,000 to be spent.

These meters work by allowing air to circulate through a space in the instrument. Infrared light with multiple wavelengths longer than 0.7 microns, out of the visible spectrum in the near infrared as explained in Chapter 1, is then shone through the air from a light emitting diode (LED) at one end. At the other end are two light filters. One only allows light with a wavelength of 4.2 microns through. The other only allows the wavelength 3.9 microns through.

Behind the filters are light detectors which produce an electric current when light is absorbed. These are connected to computers and meters to give a reading.

Carbon dioxide molecules in the air absorb light with a wavelength of 4.2 microns as explained in Chapter 1. But they do not absorb light with the wavelength 3.9 microns, nor do any of the other gases in air, nitrogen, oxygen, water vapour, argon and others as explained also in Chapter 1. So the filter allowing only 3.9 microns allows virtually all the light through but the one allowing only 4.2 microns is only affected by CO2 in the air, and, as no other gas in the air absorbs 4.2 wavelength light, the ***decrease*** in this light coming through the 4.2 filter compared to ***all*** the light coming through the 3.9 filter enable the amount of CO2 to be measured.

The NDIR meters do not measure any CO2 dissolved in water droplets in the air, just that in dry air.

For ocean and water CO2 the carbon dioxide is allowed to ***equilibrate*** with air above it then it is measured in the air and the water concentration worked out by applying ***Henry's Law*** as explained above.

NDIR measuring is fast, convenient, not affected by contaminants in the air. But the equipment does require very precise calibration.

Photoacoustic Spectography:

This similar in that the absorption of infrared light by CO2 molecules is exploited to make the measurement but it is detected by a sound - acoustic – method. When CO2 absorbs the light it heats and expands quickly; if the light pulses a sound wave results; and that can be picked by a microphone and measured on a meter. The infrared light used is made to pulsate.

The method is considered more accurate scientifically the NDIR because it does not need to be calibrated to a reference sample. But the gear is bulkier and more expensive so it is less popular.

Molecular Correlation Spectroscopy:

This is a recent high-tech method which is more complicated and expensive. It uses infrared light emitted in about one hundred very thin bands by LED equipment, the wavelengths of the bands are between 4.2 and 4.35 microns which CO_2 molecules absorb. They then emit infrared radiation in a broader band and this is picked up by sensors and measured. It can be used for tiny air samples.

Raman Spectography:

In this method the air is subjected to a powerful argon laser beam. This causes all gas molecules in the air to become excited and very unstable. They then emit light of different wavelength patterns for each gas present. This can be picked up by sensors and the patterns used to estimate the amount of each gas present including CO_2 . It is a high-tech expensive method.

Mass Spectography:

This is similar to the Accelerated Mass Spectrometer discussed above. The air sample is subjected to a beam of electrons which charge the gas molecules, they become *ions* with an electric charge. This means they will move with magnetism. They are placed in a strong magnetic field in a ***particle accelerator*** and the path each type of molecule takes varies, so they can be separated out and measured, including CO_2 . This also is very high-tech and expensive.

Heteropolysiloxane Chemical Sensors:

This relies on changes in a layer of Siloxane with CO_2 concentration changes resulting in a difference concerning currents passed through it which can be measured. The meters can be made very small but do need to be recalibrated as they tend to change over time, and have a shorter life than NDIR meters.

Metal Oxide Semiconductor:

These are used to measure hydrogen and volatile organic compounds especially in medical uses on expired air, for example the hydrogen breath test used to pick up gut bacteria. A current passed through the semiconductor changes in proportion to the gas present. CO_2 concentration can be calculated indirectly from these readings. The method is cheap but not accurate.

NASA Satellite Methods:

The first satellite measuring CO_2 was the Advanced Earth Observing (ADEOS) Satellite which was launched in 1996. It had on board a high tech spectrometer measuring infrared radiation from the Earth from 3.3 to 14 microns. From differences in the radiation detected in this range, part of the ***blackbody heat radiation*** from Earth discussed in Chapter 1, estimates of carbon dioxide air concentration, and that of other gases, could be made by sophisticated computer modelled calculation. This satellite only lasted a year. Its power supply from solar panels failed perhaps due to space debris striking them. It cost 70 billion Yen.

There was an attempt to launch a replacement Orbital Carbon Observatory OCO -1 in 2009 but there was a launch failure when the rocket exploded.

Orbiting Carbon Observatory 2 (OCO-2) was launched in 2014. This amazing craft is in orbit 705 kilometers up and weighs 449 kilograms. Although it had a design life of just two years it is still operating in 2021. It has a more sophisticated spectrometer which is designed to look especially at carbon dioxide. It measures infrared radiation at 0.75 microns, 1.61 microns and 2.06 microns. The latter two are absorbed by CO_2 molecules as explained before and the first by oxygen weakly. The radiation measured is Sun radiation reflected from the Earth's surface. It measures in a narrow band about two kilometers wide, effectively measuring in the column of air above that. This narrow focus enables CO_2 to be measured over cities, farmland, the ocean and attempts made to make conclusions about differences. It covers most of the planet by this incredible method.

The calculations to determine carbon dioxide concentration are very complex and there are errors understandably involved. Attempts are made to recognise these by comparing with data from the ground based Total Carbon Column Observing Network, whose stations use absorption patterns in incoming Sun radiation to determine gas composition including CO_2 . This ground based data is then used to correct the satellite OCO-2 data. Clouds also cause large errors. So, like carbon 14, it is far from foolproof.

Another satellite aimed at carbon dioxide measurement is the Greenhouse Gases Orbiting Laboratory (GOSAT). This was launched in 2009 as a joint American – Japanese project. It again has a spectrometer which measures radiation from the Earth and also reflected Sun radiation from the surface in 4 bands from 0.7 to 14 microns wavelength. It aims to measure oxygen, carbon dioxide, water vapour and methane concentrations. It is 666 kilometers up and has an orbit which covers a large swathe of the Earth passing over the same place every 3 days. It takes about 18,000 measurements each day. Again the calculations to produce results are extremely complicated with a large amount of computer modelling, with inevitable errors and data processing.

Another instrument is on the International Space Station. It is another joint American – Japanese project and has been operational since 2019. It was launched by a Falcon 9 retrievable rocket made by Eion Musk's Space X company. The spectrometer was made from spare parts left over from OCO-2 manufacture. It is therefore very similar, measuring especially 0.75. 1.61 and 2.05 micron bands to estimate oxygen and carbon dioxide. It can measure big "snapshots" 100 kilometers across or focus down on areas only 100 meters across. It only looks at places under the space stations orbit and does not see Polar regions.

Clearly these satellite spectrometers are a very interesting major advance. But there are large uncertainties with potential for significant errors. This is understandable when it is considered that the satellite instrument is looking down at light radiation coming up from the Earth 700 kilometers below in a column, as it were, vertically below it, ranging from 100 meters to 100 kilometers across depending on how the instrument is adjusted. The light radiation the spectrometer is analysing has been emitted from the surface, or is reflected Sun radiation from the surface, then passes up through the troposphere, stratosphere, mesosphere, thermosphere and exosphere as discussed in Chapter 1, and only then is examined by the spectrometer. The light radiation entering the instrument has passed through all those layers and been altered by most of them. It is not as though the satellite instruments examine accurately everything that is happening on the surface and the various layers of the atmosphere; rather it is just the final bit of light radiation emerging which is examined 700 kilometers up.

To produce results estimating carbon dioxide output from the surface far below there are many very complicated calculations. These require input knowledge about:

- Cloud cover. This can completely disrupt the readings

- The nature of the bit of the Earth's surface far below.

- Estimations of human use of fossil fuels in the target area from several sources.

- Details about various aerosols of dust and other atmospheric constituents which change the results considerably.

- Water vapour in the atmosphere below details. Again that can significantly alter results.

- "Background noise" information – electrical interference which might distort the signal.

Amongst many other factors. Then computer modelling is used to produce a final result.

Many of the satellite observations are rejected because they are too affected by these problems. Only about 17% of the satellite observations are considered to be of sufficient quality to be included in the final results. Also any "outlier" results, not fitting in with expectations, are rejected as being apparently inaccurate.

This results in a big reduction in the number of valid observations from the satellites for each place; for example Beijing has only one valid pass per year on average despite the satellites passing over it many times.

A study by Lei et al of OCO-2 satellite CO_2 readings over Lahore, Pakistan (reference detailed below) concluded that the increased CO_2 emissions found over the city compared to surrounding country areas were found due principally to the data fed into the computer modelling program used for the calculations about known human use of fossil fuels rather than from the OCO-2 satellite readings.

If one goes online and looks at, say, NASA websites about satellite CO_2 there is virtually no discussion about these problems; only when academic papers are studied such as that by Lei et al do they become apparent.

Efforts have been made to use the satellite data to detect CO_2 levels over cities, industrial areas, power stations and other places where humans use substantial fossil fuels. It is difficult because of the problems with the method as discussed above, understandable as the satellite sensors are examining radiation emerging from the atmosphere many kilometers above the point of interest far below. Strategies such as allowing for wind drift over a city are common, as well as rejection of data with problems. Nevertheless there are some impressive computer sites with beautifully coloured satellite tracks over cities shown. Usually red is chosen to show the increased CO_2 over a city and blue the lower level over country or the sea. But what is striking is that the CO_2 level over the cities is typically shown as about 410 parts per million, and the country as about 404. Only a difference of 6 parts per million. This is in contrast to historical ground based CO_2 measurements showing much bigger differences ranging from 10 to 70 parts per million as discussed in the next chapter. It is well known that surface CO2 levels are higher near sites with high human activity. But the satellites have trouble detecting this for the reasons given.

Interestingly a spin-off from this research has been the detection of an infrared radiation emission from plants at about 0.75 microns wavelength, just outside the red part of the visible light spectrum. This emission is called ***Solar Induced Fluorescence*** and was first discovered by Brewster 200 years ago. It is related to the photosynthetic activity of plants in that it increases when the plant is more active and decreases when less so. This wavelength is studied by the satellites as discussed, so the data from them has been used to estimate photosynthetic activity, and CO2 consumption, by plants on land and in the ocean.

But recently doubt has been thrown on this by Marrs and others from Harvard University who found that when plants were made to stop absorbing CO2 and producing oxygen by closing their openings for air to enter the leaf they still emitted this radiation the same. They suggest caution in making conclusions about world-wide photosynthesis, and therefore carbon dioxide consumption by plants, from this satellite data.

A new technology is Laser Induced Differential Induced Radar or LIDAR. This involves aiming intense laser beam pulses lasting 150 millionths of a second from an aircraft with the equipment at the ground. The beam contains radiation of 2.05 micron wavelength which is absorbed by CO2 . It is reflected back to the instrument in the aircraft, like radar. CO2 is then calculated by very complex methods. It seems to be still a research tool.

There is a temptation to be overwhelmed by the awe-inspiring nature of these immense achievements, especially as their stories are beautifully promoted on many websites and publications. Needless to say the national organisations running them have immense funding, or else could not do it. And perhaps they are keen to push views to justify more funding again.

But when it comes to interpreting the data and making conclusions from it the same care needs to be applied as for other scientific information sources: tempering with knowledge about problems, uncertainties, errors. And also the very brief time this information has been available, very short scientifically, minute in geological terms. This is important especially when interpreting trends.

Historical Methods for Carbon Dioxide Measurement:

Professor Beck's Papers:

As mentioned already historical methods have been studied by Professor Ernst-Georg Beck at the Merian-Schule, Freiburg, Germany and he published his work in 2007. He documented 180 papers published between 1855 and 1960 using earlier methods than NDIR which was invented by Keeling in 1958. In the 180 studies were many thousands of CO2 measurements.

The principal method in use from about 1812 was chemical. This involved:

- A known volume of air was dissolved in a strong alkaline solution such as sodium hydroxide, caustic soda. The CO2 in the air caused sodium carbonate to form, a salt which dropped to the bottom as it ***precipitated*** similar to the process described for the ocean and formation of carbonate rocks. This happens because sodium carbonate is not very soluble in water.

- An acid, often sulfuric acid was then added slowly until the solution was neutral, not alkaline. This meant that all the remaining sodium hydroxide had been neutralized by the acid added, which was measured. The original amount of sodium hydroxide before the air with CO2 was added being known, that used up by the CO2 could be simply calculated (no computer models !) and the amount of CO2 in the air sample also calculated.

The method is considered to be accurate within 1 to 3%.

Other chemical methods in use were volumetric, gravimetric (weight change) and manometric (pressure change) focusing on changes before and after CO2 had been removed chemically.

Before 1855 there were errors which tended to **underestimate** the amount of carbon dioxide by about 30 parts per million. These were the use of sulfuric acid to dry the air before testing which removed some CO2 ; the use of latex tubing which absorbed CO2 ; and a failure to allow for temperature.

Pettenkofer in 1855 improved these aspects as did others later including the famous Jack Haldane FRS.

So there were several very accurate methods in widespread use for the 100 years **before** 1961 when NDIR and other methods using LEDs, high tech filters, or particle accelerators and computer analysis came into use. More than 90,000 measurements after 1820 occurred.

Beck's study detailed results in the more accurate 138 papers from 1812 to 1961. Most of the measurements were done in the Northern Hemisphere, including Europe, North America including Alaska, India but there were some from Antarctica.

Some interesting results:

- Kreutz between 1939 and 1941 took 25,000 measurements at four sites near a famous meteorological station at Giessen in Germany using very accurate manometric equipment. The average result was 438 parts per million, ranging up to 550 in 1940. Some of the reading were near the city, however, and that is considered to have pushed some of the readings up by 10 to 70 parts per million. World War II is not a cause as other studies from Alaska and India showed similar results.

- Spring and Roland did 266 samples from 1883 to 1884 in Liege, Belgium. The average was 335 parts per million including one of 380.

- Hasselbarth at Dahme, Russia measured a reading of 380 parts per million in 1876. He also found that CO2 levels increase at night which is well known. His daytime measurements were about 290.

- Schulze in Rostock, Germany in 1863 – 64 found levels ranging up to 400 parts per million; Scholander 430 in 1948; Lundegardh 335 in 1924; and an earlier study in 1820 with a reading of 450 parts per million.

His main point is that there is a wealth of evidence that CO2 levels were higher than recently claimed by others including the International Panel for Climate Change (IPCC) who say they were about 280 parts per million for the last three million years.

A second point is that ice core samples of trapped air used to estimate past CO2 have inaccuracies. These sample results are used by the IPCC and others to bolster the view that atmospheric CO2 has stayed around 270 for the past 800,000 years at least. Beck points to results from 1925 to 1960 from many scientists documenting higher levels, including 410 – 440 in 1941, when the ice core samples calculated 310.

A third point is that the historical studies show a clear trend where CO2 atmospheric levels **followed** major global temperature trends, lagging behind by about 5 years for these 19th and 20th century studies. An example is that global temperature increased significantly from 1880 to the 1940s by about 0.5° C in many studies. An increase in CO2 levels was found by several groups world-wide, with a delay as said. Meaning that CO2 increase is the **result** not the cause of the temperature increase.

He also points to some of these earlier studies showing the well-known variation in CO2 levels with the Moon cycles – it increases with the full Moon – helping to confirm their accuracy.

Professor Beck's work has been strongly criticised by some others.

Ice Core Samples:

The International Panel for Climate Change (IPCC) and associates have quoted Antarctic ice core sampling where gas bubbles trapped in ice are analysed for CO2 concentration and the results used to determine earlier atmospheric levels especially for the past 800,000 years, the maximum age of the samples taken at the Vostok Station as discussed already. Other samples have been taken from the Taylor Plateau in Antarctica, Greenland, and old glaciers in the Northern Hemisphere. Greenland samples go back 130,000 years. Recently an Antarctic sample found ice 2.7 million years old, in which the CO2 in bubbles was found to be 300 parts per million. In general all the samples have found CO2 levels well less than 300 parts per million. In ice ages it falls to 180 and in warm interglacial times rises to about 280, according to these results.

But there is criticism of the validity of these measurements as mentioned by Professor Beck. This has been summarised by Dr Zbigniew Jaworoski in papers from 1994 through to 2007. In these he quotes numerous scientific studies by different authors including:

- Hurd (2006) noted CO2 in ice core samples showed evidence of decreasing CO2 due to the huge pressures up to 300 times sea level atmospheric pressure deep in the ice pushing the gas into it.

- Wagner et al (2002) documented evidence from fossil leaves showing that atmospheric CO2 was higher than the ice core level of 280 by more than 50 parts per million, that is up to 330 parts per million for the same time.

- Indermuhle (1999), Pearman (1986), and Petit (1999) have found that generally ice core sample results are always about 100 parts per million lower for CO2 than the current atmospheric level. Even for samples for ancient times when the Earth was much warmer, for example 120,000 years ago it was 5° C or more warmer, the ice core sample revealed a result of 240 parts per million CO2 . Similarly 10,000 years ago the Arctic was 5° C warmer, yet the CO2 from ice core samples was only 260. (IPCC 2007). This casts doubt on either the accuracy and/or the argument that CO2 causes the warming.

Dr Jaworowski gives more than 90 references in his 2007 review paper. He also has been strongly criticised by others.

Interestingly the ice core samples have been studied looking at which came first – the CO2 rise or the temperature rise? Obviously whichever did come might be the cause of the other increasing. There is good evidence that the temperature rose first and CO2 sometimes hundreds or thousands of years later consistent with the idea that sea temperature is the boss of atmospheric CO2 . For example:

- Caillon (2003), Fischer(1999) Idso (1988), Indermuhle (2000), Monin (2001) and Mudelsee (2001) showed that CO2 levels *followed* temperature changes in these ancient times with a lag of hundreds, even thousands of years. The temperature increase caused huge movement of carbon dioxide from the sea to the air as the sea temperature rises as discussed already.

And some studies looking at the same thing in modern times:

- Modern studies including Boden et al (1990) showing the increase in CO2 from 290 in 1885 to 440 parts per million in 1940 – higher than now - *followed* the 0.5° C temperature increase then. After that the global temperature fell by 0.3° C by 1970, and CO2 followed with a lag of about 5 years to 330 parts per million.

Other methods than analysing bubbles in old ice cores to determine earlier CO2 levels include:

- Examining the stomata on fossil leaf remnants. These are the openings used by plants to breath. Their number and size varies with the CO2 air concentration. It is less if CO2 is higher. The method involves studying fossil remains of leaves buried in old soils and sediments. Old birch tree leaves are amongst those studied.

- Isotope ratios of Carbon 13 and Boron 11, which vary in marine sediments from the deep sea bottom according to the CO2 level when the plant or sea plankton was living.

- Phytane levels in marine sediments. Phytane is a breakdown product of chlorophyll and it varies depending on the CO2 level when that plant was alive.

By these means it has been estimated that atmospheric CO2 was much higher millions of years ago. It is thought that 500 million years ago it was over 6000 parts per million, 0.6% compared to 400 today. 150 million years ago it is thought to have been over 3000 parts per million, 0.3%; and 34 million years ago 760 parts per million, 0.07%.

Wagner et al compared results from ice core samples for the period 8700 to 6800 years ago with results from fossil leaf stomata studies. The ice core samples consistently found results of 260 to 270 parts per million. But the fossil leaf stomata studies suggested higher CO2 levels ranging up to 330. Further the leaf studies showed that CO2 levels dropped to about 280 during a period when it known to have been colder from 8500 years ago to 8200 years ago, then rose again to 330 when it got warmer. The ice core studies just stayed flat at 260.

This is an argument in favour of the leaf studies being more accurate given that there is a lot of other evidence that atmospheric CO2 drops when it is colder and rises when it is warmer. Wagner again pointed to the sea temperature governing this.

But like many other aspects of these issues there are other scientists who dispute the results pointing to calibration, contamination and other potential errors.

Interestingly the temperature of the sea and air in these ancient times is estimated usually by measuring the proportion of an isotope of oxygen, oxygen 18, in the old sediments. It is higher in colder temperatures. Another method measures the ratio of magnesium to calcium in sediments. This is lower in colder times.

The age of the sediments studied is usually determined by radiocarbon 14 dating.

References and Further Reading for Chapters 8 on Carbon Dioxide:

www.wikipedia.org/wiki/Carbon_dioxide

www.wikipedia.org/wiki/Exraterrestial_atmosphere

www.science.org/doi/10.1126/sciadv.aax1420

Catling, D. and Zahnle, K. The Archean Atmosphere. Science Advances, Vol 6, Issue 9, 26 Feb 2020.

www.space.com/9599-saturn-moon-rhea-surprise-oxygen-rich-atmosphere.html

www.wikipedia.org/wiki/Chlorophyll

www.khanacademy.org/science/biology/photosynthesis-in-plants/

www.en.wikipedia.org/wiki/Carbon_dioxide_in_Earth%27s_atmosphere

Calculation: kg CO_2 in ocean and atmosphere: 39,000 gT carbon = 1.4×10^{17} kg CO_2

www.yadvindermalhi.org/blog/does-the-amazon-provide-20-of-our-oxygen (ocean photosynthesis).

Calculation ocean respiration: Ocean photosynthesis produces 240 Gt O_2 per year, virtually all used in the ocean = 330 Gt CO_2 per year.

Ice. Sir Fred Hoyle. Page 77-84.

Calculation global energy photosynthesis: oxygen production 570 Pg/year = 3.5×10^{16} moles @ 450 kcal/mole = 1.57×10^{19} kcal = approx 2×10^{16} kWh/year. CO_2 consumed about 780 Pg / year.

www.pro-oceanus.com/images/pdf/PSITechnicalNote1.1-DissolvedCO2and UnitsofMeasurement2019.pdf

www.wikipedia.org/wiki/Carbon_dioxide_in_Earth%27s_atmosphere (ancient CO_2).

Luthi et al. Nature Vol 453, pp 379 – 382 2008.

www.en.wkipedia.org (Ice age CO_2).

Emiliani, C. Scientific American Vol 198 1958. Ice. Sir Fred Hoyle, page 103. (deep sea temperature 40 million years ago).

www,en.wikipedia.org/ (proportion of CO2 dissociating to carbonic acid).

www.rwu.pressbooks.pub/webboceanography/chapter/5-5dissolved-gases-carbon-dioxide-ph-and-ocean-acidification/

Wiebe, R. and Gaddy, L. The solubility of carbon dioxide in water at various temperatures from 12 to 40° C and at pressures to 500 atmospheres. J.Am.Chem.Soc. 1940, 62, 4, 815-817

https://reefs.com>magazine>chenmistry-and-the-aquarium

www.aslopubs.onlinelibrary.wiley.com/doi/pdf/10.4319/lo.1980.25.2.0367 (concentration Ca ions in seawater)

www.web.stanford.edu/group/Urchin/mineral.html (ions in seawater).

Broeckner,W. et al. The fate of fossil fuel carbon and the global carbon budget. Science 206, 409-418. (saturation calcium carbonate)

Ice. Sir Fred Hoyle. Page 87. (Blue-green algae rock formation)

www.en.wikipedia.org/wiki/Geochemistry_of_carbon (amount carbon in crust, mantle, core).

www.physicalgeography.net/fundamentals/9r.html (carbon stores)

Pidwirny,M. The Carbon Cycle. Fundamentals of Physical Geography. 2nd Ed. (2006)

Lui, Z. et al. Breakpoint Lead-Lag analysis of the last deglacial climate change and atmospheric CO2 concentration on global and hemispheric scales. Quat. Int. 2018, 490, 50-59.

Peti, J.R. et al. Climate and the atmospheric history of the past 420,000 years from the Vostol Ice Core Anarctica. Nature, 399, 429-436. 1999.

Roe, G. et al. In Defence of Milankovitch. Geophys. Res. Lett. 2006. 33.

Koutsoyiannis, D. and Kundzewicz, Z. Atmospheric Temperature and CO2 . Hen-or-Egg Causality? Sci-02-000 83-v2.pdf. 25/11/20.

La Quere, C. et al. Temporary reduction in daily CO2 emissions during the Covid 19 forced confinement. Nat.Clim.Chang. 10, 647-653. 2020.

Beck, E-G. 180 years of atmospheric CO2 measurement by chemical methods. Jn. En. & Environ. Vol 18, No. 2, 2007.

www.gml.noaa.gov/ccgg/trends/gl_trend.html (Global CO2 levels).

https://e360.yale.edu>features>soil_as_carbon_storehou...

Liu, C. et al. An estimate of rainout of atmospheric CO2 . J Environ Sci (China) 2004; 16(1): 86 -9

www.chemistry.stackschange.com (raindrops falling)

Calculation: CO2 and rain: Earth surface area 5.1×10^{14} square meters, one meter precipitation assumed to be all rain, = 5.1×10^{14} m^3 surface water, @ 10kg CO2/ kg = 5.1×10^{15} kg = 5×10^{12} tonnes = 5000 billion tonnes.

Fischer, T. et al. The emissions of CO2 and other volatiles from the world's subaerial volcanoes. Scientific Report 9, No. 18716 (2019) www.nature.com/articles/s41598-019-54682-1

Shinohara, H. et al. Excessive degassing from volcanoes and its role on eruptive and intrusive activity. Rev. Geophysics. 46 RG400S https://doi.org/10.1029/2007RG000244 (2008)

Hauri, E. et al. Carbon in the convective mantle. https://doi.org/10.1017/9781108677950 (2019) (Mid Ocean Ridge CO2 emissions)

www.usgs.gov/programs/VHP/volcanoes-can-affect-climate (US Geological Survey re volcanic gases).

www.modis.gsfc.nasa.gov/data/dataprod/nontech/MOD14.php (MODIS fire detecting satellites).

Tang, W. et al. Widespread phytoplankton blooms triggered by2019 – 2020 Australian bushfires. Nature 597, 370-375 (2021). www.nature.com/articles/S41586-021-03805-8

Grace, K.S. et al. Towards a better understanding of fish-based contribution to ocean carbon flux. Limnology and Oceanography, 2021; doi.10.1002/no.11709 . (marine decay carbon)

Green, C. and Byrne, K. www.sciencedirect.com/topics/earth-and-planetary-sciences/globa-carbon-cycle Encyclopedia of Energy 2004. (Respiration and decay CO2).

Sarmiento, J. et al. Sinks for anthropogenic carbon. Physics Today 55(8) 30. https://doi.org/10.1063/1.1510279 (Ocean CO2).

Tanhua, T. et al. The marine carbon cycle and ocean anthropogenic CO2 inventories. Ocean civilization and climate. J. Church et al pp 103 https://doi.org/10.1016/B978-0-12-391851-2.00030-1

Cai, Q. et al. Global patterns of human and livestock respiration. Scientific Reports 8 article 9278 (2018). www.nature.com/articles/s41598-018-27631-7

www.ucusa.org/resources/each-countrys-share-CO^2-emissions

www.iea.org/reports/global-energy-review-2021/co2-emissions

www.irena.org/energy

www.sciencedirect.com/science/article/abs/pii/S0301679X14001820

Global Energy Consumption due to friction in trucks and buses. Holmberg, K et al. https://doi.org/10.1016/j.triboint.2014.05.004

www.iea.org/energy

www.eia.gov/outlooks/ieo.pdf/industrial.pdf

www.c2es.org/content/international-emissions/

www.ourworldindata.org/emissions-by-sector

Davis, W. Carbon-14 Production in Nuclear Reactors. 1977. www.osti.gov/servelts./purl/7114972

www.en.wikipedia.org/wiki/Carbon-14

www.gml.noaa.gov/ccgg/isotopes/decay.html (Carbon 14 and carbon 13 measurement).

www.pnas.org/content/117/24/13300 (difficulties with C^{14}).

www.capnography.com/physics/physical-methods-of-CO2-measurement (NDIR etc)

www.en.wikipedia.org/wiki/Carbon_dioxide_sensor (Chemical).

www.controlglobal.com/articles/2019/measuring-atmospheric-carbon-dioxide (NASA satellites).

www.directory.eoportal.org/web/eoportal/satellite-missions/o/oco-2 (OCO-2 satellite)

www.gosat.nies.go.jp/en/about_2_observe.html (GOSAT satellite).

Marrs, J.K. et al. Geophysical Research Letters. 24 June 2020. https://doi.org/10.1029/2020GLO87956 (Solar Induced Fluorescence).

Jaworowski, Z. Ancient Atmosphere – Validity of Ice Records. Environ Sci & Pollution Res. Vol 1 No. 3pp 161 – 171, 1994. www,aph.gov.au>committee>submissions pdf

Jaworowski, Z. EIRScience 16 March 2007 pages 38 – 53.

www.science.org/content/article/record-shattering-2.7-million-tear-old-ice-sample

Wagner, F. et al. Rapid atmospheric CO2 changes associated with the 8200-years-B.P. cooling event. Proc. Natl.Acad. Sci. U.S.A. 2002 Sep 17; 99(19): 12011-12014. www.ncbi.nlm.nih.gov/pmc/articles/PMC129389

Lei, R. et al. Remote sensing of the environment. Vol 264, Oct 2021, 112625. www.sciencedirect.com/science/article/pii/S003442572100345X (errors estimating human CO2 output and OCO-2 satellite limitations).

Chapter 9

Carbon Dioxide Heating of the Atmosphere

Concentration:

At present one in every 2500 molecules in the air is CO2 ,400 parts per million, 0.04%. If the CO2 molecule absorbs heat energy it will clearly get hotter, its temperature increases. If its temperature is greater than surrounding air molecules it will quickly loose that to adjacent molecules either by conduction, that is directly transferring the extra heat vibration to the next molecules, or by **blackbody radiation**, as explained in Chapter 1. As a result the hot CO2 molecule drops to the same temperature as surrounding air molecules, the extra heat energy it had is *spread evenly* through the air. This is aided by **convection**.

It is easy to understand that **direct conduction** and **convection** will quickly even out the extra heat energy through the air. How **blackbody radiation** does so is more complicated.

The blackbody heat radiation emitted from the hot CO2 molecule has wavelengths from 4 to 90 microns with a peak at about 10 as explained before, for usual air temperatures on Earth. The wavelength is a little less for hotter air temperatures and longer for colder ones. The process from molecule heat energy absorption to blackbody heat radiation emission is very fast, 80 **femtoseconds.** A femtosecond is 10^{-15} seconds, a million billionth of a second.

The radiation, which is put out in all directions by the molecules, will travel through the air until it strikes a molecule which will absorb it, as discussed in Chapter 1. Most likely this will be a water vapour molecule as explained. These are very common, usually from 2 to 4% of air molecules, or one in 50 to one in 25 of air molecules. The other common air molecules, nitrogen, oxygen, argon, all allow the heat radiation in these wavelengths from 4 to 90 microns to pass straight through without heating them as also explained in Chapter 1. But water vapour molecules absorb thoroughly for these wavelengths.

The heat radiation from the CO2 molecule ends up heating surrounding water vapour molecules, aiding the spreading and evening out process mentioned.

As the CO2 molecules are far less common, one in 2500 air molecules as said compared to one in 25 or 50 of water vapour molecules in the air, the chances of the blackbody radiation from a hotter CO2 molecule reaching another CO2 are much less; and even if it did so most of the radiation would pass straight through the CO2 molecule as explained in Chapter 1. Only the narrow band from 13 to 17 microns would be absorbed, the rest of the 4 to 90 blackbody radiation would pass through in contrast to water vapour molecules where a great deal of it would be absorbed.

Some have postulated that CO2 at these low concentrations, one in 2500 air molecules, can cause significant atmospheric heating. But a problem with this idea is the **rarity** of the CO2 molecules. Even if a CO2 molecule got very hot it would quickly lose its heat to surroundings. And that occurs mostly by far to water vapour molecules which are, as discussed already, very able to deal with heat energy by convection, phase changes and radiation in the troposphere, then the stratosphere to space. There is such huge capacity for atmospheric water to transfer massive amounts of heat energy to space, as has been explained, that it is unreal to assert that this can be overwhelmed by far less common CO2 molecules.

Even if the CO2 concentration were to triple the situation would be essentially the same. Then one in every 800 air molecules would be CO2 , still comparatively rare. Historically CO2 has been as high as 5000 parts per million, 0.5%, 700 million years ago. This is still only one in 200 air molecules, minor compared to water vapour at least one in 25 in those times.

Concerning this concentration question it is necessary to go back 2.5 to 4 billion years ago when CO2 was perhaps 25% or more of the atmosphere, one in 4 molecules, to understand it could have a significant heating effect then. Or to consider Mars and Venus where CO2 is 95% of the atmosphere.

But to postulate that a gas comprising one in 2500 atmospheric molecules can significantly heat the atmosphere in the face of the evening out processes and the powerful effects of atmospheric water requires, to be polite, imagination.

Heat Radiation Absorption of Carbon Dioxide:

For convenience the earlier sections on carbon dioxide heat radiation absorption in Chapter 1 are copied here:

"Regarding the effect on the Earth's heat radiation from 4 microns to 90 microns as shown in Figure 7 Carbon Dioxide has no absorption effect above 17 microns, unlike water vapour which powerfully absorbs radiation above 20 microns. So heat radiation with longer wavelengths the 17 microns passes through carbon dioxide unaffected."

"For wavelength from 4 to 17 microns there are some areas of absorption. There is a narrow but 100% area at 4.3 microns, this goes from 4 to 4.4 microns. To assess the importance of this on Earth's heat radiation Figure 7 shows there is not much to affect for that wavelength, so it is of little importance. There are weak areas of absorption at 9.2 and 10 microns as shown, again of little importance."

"Then there is an area on intense absorption from 13 to 17 microns, it reaches 100% absorption from 14 to 16 microns. It is very intense absorption. Indeed, at sea level, even though carbon dioxide is only 0.04% of the air there, one in every 2500 air molecules, *all* of the Earth's heat radiation in that wavelength band from 14 to 16 microns is completely absorbed by CO2 molecules in just 10 meters of air."

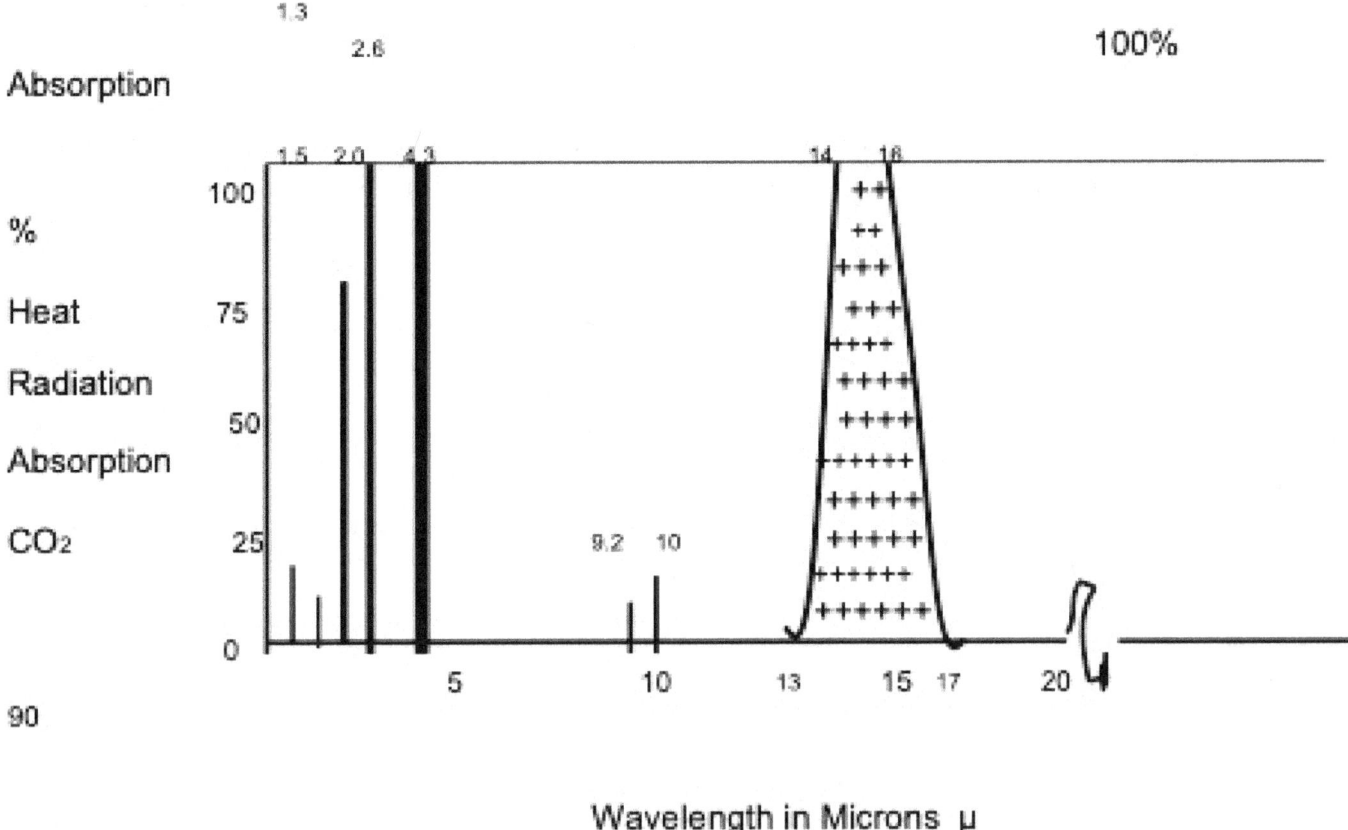

Figure 10: Radiation Absorption by Carbon Dioxide Molecules CO₂

"As a consequence of this intense absorption in this narrow wavelength band increasing the concentration of CO2 in the air does not have any more effect on the heat radiation from the Earth's surface because all the radiation in that narrow band which CO2 affects has already been absorbed at lower concentrations of CO2. It is like a car going flat out – it cannot go faster."

This effect of carbon dioxide molecules to intensely absorb 100% of the heat radiation in the narrow band 14 to 16 microns wavelength is true for very low concentrations of CO2 - as low as one part per million, 0.0001%. But the distance the radiation travels in the air gets longer before complete absorption in this narrow band is complete – up to 300 meters for such low concentrations at sea level, longer in thinner higher air. That is clearly a very short distance when one considers the height even just of the troposphere from 8 to 17 kilometers as discussed.

For current atmospheric CO2 concentrations of 0.04% the distance is about 10 meters in sea level atmosphere for complete absorption in this band.

The Earth loses heat energy by water vapour processes in the ocean and troposphere, then radiation to space, as has been discussed. How can CO2 interfere with this process, especially, what is the effect of increasing the CO2 concentration?

Consider heat radiation just leaving the Earth's surface with wavelengths of 4 to 90 microns and a peak of 10 microns for the average surface temperature of 14° C as discussed as shown in Figure 7 repeated here.

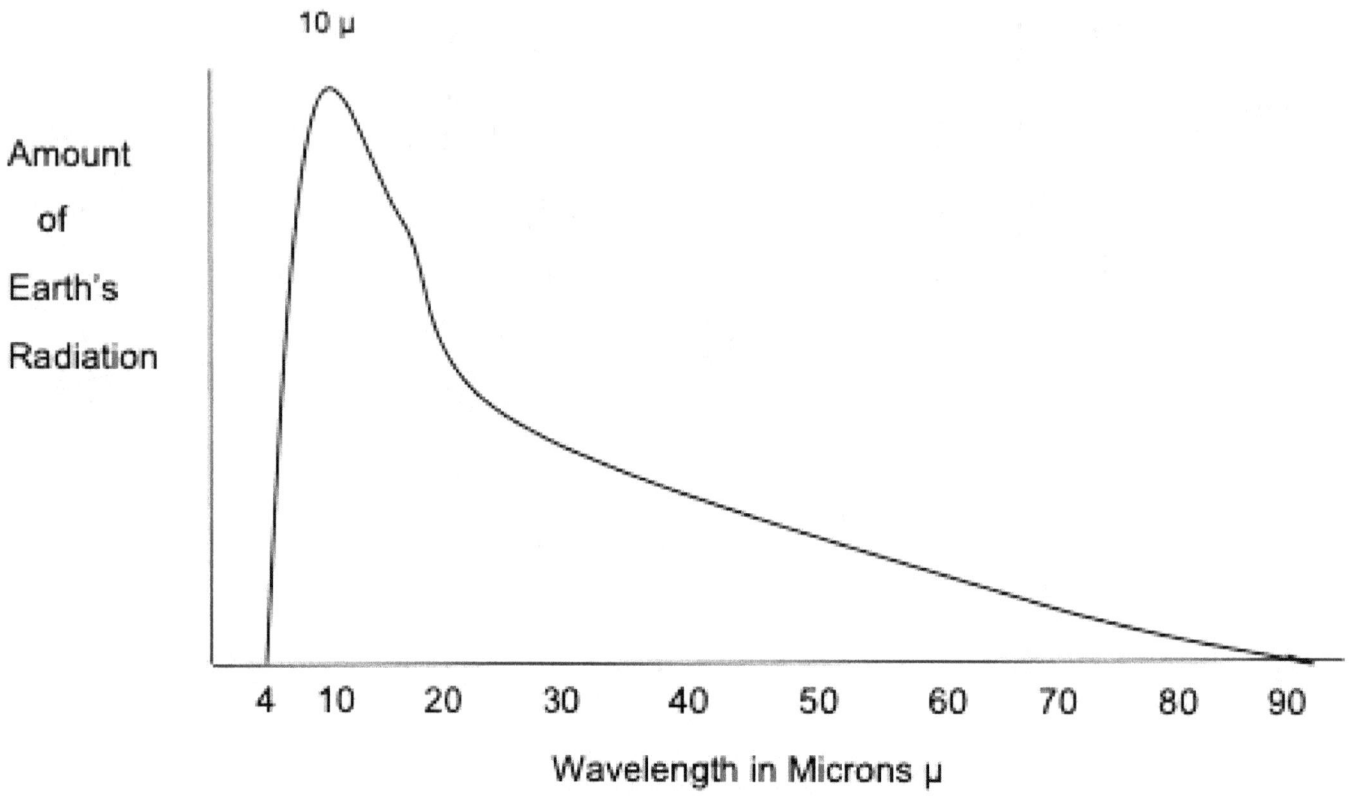

Say the radiation strikes a CO2 molecule just above the surface. All of the heat radiation except for that between 13 and 17 microns will pass straight through the molecule not heating it. Only the smaller amount of heat radiation, about 7% as already discussed, will be absorbed by the molecule heating it. Radiation between 14 and 16 microns wavelength will be quickly completely absorbed. Some of that between 13 and 14 microns, and between 16 and 17 microns, the so called *"shoulders"* of this absorption curve, will be absorbed, some will pass by.

There has been much debate about the proportion absorbed by the shoulders. The reason for this debate is that there is broad agreement that the absorption in the band 14 to 16 microns is complete; which means, as said above, that changing the concentration of CO2 does not alter the amount of heat radiation absorption in this band because it is already complete – the accelerator is flat out.

But it is argued that changing the CO2 concentration does change the amount of heat radiation absorption concerning the shoulder wavelengths; because the absorption is not complete there. So there is potential for the heating effect of CO2 in the atmosphere to increase and decrease with changes up and down in the level of CO2 *but only concerning the wavelengths in the shoulders.*

Some say the shoulders are important; others that there is practically no heat radiation absorption in the shoulders because the absorption curve is very steep, the shoulders hardly exist. The broad consensus is that there is a small amount in the shoulders, about 5% of the heat radiation absorbed by CO2 in this band from 13 to 17 microns; it is agreed by most that 90 – 95% of the heat absorption by CO2 molecules is in the intense 14 to 16 micron band, not able to change with changes in CO2 concentration.

To put the shoulders in more perspective CO2 molecules absorb about 7% of the Earth's heat radiation passing through the troposphere, water vapour absorbs far more The shoulders affect maybe 5% of this 7%, 0.35% of the total, some would say less.

Returning to the molecule of CO2 just above the surface which has now absorbed what heat energy it can and increased its temperature; as discussed above it will now transfer that heat energy to its surroundings to bring its temperature down to the level of the surrounding air molecules. It does this by direct conduction aided by convection and by radiation as mentioned above. The radiation is of the *blackbody* type already discussed from 4 to 90 microns wavelength. This will include heat radiation with wavelengths between 13 and 17 microns, the absorption band of CO2 because all matter with heat energy, temperature above absolute zero minus 273° C, molecules vibrating as a result, emit *blackbody radiation in wavelengths appropriate to their temperature* as discussed in detail in Chapter 1.

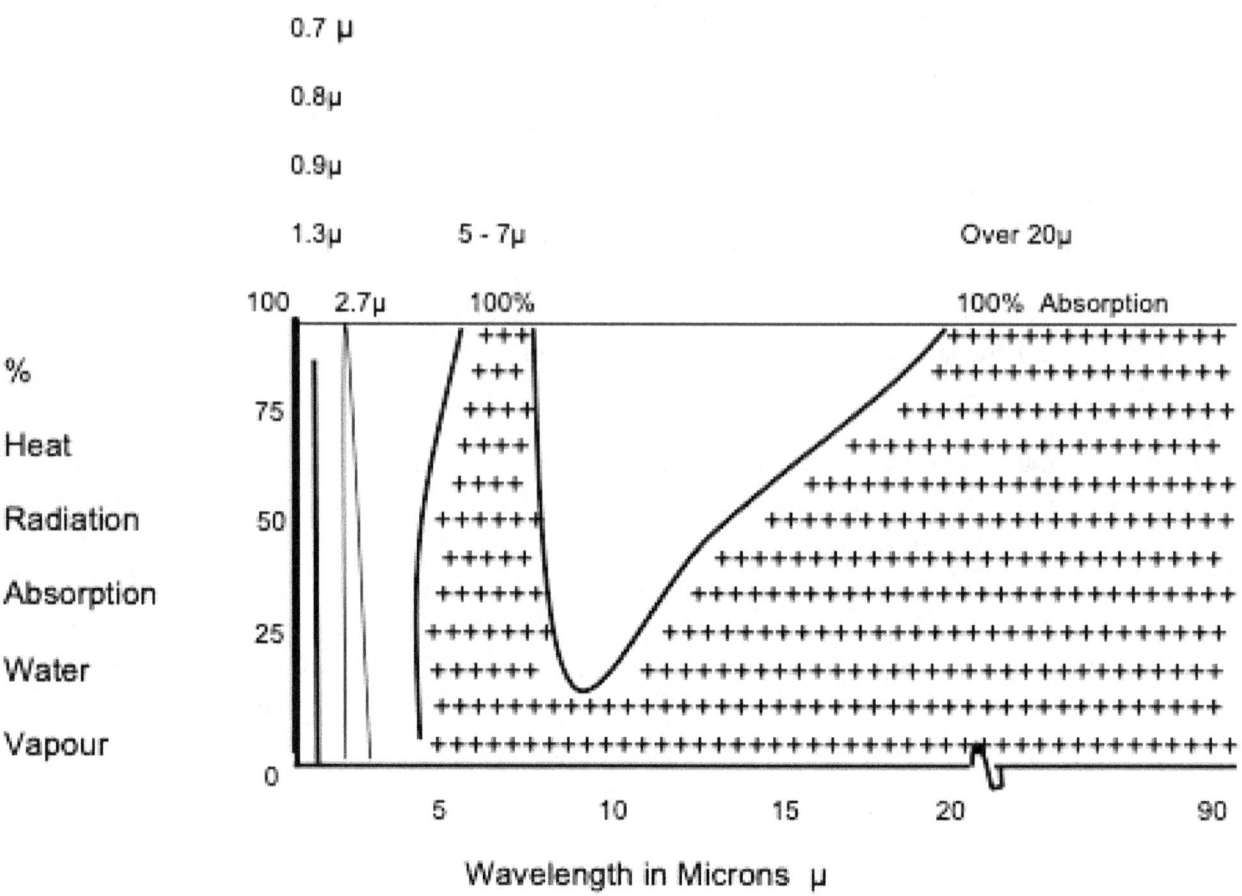

Figure 9: Radiation Absorption by Water Vapour Molecules H_2O

CO_2 is present right through the atmosphere and all the way up the CO_2 molecules follow the same pattern of absorbing intensely in the 14 to 16 micron band, less so in the shoulders. Each molecule heats up, then loses its heat energy as said including blackbody radiation in the band 4 to 90 microns, and direct conduction aided by convection. Then the next CO_2 molecule absorbs the band wavelengths and so on. And the satellites looking at this find a graph showing **virtual absence** of heat radiation wavelengths in the 14 to 16 micron band, as well as decreases reflecting the more widespread absorption by water vapour shown again in Figure 9 repeated here.

It has already been said that each atmospheric molecule emitting heat, whether by conduction or radiation, does so in **all directions**; up, down, sideways. But because the surface is warmer than the atmosphere generally, and surrounding space is so cold at minus 270° C the heat energy gradually moves up through the atmosphere to escape by radiation to space as already discussed. And the dominant method of this heat movement is by convection in the atmosphere, aided very substantially by water phase changes as already mentioned. These processes go up not down.

Considering the very low concentration of CO2 and the nature of the heat energy absorption band it is very difficult to conclude that changing carbon dioxide levels affects air temperature significantly. Especially when the massive heat energy transfers by atmospheric water described in previous sections are considered.

The History of Carbon Dioxide and Water Vapour Heat Energy Absorption Discoveries:

Some Prominent Scientists:

John Tyndall:

The first scientist to detail that the atmosphere absorbed heat energy was **John Tyndall**, a brilliant 19th century British physicist who followed Michael Faraday with many discoveries. In 1859 he reported that, while the atmosphere freely allowed the Sun's radiation to pass through it quickly, it was much slower to allow the returning heat radiation from the Earth's surface to pass to space, essential for the Earth to stay at the same temperature. He was the first to distinguish between the shorter wavelengths of the incoming Sun radiation and the longer ones of the outgoing heat radiation, which he called "Radiant Heat" and later "Ultra-red" recognizing that this radiation was of longer wavelength than visible red light. He explained that the Earth was warmer due to this slowing effect, which we now know raises the Earth's temperature by 33°C.

Further he studied the various atmospheric gases measuring their fingerprint heat radiation absorption patterns in the "ultra-red" and concluded that water vapour was by far the most important slower of heat loss, the other gases he considered minor by comparison. He studied nitrogen, oxygen, carbon dioxide and methane as well as water vapour.

In the 1880 "ultra-red" was superseded by "infrared' the modern term for this part of radiation spectrum.

Tyndall had been preceded by Joseph Fourier who in 1820 observed the atmosphere had a warming effect for Earth compared to if it had been a naked rock in space in relation to its distance from the Sun.

Svante Arrhenius:

A controversial figure in his lifetime as well as now was **Svante Arrhenius**, a Swedish scientist born in 1859 the same year Tyndall made his announcements. He initially worked in chemistry and discovered that salts when dissolved in water do so by forming ions, positively and negatively charged. Faraday before him had named ions and felt they were only present when electricity was passes through a solution but Arrhenius showed they were present naturally. He aggravated his professor however and only achieved a 4th class degree in 1884 at age 25. But he was awarded the Nobel Prize for chemistry for this work in 1903.

In the 1890s he became interested in the reasons for ice ages. This led him to study atmospheric warming effects including CO2 . This included studying the reflected light from the Moon and by complicated calculations making conclusions about its effect on warming and cooling of Earth. He was made a professor but the appointment was opposed by many colleagues.

He concluded that CO2 was the *regulator* of the Earth's temperature. He considered that water vapour was the most powerful atmospheric gas by far influencing temperature but that CO2 changes came first. They resulted in the air heating, then more water vapour was drawn in by more evaporation and the extra water vapour caused the temperature to go higher. And the reverse happened when it got colder, he said. A kind of servo switch mechanism. His theory did not disagree with the clear evidence then that water vapour was dominant.

He did very complicated calculations and devised mathematical formulas to predict the Earth's temperature as a result of CO_2 changes. He predicted that a halving of the level would result in a temperature drop of 4° C world-wide. He also calculated that an increase in CO_2 4 times, say from 300 to 1200 parts per million, would result in a temperature increase of 8° C. He considered that human use of coal and fossil fuels would increase the CO_2 levels substantially over centuries and result ultimately in no ice age, a warmer world including Sweden by 5° C which he considered to be a very good outcome. He encouraged lots of coal burning! He was confident he had found the cause of ice ages and saved humanity from them.

But he made several important errors in his calculations. Especially he ignored convection in the atmosphere, the dominant method of heat transfer in gases. He also ignored the effect of clouds due to more water vapour, and the effects of water atmospheric phase changes. He also considered the oceans only contained 5 times the amount of CO_2 as the atmosphere whereas we know this to be 50 times today. He felt that the transfer of CO_2 to and from the oceans was very slow, thousands of years, increasing the importance of atmospheric level changes. For these and other errors many of his contemporaries vigorously disagreed with him about CO_2 being the lead cause.

Knut Angstrom and Herr J. Koch:

Amongst these was **Knut Angstrom**, another Swedish physicist. In 1900 he got an assistant, Herr J. Koch, to measure the heating effect of shining infrared heat radiation through a tube filled with varying amounts of CO_2 . It was found that the heat absorption barely changed with big differences in CO_2 , the first time this had been noted, consistent with the virtually total absorption even at low concentrations of CO_2 in the relevant wavelength band as has been discussed. Angstrom concluded that CO_2 could not be the controller Arrhenius said it was and a major feud erupted. Most agreed with Angstrom.

The dominant scientific view from then until the 1960s and 1970s was that CO_2 was not important concerning the Earth's surface temperature; that water vapour, the oceans, small variations in the Sun's activity, other factors such as volcanic eruptions, large meteorite strikes, changes in the Earth's orbit, which have been discussed above, were the important factors.

Guy Callendar:

But, as is common in human affairs, not everyone agreed. Some still adhered to the "carbon dioxide is the leader" argument. Amongst them was a British steam engine expert, **Guy Callendar**, who became interested in this in the 1930s and published his calculations and thoughts. He concluded that a doubling or halving of CO_2 levels would cause a 2° C change. Essentially he agreed with Arrhenius. But he made the same mistakes, omitting from his calculations the effects of clouds, water vapour phase changes, convection especially. Nevertheless he attracted a following of scientists of a similar view. He also felt, like Arrhenius, that human use of coal and fossil fuels was a very good idea. The key plank of his arguments was that warming had occurred from the 19th century to the 1940s as has been discussed above.

Edward O. Hulburt:

Another scientist, a geophysicist, who considered CO2 was the main controller was **Edward O. Hulburt** born in 1890. He worked at the U.S. Naval Research Laboratory in San Diego for most of his career and was professor from 1926. His main interest was the **Ionosphere,** the very high atmospheric layer which helps protect Earth from harmful Sun radiation as already discussed. He discovered that the Sun's radiation broke up molecules and atoms in this layer to turn them into charged particles, ions, which block some dangerous Sun radiation. But he later became interested in the cause of ice ages and made calculations based on Arrhenius's theories. He calculated that doubling CO2 concentration would result in a 4°C world-wide temperature increase and halving it a similar decrease. He also did not oppose humans using fossil fuels He published this in 1931. Again in retrospect he did not allow for clouds, convection and water vapour effects.

Some Prominent Modern Scientists Convinced that Human Fossil Fuel Use and Subsequent Increased Carbon Dioxide Production is the Dangerous Cause of Global Warming – *"Alarmists".*

Gilbert N. Plass:

In the 1950s scientists who believed human use of fossil fuels was risking a hazardous global temperature rise emerged. By this time some early computers allowed more complex calculations based on models. One was **Gilbert Plass.** He was born in Toronto, Canada but spent his career in the U.S. He gained a PhD in physics at Princeton and after various research roles was professor of physics at Texas A and M University, one of the largest and most prestigious American universities.

In his early career he studied infrared radiation absorption in the atmosphere. He became the editor of a research journal "Infrared Physics and Technology". He was one of the first scientists to use the early computers for complex calculations. He became concerned about CO2 heat absorption. He studied the *shoulders* at the edge of the 14 to 16 micron wavelength intense absorption band which have been discussed above and considered them important as causing more heat absorption if the CO2 level increases, and less if it dropped. He was also interested in the high atmosphere up to 75 kilometers, including the stratosphere and part of the mesosphere where there is CO2 . Above that CO2 are largely broken up by Sun radiation. He calculated that doubling CO2 would result in a 3.6° C increase, that in 2000 global temperature would be 1°C higher and CO2 30% higher than in 1900. He also was later criticized for not allowing sufficiently for clouds, convection and water vapour in his calculations.

In 1953 he was featured in "Time" magazine in an article expressing concerns about carbon dioxide as a greenhouse gas, human use of fossil fuels and rising temperature risks. In 1956 he published his findings scientifically. He was also a prominent philatelist (stamp collector) and intensely interested in classical music including compering radio programs. He died in 2004

Charles David Keeling:

Charles David Keeling was another American scientist who became a giant in this field. He was born in 1928 and after a Bachelor of Arts degree obtained a PhD in chemistry from the California Institute of Technology (CIT). He was based at the Scripps Institute of Oceanography at the University College of San Diego for 50 years, the Head of Department for many.

In the 1950s he worked on new methods to accurately measure atmospheric CO2 levels including using the new **mass spectrometer** discussed already. He also worked on the new **Non Dispersive Infrared Meters (NDIR)** which had just developed. These offered faster more accurate and convenient methods of CO2 level measurement than the previous methods. Keeling and his team did thousands of measurements in many places ranging from urban cities to dense forests from sea to mountain top. He observed the daily cycle which occurs in forests, the CO2 rising at night when photosynthesis of plants stops, and dropping during the day when it restarts with sunlight. He also observed the seasonal cycle from northern hemisphere winter to summer; in the northern hemisphere winter the global CO2 increases and in the northern hemisphere summer it decreases. This is because most land is in the northern hemisphere, the southern is largely ocean. In the northern winter a large amount of the land is covered with snow and ice, very cold, so land plant photosynthesis stops and world-wide CO2 rises.

Keeling also found that while CO2 levels varied from place to place, urban to forest, there was a certain *general world-wide background level* which could be calculated by computer modelling from many individual measurements. And especially from measurements at some remote places including the South Pole and high up the Hawaiian **Mauna Loa Volcano** where he started doing measurements in 1958. These results were, and are now, **adjusted** for factors including urban pollution and CO2 production by the volcano.

Keeling documented a rise in the general background level from when they started measurements in 1958. The rising graph is known as the **Keeling Curve**. It has steadily gone up from 320 parts per million in 1958 to 410 now, 0.032% to 0.041%. Keeling considered that the rise started with the **Industrial Revolution** and widespread use of coal 150 years before, when he considered levels were around 280 (0.028%). He became very alarmed about the risk of dangerous global warming arguing very forcefully and with great political effect for cessation or least a decrease in human use of fossil fuels.

Keeling also discovered the difference in carbon isotope uptake by land plants and ocean plants. There is a difference in the C^{12}/C^{13} ratio of the CO2 air and sea molecules taken up by these plants in photosynthesis related to a slight difference in the process itself. Keeling tried to use this to determine how much atmospheric CO2 was taken up by land versus ocean plants. He also studied so-called CO2 sinks, the uptake by land and the oceans already discussed. He tried to measure the rate of CO2 transfer to and from the air and ocean and tried to develop computer models for the massive movements of CO2 through the oceans already discussed. He had difficulties with these aspects and in response started research organizations to do that and other work in this area, and these continue.

Keeling achieved great eminence and many prestigious awards including the National Medal of Science, the nation's highest for science, in 2002 and the Tyler Prize, the world's most distinguished environmental science award in 2005. He was undoubtedly a very fine man with a strong family, interested in playing the grand piano, choir member, and hometown politics as well as his global advocacy concerning climate change.

Looking back it is argued that he underestimated the amount of CO2 in the oceans and the transfers to and from the oceans and atmosphere. He considered the rate of transfer from the atmosphere to the ocean to very slow taking hundreds of years. This would have added to his concerns about fossil fuel use.

James Hansen:

Another very prominent brilliant American scientist concerning the role of CO2 is **James Hansen.** He was born in 1941, 13 years after Keeling. At the University of Iowa he gained a Bachelor of Arts degree in physics and mathematics with distinction: he went on to gain Master of Science degree in Astronomy and a PhD in Physics. He worked at the Goddard Institute of Space Studies from 1981 to 2013, most of that time as Director. His early years were spent studying computer radiative (heat energy radiation) transfer models imitating the atmosphere of Venus. Venus to us is an extremely hostile place with temperatures around 400°C, an atmosphere 95% of which is carbon dioxide with clouds of sulfuric acid which Hansen discovered.

There is a theory that billions of years ago when the Sun had one third less heat radiation than now Venus was cooler and had liquid water. But that is just a theory. Another theory, as already discussed, is that if Earth were only 5 million kilometers closer to the Sun liquid water could not exist here and Earth would be similar to Venus.

Hansen later became the professor of the Program on Climate Science, Awareness and Solutions at the Earth Institute of Columbia University, New York.

Hansen became concerned, undoubtedly partly at least due the above studies, that Earth was at risk from overheating due to a build-up of CO2 . At this time, 1980s, Keeling and others' work was well known. He started the Global Temperature Record in 1980 and the later Global Historical Climatology Network in the 1990s. These networks took data, including temperature from multiple world-wide stations. These tended to be located where the scientists were mostly in northern hemisphere continents but many from other parts also. By the 1990s there were more than 280 stations, and 3800 readings extending over a 50 year period, 1600 over 100 years, and 220 over 150 years. Hansen was confident that his computer modelling accurately predicted the future.

By 1988 he had formed the view that the global temperature had increased by up to 0.7°C from the previous century. He was sure this was due to human use of fossil fuels especially coal. He made a confident submission to the U.S. Senate hearings on this subject in 1988 advocating a move away from fossil fuels. He predicted the temperature would increase in the 1990s.

In 1999 in a further report he found that 1998 was the warmest year since 1980 according to his results. But his data showed that the Eastern U.S. and the Atlantic Ocean were cooler.

Also in 1999 other researchers, Spencer and Christy, found that satellite temperature measurements of the troposphere upper layers showed cooling in recent years in contrast to computer model predictions. But the satellite systems had their problems as has already been discussed especially decay of their satellite orbits and calibration issues.

He became interested in the influence of atmospheric aerosols, particles of dust and other substances on the climate and in 2000 published results on **black carbon,** a common pollutant in the air over industrial areas especially China. He felt this was masking the influence of CO2 explaining why his earlier heating predictions were not occurring. He also said then: "---present knowledge does not permit accurate specifications of the dangerous level of human-made greenhouse gases." Admitting understandable limitations in this difficult field.

Also in 2000 he advocated stopping the use of coal by 2030.

In 2007 he found the climate system was heating faster with rising sea levels, faster than his earlier models predicted. In 2009 he advocated nuclear power are the best option, especially ***fast reactors and thorium reactors*** which consume their waste or do not produce any. These are still at the research stage.

Also in 2007 Stephen McIntyre and others reported errors in Hansen's group's work with the computer processing which increased the reported temperature by 0.15°C. The critics considered that 1934 remained the hottest year of the century. There was criticism of the "adjustments" in Hansen's group's results. The criticisms were accepted including by the IPCC.

Also in 2007 he stated that a CO2 level of 450 parts per million (0.045%) was dangerous, and that a doubling (to 600 parts per million) would result in a 3°C global temperature increase.

In 2009 he published a study finding that at least 1.8 million deaths due to world-wide air pollution from fossil fuel and coal use could have been prevented from 1971 if the world had shifted to nuclear. Critics, especially antinuclear groups, commented that he had underestimated considerably deaths from nuclear accidents including Chernobyl.

Also in 2009 more problems with computer modelling surfaced, mainly due to difficulties with clouds and sea ice modelling.

He endeavoured to encourage movement away from fossil fuels especially coal by participating in protests including one at the White House in 2013 against the Keystone Pipeline planned to take oil from Canada to the Gulf of Mexico. He was also in protests against coal mining.

In his distinguished career he has received many prestigious awards and prizes including a Taiwanese one worth $25 million for his department. And an award from the American Meteorological Society. This was criticised by some dissenting members of the Society.

He has been at the forefront of research in this field with special interests in information from satellites and computer modelling which have developed rapidly in his time.

Hubert Lamb:

Hubert Lamb was an English meteorologist who was born in 1913 and spent many years working for the United Kingdom Meteorological Office as a practical forecasting meteorologist and Technical Officer. As a Quaker he was a conscientious objector to some aspects of wartime work in the U.K. so spent time at the Irish Meteorological Office. Later he served in West Germany and Malta.

He was clearly a very brilliant researcher and was promoted on merit, eventually founding the Climate Research Unit of the University of East Anglia in 1972. He did early research on the Medieval Warm Period and Little Ice Age, being one of the first to detail these. For this he reviewed paleontological evidence extensively including fossil pollen data.

He was interested also in the likelihood of the ice age resuming, along with others in the 1960s, a cool period. He became interested in carbon dioxide warming effects but commented in a book written in 1977 that CO2 effects were: "much smaller than suggested.".

He produced an early version of a global average temperature graph showing the Medieval Warm Period and the Little Ice Age.

In 1984 he predicted that the ice age was 3000 to 7000 years away. And that over the next few hundred years there would be warming due to CO2 from human fossil fuel use, but that would stop when the fuels ran out. There would then be cooling leading up to the ice age resuming.

He was awarded the Symons Gold Medal of the Royal Meteorological Society, and received an Honourary Doctorate. He died in 1997.

Jerry D. Mahlman:

Another prominent American scientist in this field was *Jerry Mahlman.* He was born in 1940 and later obtained a PhD at Colorado State University in geophysics especially fluid dynamics. He became Director of the Geophysical Fluid Dynamic Laboratory at Princeton University in 1984, a post he held until 2000. He worked on the development of computer models for atmospheric research. He was involved in research into the effects of atmospheric nuclear bomb explosions earlier.

He later worked on *chlorofluorocarbons* escaping from refrigeration equipment into the atmosphere and destroying ozone as has been discussed already. He was one of those instrumental in the Montreal Protocol to ban them.

He became concerned about the risk of global warming due to human use of fossil fuels. He was the first to describe the graph of northern hemisphere average temperature produced by the work of Mann, Bradley and Hughes in 1999 as a *"Hockey Stick".* This was because it showed the global average temperature had remained low for the last 1000 years then risen over the 100 producing a shape like a hockey stick. There are many online examples of the graph.

Mahlman continued to work with computer modelling especially for the stratosphere. He became concerned about carbon dioxide from human use of fossil fuels. He was awarded the American Meteorological Society Medal, and also a Presidential Award for his work, especially on chlorofluorocarbons and ozone destruction. He died in 2012.

Michael E. Mann

Michael Mann is an American climatologist who is widely known for his work with Raymond Bradley and Malcolm Hughes on computer modelling of multiple data to estimate the northern hemisphere temperatures from 1000 A.D. producing the "Hockey Stick" graph.

He was born in 1965, his father was a professor of mathematics. His early academic work was on computer simulations of randomness in crystals, then later on seismology. He gained his PhD from Yale University in 1998, his thesis on ocean-atmosphere interactions.

He teamed up with Bradley and Hughes. He became Professor of Meteorology at Pennsylvania State University and Director of the Earth System Science Centre there.

In commenting on the "Hockey Stick" paper "MBH99" he said: "---20th century exceptionally warm in the last 900 years ---" and: "---though substantial uncertainties exist ----" referring to the difficulties determining surface temperature hundreds of years ago and the inevitable errors.

Nevertheless he was convinced there was a problem with human use of fossil fuels and he has been at the forefront of advocacy about that. He has written several books including: "The Madhouse Denial: How climate change denial is threatening our planet." and: "The New Climate War: the fight to take back our planet.". He runs a website with Gavin Schmidt and others: "RealClimate."

In 2018 he predicted that the Antarctic may lose double the expected ice before 2100 and sea levels rise by 2 meters world-wide due to human fossil fuel use.

He has received many awards including Fellow of the American Meteorological Society and the Medal of the European Geosciences Union.

Raymond Bradley:

Raymond Bradley is a climatologist and paleoclimatologist (the study of ancient climate by fossil records, fossil tree ring studies, coral fossils, ice core studies and other methods). He is the Professor of Geosciences at the University of Massachusetts Amherst and Director of the Climate System Research Centre there. He is closely associated with the National Oceanic and Atmospheric Administration (NOAA) and the International Panel on Climate Change (IPCC).

He obtained his PhD at the Institute of Arctic and Alpine Research University of Colorado and also a DSc from Southampton University U.K. in paleoclimatology. He has done a lot of field work in the Arctic and North Atlantic.

He was one of the lead researchers with Mann and Hughes for the "Hockey Stick" study.

He has a special interest in climate changes since the last ice age. He is an advocate for stopping human use of fossil fuels and has had many media interviews concerning that. He has written a number of academic books and has received many awards including the Medal of the European Geoscience Union and an honourary doctorate from Liverpool University.

Malcolm Hughes:

Malcolm Hughes is a climatologist and dendrochronologist (tree ring studies used to determine the ancient climate). He is the Professor of Dendrochronology at the University of Arizona.

He is British and obtained his PhD in ecology at the University of Durham.

He is interested in using fossil tree ring, sedimentary rock, ice core and historical data to estimate ancient climates. He participated with Mann and Bradley in the research leading to the "Hockey Stick" papers, MHB99.

Gavin A. Schmidt:

Gavin Schmidt is the Director of the Goddard Institute of Space Studies (GISS) in New York, which is part of the National Oceanic and Atmospheric Administration (NOAA) there. He succeeded James Hansen in this role.

He is British. He gained his PhD in Applied Mathematics at University College, London. He is especially interested in developing computer models concerning ocean circulation and climate forcing. He continues the work of Hansen advocating for cessation of human use of fossil fuels.

Susan Solomon:

Susan Solomon is an atmospheric chemist. She is professor of Atmospheric Chemistry and Climate Science at the Massachusetts Institute of Technology, a premier American research institution. She gained her PhD in atmospheric chemistry at the University of California at Berkely in 1981.

She has been Head of Department for the Chemistry and Climate Processes Group at the National Oceanic and Atmospheric Administration (NOAA) also.

She was the first to describe the ozone hole due to the effect of chlorofluorocarbons from refrigeration and active concerning the Montreal Protocol and its banning of these.

She also has been a contributing author for the International Panel on Climate Change (IPCC) reports. She was awarded the "Future of Life Award" for her work on ozone.

Kate Marvel:

Kate Marvel is an associate research scientist at the Goddard Institute for Space Studies (GISS) part of the National Oceanic and Atmospheric Administration (NOAA) in New York.

She gained her PhD in 2008 at the University of Cambridge in theoretical physics. She did later studies at Stanford University especially on computer modelling to predict temperature including the difficult area of cloud modelling. She has also worked on the effects of aerosols and atmospheric pollutants, and the effects of drought.

She regularly writes a section called "Hot Planet" for Scientific American magazine, and is active advocating for cessation of human fossil fuel use.

Mauro Facchini:

Mauro Facchini is Italian and is the Head of Copernicus, which is the European Earth Observation Program of the European Commission.

He obtained his PhD in mechanical engineering, then had a number of other academic posts before this current one concerned with the Earth's climate.

He has commented on the Copernicus Report for 2021 highlights of which are:

- CO2 rose to 414 parts per million (0.04%) from their data.

- Methane levels rose to 1876 parts per billion (1.8 parts per million, 0.00018%) For reasons that are unknown as there are many natural causes for methane increasing.

- The 2021 global average temperature from their data was 0.3° C above the 1990 – 2020 average, and 1.1° C above the 1850 -1900 records.

- The last seven years are the warmest on their records for global average .But some places were cooler especially East Australia, Siberia, Alaska, the Pacific probably due to the La Nina ocean cycle.

- Their global record starts in 1970 (the end of the cool period as discussed) and the global temperature is 1°C higher than then.

- The European temperature is 0.1° C higher than the 1991 -2020 average for Europe. April was colder and the summer warmer than the average.

Vincent-Henri Peuch:

Vincent-Henri Peuch is the Head of the Copernicus Atmospheric Monitoring Service (CAMS) of the European Union and joined that organization in 2011. He has had a very distinguished academic career and before the CAMS association spent 15 years researching a number of atmospheric subjects including:

- Air quality modelling.

- Pollutants.

- Ozone.

- Climate.

- Satellite data.

- Cloud computer modelling.

He obtained his PhD from Ecole Normale Superieure de Lyon in 1996 on "Numerical Modelling of elementary chemical processes in catalytic processes." He subsequently directed 12 students through PhD theses and has 80 peer reviewed publications in the scientific literature. From 1997 to 2005 he was a scientist researching atmospheric modelling at the National Centre for Meteorological Research, Toulouse and from 2005 to 2011 did further research in Toulouse.

Jim Salinger:

Jim Salinger was born in New Zealand in 1947. In 1971 he gained a Bachelor of Science degree in geography at the University of Otago and in 1981 a PhD also in geography at Victoria University of Wellington. The thesis for this was on: "The New Zealand climate: the instrumental record." In 1999 he gained Master of Philosophy in Environmental Law with first class honours from the University of Auckland, his dissertation on New Zealand signing up to the Convention to combat drought and desertification, which the country subsequently did.

After a period as a research associate at the University of East Anglia he worked at the Meteorological Service of New Zealand initially as the senior agricultural scientist then later as principal scientist. He had a high public profile in New Zealand often appearing on TV regarding weather matters. In 2006 he became president of the World Meteorological Association for 4 years winning an award for exceptional service concerning agricultural meteorology. Other awards included companion of the Royal Society of New Zealand for the promotion of climate science to the public, and in 2016 he was a semi-finalist in the New Zealander of the year awards.

Later appointments included as a visiting professor at Stanford University, California, another at the University of Florence, another at the University of Haifa in Israel, another at the Pennsylvania State University, and another as research fellow at the University of Tasmania. These were all in climate science. He is now retired but is still active concerning climate change.

Concerning climate he is a passionate advocate for cessation of fossil fuels. In 2016 he stated that up to 90% of corals would die if temperature increased 1.5° and 99% would if the increase was 2°. There was a heatwave in New Zealand in 2018, which was an El Nino year, and he maintained the heatwave was caused by human fossil fuel use. Also in 2018 he criticised a New Zealand government tree planting program as it was mostly pine trees, not natives which have a much longer lifespan.

In 2019 he was in a deputation to the government stating that "time was running out" and advocating immediate severe action to reduce CO2 emissions, including for farmers. He wrote an impassioned publicised letter to his grandchildren apologising for the disastrous world left to them by his generation. In 2020 he linked a major forest fire in the South Island to human fossil fuel use, although others considered a spark from power lines was the cause.

He has written several books including: "Climate Change in New Zealand: Scientific and Legal Assessments."

Kevin Trenberth:

Kevin Trenberth was born in Christchurch, New Zealand in 1944. He was in the New Zealand Army for a time and very good at rugby. He initially studied mathematics before meteorology. He won a New Zealand Government scholarship to study at the Massachusetts Institute of Technology (MIT) and gained a Doctor of Science degree there in meteorology.

In 1984 he moved to the National Centre for Atmospheric Research (NCAR). He also has had appointments at the University of Auckland, the New Zealand Meteorological Service, the University of Illinois. He has served on a number of important committees including the Joint Scientific Committee of the World Climate Research Association.

He has been heavily involved in the Intergovernmental Panel on Climate Change (IPCC) and was a lead author for the 2001 and 2007 reports. His research has included especially the water cycle in the environment, the Earth's energy budget and global climate change. He has been a powerful advocate for action to reduce human use of fossil fuels. He has frequently been in debates with sceptics including Richard Lindzen, Roy Spencer and John Christy.

He has over 270 papers published in the peer reviewed scientific literature and has contributed to many books and reports on human induced global warming due to fossil fuels.

In the Climate Research Unit at the Hadley Centre email hacking event an email from him was publicized widely. It said: "The fact is we can't account for the lack of warming at the moment and it is a travesty that we can't." The lack referred to was a cooling period from 2004 to 2008.

In 2013 he maintained that warming in the oceans was occurring deep below 700 meters, not on the surface.

He has received many awards including the Roger Revelle Medal from the American Geophysical Union, an NCAR Distinguished Achievement Award, is a Fellow of the Royal Society of New Zealand, and a Fellow of the American Meteorological Society.

Clearly there are many more scientists concerned about this issue.

Some Prominent Scientists who do not consider human fossil fuel use and consequent Carbon Dioxide production is a significant cause of dangerous Global Warming – *"Naturalists" or "Sceptics", or, unkindly by Alarmists – "Deniers"*:

Sir Fred Hoyle:

Fred Hoyle was born in 1915 in Yorkshire. His mother was a pianist including in cinemas in the silent movie era. His father was a violinist, worked in the wool trade and served as a machine gunner in the first world war. The family was poor but he won a scholarship to Cambridge University and in 1936 at age 21 gained a Bachelor of Mathematics degree with distinction, winning the Applied Mathematics prize. He became interested in nuclear physics and in 1939 fulfilled the requirements for PhD but declined to accept it as it would affect his tax status! He later did research at Cambridge in astrophysics especially concerning star composition and the origins of the universe.

During World War II he was involved in research on radar. After the war he pioneered work that showed that stars had massive temperatures and pressures inside them that caused atoms to join making heavier elements. In 1957 he and others published a famous paper: "Synthesis of the Elements in Stars". This led to him becoming Professor of Astronomy at Cambridge. He was knighted in 1972.

He received many awards and honours including the Gold Medal of the Royal Astronomical Society of which he was President for a term; made Fellow of the Royal Society in 1957; elected a member of the US National Academy of Science in 1969; and the Crafoord Prize of the Royal Swedish Academy of Science, amongst many others.

He was involved in several controversies including amongst staff at Cambridge causing him to resign from there in 1972. He subsequently wrote many books both fiction and non-fiction. Other controversies including disputing the "Big Bang" theory of the origin of the universe, and the Earth based origin of life, favouring instead a theory of seeding from space.

He was a proponent of the resumption of the ice age and in 1981 his book: "Ice – a chilling scientific forecast of a new ice age." was published. In this he makes it clear he disagreed with those stating human fossil fuel use was leading to increased global temperatures. In particular he points to Carbon Dioxide intensely absorbing heat infrared radiation in the wavelength band 14 to 16 microns so that increasing or decreasing the CO_2 concentration makes no difference because the effect is maximal at very low concentrations, and therefore not increased by more CO_2 in the atmosphere, it is at full song as discussed already. He discusses that 550 million years ago CO_2 levels were more than 10 times higher than now, about 5000 parts per million. 0.5%. Yet there was no runaway greenhouse effect on global temperature then.

An additional argument he put is that if CO_2 was the lead cause of global temperature changes then ice ages would be preceded by very low levels, less than 180 parts per million, 0.018%. That would result in a failure of photosynthesis and plant life, with consequent mass starvation for all life on Earth. But that did not happen. Rather life continued thriving in tropical areas especially suggesting CO_2 levels were not unduly low.

He also cites evidence already discussed that CO_2 levels follow by several hundred years global temperature changes.

He considered the most likely explanation for ice ages was large meteors striking Earth resulting in massive cooling due to atmospheric dust and particles blocking the Sun. He also discussed "diamond dust", tiny ice particles forming in the upper atmosphere reflecting the Sun's heat resulting in more cooling, a vicious circle leading to an ice age.

This book was published when James Hansen and others were promoting their arguments as discussed.

Sir Fred Hoyle was a very independent brilliant thinker. He suffered a severe accident walking on the Yorkshire Moors near his home in 1997 and died in 2001.

Joseph D'Aleo:

Joseph D'Aleo is a prominent American meteorologist with 30 years' experience in that science. He graduated Master of Science from the University of Wisconsin and subsequently was professor of meteorology and climatology at Lyndon State College in Vermont, USA. He has served as chair of American Meteorological Society committees on weather forecasting. He has been very active in public broadcasting on weather issues and is currently involved in one of these, Weatherbell Analytics. There is a religious aspect to his believe that the climate is stable "created by God's intelligent design ---".

He became a leading public figure in the controversy about hacked emails from the Climatic Research Unit at the University of East Anglia in 2009. He alleged that temperature data there and at other centres promoting climate alarmism had been "manipulated". In particular he criticised views that 2005 was the warmest year on record, pointing to the 1930's and 1940's as warmer, and also to the Medieval Warm Period as being warmer also. He with others runs a blog called "Icecap".

Points he makes include:

- That global temperatures have been cooling since 2002 despite CO_2 increasing.

- CO_2 is a trace gas and causes little warming.

- There is no correlation between CO_2 levels and temperature.

- CO_2 is a natural plant food.

- CO_2 levels now are near the lowest levels they have been for 550 million Years when they were more than ten times as high without causing a "runaway greenhouse effect."

- The evidence is clear, as already discussed, that temperature increases come first before CO_2 changes by usually several hundred years and not the other way round.

With other like-minded scientists including Sallie Balilunas, Robert Carter, Robert Baling he continues to state that water vapour is the dominant heat trapping gas in our atmosphere and that alarmism about human use of fossil fuels is not justified.

His other work includes research on solar and oceanic weather cycles, and improved methods of forecasting.

Sallie Balilunas:

Sallie Balilunas was born in New York and gained a Bachelor of Science degree in astrophysics at Villanova University in 1974 and later a PhD from Harvard in astrophysics in 1980. Her thesis was on: "Optical and ultraviolet studies of stellar chromospheres of Lamda Andromeda and other late-type stars." She was later a research associate at Harvard, then astrophysicist at the Smithsonian Astrophysical Observatory in 1989. She became a professor at Tennessee State University and deputy director at the Mount Wilson Observatory. She has received many awards including the Bok Prize from Harvard, the Newton Lacy Pierce Prize from the American Astronomical Society, the Donald E Billings Award in Astrophysics from the University of Colorado and the Amelia Earhart Fellowship from Zonta international. She was named one of America's outstanding women scientists by "Discover" magazine in 1991.

In 1992 she was a co-author of a paper concluding that the Sun could vary its output more than had been thought based on studies of other stars. In 1995 she publicized her view that fears of global warming were greatly exaggerated. In 2003 she and Willi Soon, another astrophysicist, published a review paper in the journal "Climate Research" which concluded that the Medieval Warm Period was probably at least as warm as now, and that variations in solar output were the likely explanation for warm and cool periods, like that and the later Little Ice Age. The paper caused an outcry from alarmists as discussed below. She is now retired.

Willie Soon:

Willie Soon was born in 1955 in Malaysia. He emigrated to the US in 1980 and obtained a Bachelor of Science degree in 1985, a Master of Science in 1987, then a PhD in aerospace engineering with distinction in 1991 from the University of Southern California. His thesis was on: "Non-equilibrium kinetics in high temperature gases." He then had a research position at the Smithsonian Centre for Astrophysics at Harvard where he remains. He was also an astronomer at the Mount Wilson Observatory.

In 2003 he and Sallie Balilunas published their paper in "Climate Research" mentioned above. The paper was a response to that by Mann, Bradley and Hughes already mentioned where the "Hockey Stick" graph featured, and which alleged that current temperatures where the warmest for a thousand years due to human use of fossil fuels. Soon and Balilunass' paper reviewed evidence showing that the Medieval Warm Period was probably warmer, and that variations in solar output were the likely explanation for warm and cooler periods, not carbon dioxide changes. Their paper, especially its publication in the prestigious "Climate Research" journal, caused an outcry amongst alarmists. No less than 10 of the journal's editors resigned. It was alleged the peer review process had been compromised before publication, and that the evidence quoted did not prove the points made. In addition Soon was alleged to have received funding from the oil and coal industries. He denied that he had been influenced by that.

The managing director of the journal, Otto Kinne, a German Marine biologist and professor at the University of Kiel, said: "while these statements may be true (that is about the Medieval Warm Period being warmer and the Little Ice Age being colder due to solar activity changes) the critics point out that they cannot be concluded from the evidence provided in the paper."

When the paper is read it cites evidence including from fossil sediments, coral fossils, ice cores, tree ring studies, and glacial retreat studies that the Medieval Warm Period and Little Ice Age were global events. It points to difficulties accurately determining what the temperatures were in those times, estimating errors at between 1.3 and 1.8°C. It states that the conclusions in the paper are provisional and that some of the data is poor.

It is concluded that the coldest time in the last 1000 years was from 1530 to 1730, and the next coldest the 19th century. They conclude that 20th century winters may be 0.5°C warmer than the Medieval Warm Period, and the summer temperatures now similar to then, although it might have been warmer. The Vikings settled Iceland in 986 AD and abandoned it due to cooling in 1400 AD. Other evidence cited in the paper was :

- Evidence of more warming than now by 0.8°C from Asian tree ring studies from 1000 to 1300 AD.

- Of a warm period in Japan from 900 to 1300 AD both from historical information and late freezing dates from cherry and cedar tree fossil data of about 1°C more than now.

- Evidence of droughts in North America from 800 to 1200 suggesting warmth.

- Stalagmite cave records from Nelson in New Zealand suggesting a warm period from 1200 to 1400 and a cold period from 1600 to 1700.

- A great deal of other evidence from Africa, South America, Antarctica, South Georgia where glaciers peaked from 1700 to 1900 but had retreated in 1000 AD.

In their discussion they argue that Mann, Bradley and Hughes were "premature" in their conclusions because of the difficulty correlating old data from tree rings, fossils, ice cores etc with modern instrument readings, which are comparatively of very short duration. They point to tree ring data especially suggesting the Medieval Warm Period was at least as warm as now. Some data from South Africa suggests it may have been 3 or 4°C warmer there then.

They also point out that choosing the last 1000 years, as Mann, Bradley and Hughes did, is: "strictly a convenience that merits little scientific weight." Meaning that the Earth is 4.5 billion years old and complex life has existed for over 500 million years, a thousand years is nothing.

Given all this, and that it was a review paper discussing existing research and not pretending to discuss new research , the intense reaction against it is interesting, as are the many attempts to discredit the authors. A number of scientists defended the paper. It could be argued that in a more tolerant world there would have been a gentler more academic approach; instead there was intense condemnation.

Willie Soon continues to work at the Smithsonian and as a professor at the University of Putra in Malaysia. He is also active against climate alarmism including with the "Heartland Institute", a think tank.

Robert M Carter:

Robert Carter was born in 1942 in Reading, England. He emigrated to New Zealand in 1956, attending Lindisfarne College in Hawkes Bay. He then gained a Bachelor of Science degree in geology at the University of Otago in 1963. He returned to England and gained a PhD in Palaeontology from Cambridge in 1968. His thesis was on: "The Functional Morphology of Bivalved Mollusca."

He then became professor and head of the School of Earth Sciences at James Cook University, Queensland, where he remained until 2013. He held many other positions including chair of the Earth Sciences Panel of the Australian Research Council, director of the Australian Ocean Drilling Program. He received many awards including Honorary Fellow of the Royal Society of New Zealand, Special Investigator Research Award from the Australian Research Council, the Hochstetter Lecturer from the Geological Society of New Zealand.

In 2006 he publicized his view that changes in climate were not related to human use of fossil fuels, but rather natural causes including El Nino and other ocean cycles. He pointed to no warming having occurred since 1998 despite CO_2 levels having increased. He published widely in journals including the Journal of Geophysical Research. He testified to a Select Committee on Climate Policy of the Australian Parliament in 2009 and also before the United States Senate. He became active in the "Heartland Institute", along with Sallie Balilunas, Willie Soon and others who did not consider human fossil fuel use harmful.

He was a founding member of the New Zealand Climate Science Coalition with Brian Leyland, a prominent electrical engineer, and Augie Auer, the Chief Meteorologist of the New Zealand Meteorological Service. The Coalition promoted its view that human use of fossil fuels was not causing changes in the climate.

In 2007 he appeared in a UK Channel 4 documentary: "The Great Global Warming Swindle." Points he made included:

- Global temperatures decreased from 1940 to 1970, then increased to 1997 and have remained stable since then despite CO_2 levels still increasing.

- The Vostok ice core studies show atmospheric CO_2 levels follow ocean temperatures changes with a lag period of 800 years as already discussed.

- Water vapour is the most important warming gas, but computer models cannot deal with cloud formation, a crucial error. The Sun is the biggest influence on global temperature.

- The Medieval Warm Period was warmer than now but CO2 levels were lower.

He was very highly respected as a principled scientist. He died suddenly in 2016 of a heart attack at age 73.

Nicola Scafetta:

Nicola Scafetta is Italian and after studying physics at the University of Pisa emigrated to the US and obtained a PhD at the University of North Texas in 2001. His thesis was on: "An entropic approach to the analysis of time series."

He then went to Duke University in North Carolina as a research scientist in physics. He became an assistant professor there. His work included developing the ACRIM satellite which measures solar radiation. One of these satellites was destroyed in the Challenger disaster.

Concerning climate change he considers that natural cycles in the solar system including the 11 year Schwabe and 22 year Hale Sun cycles, the 11 year Jupiter cycle, a 29 year Saturn cycle, and lunar cycles, explain climate changes, He does not consider human use of fossil fuels is significant given this.

He is currently at the University in Naples.

Vincent R Gray:

Vincent Gray was born in London in 1922 and gained a PhD in physical chemistry from Cambridge in studies on incendiary bombs. He moved to New Zealand in 1970 and became the chief chemist for the Coal Research Association. After retirement he spent 4 years in China and became concerned about the views that human use of fossil fuels was causing serious warming, disagreeing with that. He commented on more than 1800 aspects of the 2007 Intergovernmental Panel on Climate Change report, disagreeing with it widely. He considered the methods of the IPCC were "unsound" and called for it to be abolished. In 2002 he published a book: "The Greenhouse Delusion: a Critique of Climate Change 2001."

In 2006 he appeared on "Counterpoint" an Australian ABC program with Dr Michael Manton, who had been chief of the Meteorological Research Centre and Professor of Mathematics at Monash University. The program was called: "Nine Lies about Global Warming." Dr Manton took a neutral position. The "Lies" included:

- That the "Hockey Stick" graph and paper by Mann, Bradley and Hughes was "fraudulent". Vincent Gray maintained the IPCC has withdrawn the graph for that reason.

- The instrument record of the last 100 years was influenced by "urban drift" and when that was corrected there was hardly any warming.

- Chinese research showed only minor warming.

- The warm period leading up 1940 occurred before any significant rise in CO2 levels. There was then a cold period to 1970, then warming not related to CO2 levels.

- At the Kyoto United Nations COP-11 climate meeting in Montreal in 2005 there was a severe blizzard and that winter was the coldest in that region for many years.

- Data had shown 1998 was the warmest year and that coincided with a major El Nino as the major cause.

- Ocean temperature readings by ships have changed because the ships are much bigger. Samples are taken from engine water intakes not buckets over the side arguably causing a warming artifact or "difference" compared to the previous method.

- Satellite data does not show any change in global temperature when the "raw data" not "corrected data" is examined.

- Weather balloon studies have not shown any lower atmospheric change in temperature since 1958.

Vincent Gray died aged 96 in Petone, New Zealand in 2018.

Roy Spencer:

Roy Spencer was born in the US in 1955. He gained a Bachelor of Science degree in Meteorology in 1978 from the University of Michigan and a PhD in Meteorology in 1982 from the University of Wisconsin-Madison. His thesis was on: "A Case Study of African Wave Structure and Energetics during Atlantic Transit."

Subsequently he worked at NASA's Marshall Space Flight Centre and became senior scientist for climate studies there. Since 2001 he has been principal research scientist at the University of Alabama. He is also the US Science Team Leader for the Advanced Microwave Scanning Radiometer, an instrument on NASA's Aqua satellite and has had that role since 1994. He also is on several other scientific bodies connected with weather instruments on satellites including the NASA Headquarters Earth Science and Applications Advisory Committee, and National Research Council study panels.

In 2007 he and others published a paper showing that tropical clouds, especially high altitude cirrus clouds, decreased as temperature increased allowing more heat -infrared radiation – to escape to space and cooling the atmosphere as a result. He considered this would reduce the amount of warming predicted by computer models, which did not take this factor into account.

In 2008 he and William Braswell, another meteorologist, published a paper in the Journal of Climate showing that clouds could cause changes in temperature, and not just responding to temperature changes already there. He felt this showed the climate system stabilized itself concerning temperature and other factors.

In 2011 they published another paper showing that satellite observations showed that much more heat is lost to space from the atmosphere than computer models show, especially over the oceans. The paper was criticised by alarmists, including statements that it should not have been published by the journal.

In 2006 he publicly disagreed with the documentary by Vice President Al Gore "An Inconvenient Truth." He considers that climate change is due to natural factors not human use of fossil fuels. He thinks the

Medieval Warm Period and the Roman Warm Period were as warm as the present day before fossil fuel use. He publicly disagrees with the concept of catastrophic global warming caused by humans. He has written a number of books on this including: "The Great Global Warming Blunder: how Mother Nature Fooled the World's top Climate Scientists." in 2010.

He agrees with James Hansen's statement that: "It is impossible to find the average global temperature". He considers that published global temperatures by alarmists are "adjusted" for solar variation, aerosols and atmospheric dust and other factors; and that the raw unadjusted figures show no significant warming. He points to satellite temperature records showing variations from minus 0.7°C to plus 0.7°C over the years from 1979 when they started, with the hottest years in 1998, 2010 and 2016 during El Nino's. The same is shown by land based measurements.

Awards include the American Meteorological Society Special Award for developing a precise record of global temperature from polar orbiting satellites. Articles in the press include: "More Carbon Dioxide Please." in the "National Review".

Richard Lindzen:

Richard Lindzen was born in Massachusetts in 1940. His father was a German Jewish shoemaker and fled Nazi Germany with his wife. The family moved to the Bronx in New York. After early education there Richard gained a PhD from Harvard in 1964. His thesis was on: "Radiative and photochemical processes in stratospheric and mesospheric dynamics."

After positions at the University of Washington, the Institute for Theoretical Meteorology at the University of Oslo, the National Centre for Atmospheric Research, the University of Chicago and Harvard University he became Professor of Meteorology at the prestigious Massachusetts Institute of Technology (MIT) from1983 until he retired in 2013.

During his career he published 230 papers in peer reviewed journals. His early work was on ozone photochemistry in the stratosphere and mesosphere high atmosphere. Later work was on atmospheric tides, the action of high carbon dioxide radiating heat into space, and "super rotation" of the upper atmosphere of Venus, which rotates 50 times faster than the Earth. He also studied the Hadley Circulation which has been discussed already.

Lindzen proposed an "Infrared Iris" theory. This states that as tropical sea surface temperatures increase high cirrus clouds in the upper troposphere decrease allowing more heat, infrared radiation, to escape into space, thus cooling the surface. He considered this an important stabilizing negative feedback mechanism controlling surface temperature, noting that this was not in computer models used by the IPCC and other alarmists.

Lindzen also considered that water vapour's ability to lose heat from the surface was not modelled correctly by computer models in use and greatly underestimated. He also has criticised the way clouds are handled by the models.

He served on the American National Academy of Sciences panel on "Climate Change Science: An Analysis of Some Key Questions." He later criticised the panel's report because it did not acknowledge that just 20 years of data was too short to make conclusions.

Lindzen was a member of the scientific group for the 1995 Intergovernmental Panel on Climate Change (IPCC) report, and also the 2001 report. He criticised the latter saying it had been amended by non-scientists to make more definite conclusions about human use of fossil fuels on climate than the scientists concluded.

In 2007 on the "Larry King Show" on TV Lindzen said: "we're talking a few tenths of a degree change in temperature. None in the last 8 years by the way (since 1998). ---- I think it it's mainly like little kids locking themselves in dark closets to see how much they can scare themselves."

In 2001 he urged the Bush administration not to sign the Kyoto Protocol. In 2017 he urged the Trump administration to withdraw from the United Nations Convention of Climate Change, which they did.

He has been a long term severe critic of computer modelling regarding climate changes. In 2009 he pointed out that the models had failed to predict warming trends from major ocean cycles such as El Nino, the Pacific Decadal Oscillation, the Atlantic Oscillation. He noted that these warming events had been falsely attributed to human use of fossil fuels.

He disagrees with the IPCC prediction of 3°C increase in temperature if CO_2 levels double to about 800 parts per million (0.08%). Rather he considers the increase would be 0.6°C. He thinks the current data time period is: "too short for estimating long term trends."

He is truly eminent and has many awards and honours. These include: Fellow of the American Academy of Arts and Sciences; Fellow of the American Meteorological Society; Distinguished Visiting Scientist at the California Institute of Technology Jet Propulsion Laboratory. He has interested in amateur radio, photography and oriental rugs.

William Gray:

William Gray was born in 1929 in Detroit and after obtaining a Bachelor of Science degree from George Washington University joined the US Air Force as a weather forecast officer. After 4 years of that he was a research assistant at the University of Chicago Department of Meteorology where he gained Master of Science in meteorology . He then obtained a PhD in geophysical science.

Later he went to Colorado State University where he was professor of Atmospheric Science. He was very interested in tropical meteorology and did considerable research on hurricanes. He ran the Hurricane Centre at that university. He was very active with the World Meteorological Organization and travelled widely internationally promoting international cooperation amongst meteorologists. He was also an enthusiastic teacher and many of his students won awards.

Concerning hurricane research he established hurricane forecasts and found that they were cyclical with ocean cycles. He published 80 papers and 60 major reports. He also was very interested in water vapour heat transfer in the atmosphere. He was sceptical about the value of computer modelling in meteorology, advising students to exam by direct observation instead. He said: "Thou shalt not bow before computer terminals nor involve thyself with numerical models."

After retiring from the university he became active opposing alarmists concerning human use of fossil fuels. He considered any human effect was tiny compared to natural variations. He stated that many scientists only went along with alarmism to get funding and grants, which they otherwise would lose. He considered that government leaders and environmentalists were using the issue to gain aid and finances, and to run world affairs, that it was not scientifically based. In 2011 he criticised the American Meteorological Society joining the alarmist position in a paper.

He won many awards including Fellow of the American Meteorological Society, that society's Miller Banner Award and also it's Jule G Charney Award. He won an Honorary Lifetime Achievement Award from the World Meteorological Organization. In 1995 he was ABC (American) TV "Person of the Week" for his hurricane work.

He remained a Lieutenant Colonel in the reserves for the US Air force for many years. He was good at baseball and tried to become professional when younger but a knee injury prevented that and he became an eminent meteorologist instead. He was a strong family person with four children.

He died in 2016 aged 86.

Siegfried Fred Singer:

S. Fred Singer was born in Vienna in 1924. His father was a jeweller and the family fled from the Nazis as they were Jewish. After a time in England he emigrated to the U.S. and obtained a Bachelor of Electrical Engineering degree from Ohio State University in 1944. During WWII he worked on mine development for the U.S. Navy. Subsequently he was a lecturer in physics at Princeton University and in 1948 gained a PhD there in Physics. His thesis was on: "The density spectrum and latitude dependence of extensive cosmic ray showers."

He worked for a large number of academic and government entities. He pioneered the development of early Earth observation satellites. He established the National Weather Bureau's Satellite Service Centre in 1962. He was founding Dean of the University of Miami's School of Environmental and Planetary Sciences in 1964. He was a professor in environmental science at the University of Virginia and held positions at the Environmental Protection Agency.

He was interested especially in satellite development for weather observation, radiation belts, the magnetosphere and meteorites. He also was very interested in the Martian moons, Phobos and Deimos, considering space craft could land on them with advantages. He frequently wrote about issues in the mainstream press, often contrary to accepted thinking. For example in 1994 he wrote articles criticising the Environmental Protection Agency's condemnation of second hand tobacco smoke, "passive smoking." He felt the quoted science was not rigorous –"junk science." He also was critical of accepted ideas about sunlight and melanoma, and also about chlorofluorocarbons and ozone loss for the same reason – considering the science not rigorous enough.

He was the author of more than 400 articles in journals and more than 400 in popular publications including the Wall Street Journal, New York Times, Washington Post.

He widely promoted his view that there was "no convincing evidence that the global climate is actually warming." In 2003 he wrote to that effect in the "Financial Times" and in 2004 was on a British Channel 4 documentary: "The Great Global Warming Swindle." He believed temperatures were higher 1000 years ago during the Medieval Warm Period pointing to the Vikings settling Greenland, including pastoral farming and crops. He felt that if warming did occur it would be good for life on the planet. He agreed humans were putting more CO_2 into the atmosphere but did not think it was making any difference. He considered that the atmosphere was very complicated with many feedback mechanisms controlling temperature which were not allowed for in computer modelling by alarmists.

He was critical of the IPCC from it's first report in 1995, saying it had been "selective" in including evidence and made a statement that the IPCC had ignored satellite weather data showing a slight global cooling trend. He quoted raw satellite data, his critics pointed to "adjusted" data as showing slight warming.

He called the 1995 Kyoto Protocol where nations agreed to cut CO_2 emissions: "dangerously simplistic, quite ineffective, and economically destructive to jobs and living." He set up the "Nongovernmental Panel on Climate Change" in 2004 and this continues as a focus for scientists of this view. He wrote the book: "Hot Talk, Cold Science" in 1997 which was and is very successful. A second edition was published in 1999 and a third edition in 2021. This was co-written with Dr David Legates PhD (in climatology) and Dr Anthony Lupo PhD (in atmospheric science). The book focuses on scientific data widely. The authors summarised their conclusions in a statement in "Cosmos" magazine: "The scientific base for greenhouse warming is too uncertain to justify drastic action at this time." The book has been widely praised by many scientists and public figures including:

Dr Richard Lindzen, discussed above.

Dr John Christy PhD, Distinguished Professor of Atmospheric Science, University of Alabama.

Dr Elliott Bloom PhD, Professor Emeritus for Particle Astrophysics, Stanford University.

Dr Ian Clark PhD, Professor of Earth and Environmental Sciences, University Of Ottawa, Canada.

Dr Larry Bell PhD, Professor Sasakawa International Centre for Space Architecture, University of Houston.

Dr Willie Soon PhD, already discussed.

Dr Nils-Axel Morner, Emeritus Professor of Paleogeophysics and Geodynamics, Stockholm University, Sweden.

Dr Patrick J Michaels PhD, Former Professor of Environmental Sciences, University of Virginia and Former President American Association of State Climatologists.

Dr Sallie Balilunas PhD, already discussed.

Dr L. Graham Smith PhD, Professor Emeritus of Geography, University of western Ontario, Canada.

Dr Peter D Friedman PhD, Professor of Mechanical Engineering, University of Massachusetts Dartmouth.

Dr R Timothy Patterson PhD, Professor of Geology, Carleton University, Canada.

Dr Randy Simmons PhD, Professor of Political Economy, Utah State University.

Dr Curt G Rose PhD, Emeritus Professor of Environmental Studies and Geography, Bishop's University, Canada.

Dr Jan-Erik Solheim, Professor Emeritus, Institute of Theoretical Astrophysics, University of Oslo, Norway.

Dr Peter Stilb PhD, Emeritus Professor of Chemistry, Royal Institute of Technology, Sweden.

Dr Lawrence E Gould PhD, Professor of Physics, University of Hartford, Connecticut, U.S.A.

Dr Clifford D Ollier PhD, Professor Emeritus of Geology, University of Western Australia.

----- and many others. All these agree with the authors' statement that the data is too uncertain to justify drastic action. Many state in their comments that the scientific debate on climate issues has been distorted badly by non-scientific forces, and many use passionate language to describe their disquiet about this issue.

Concerning awards and honours Singer received the White House Special Commendation Gold Medal Award from the U.S. Department of Commerce, and the first Science Award from the British Interplanetary Society. He was made a Fellow of the American Association for the Advancement of Science and received an Honorary Doctorate from Ohio State University.

He died in 2020 aged 95.

Ian Plimer:

Ian Plimer was born in 1946 and grew up in Sydney, Australia. After obtaining a Bachelor of Science degree in mining engineering at the University of New South Wales he gained a PhD in geology at Macquarie University. His thesis was on: "The pipe deposits of tungsten-molybdenum-bismuth in eastern Australia."

He then worked in Broken Hill as a lecturer in geology at the University of New South Wales there, as well as being Chief Research Geologist for North Broken Hill Ltd. In due course he became professor of geology at the University of Newcastle, then as professor and Head of Geology at the School of Earth Sciences, University of Melbourne from 1991 to 2005 when he became professor emeritus. He also was professor of mining geology at the University of Adelaide.

He has published more than 120 papers on mining geology.

In addition he has had an extensive business career in mining in Australia.

He has been very active publicly concerning climate change, considering that the alarmists are "irrational, and prejudiced by grants and funding." He is critical of computer models used by alarmists, saying they emphasise unduly CO_2 levels, and underestimate other factors such as solar variations, ocean cycles such as El Nino. He also considers that volcanic production of CO_2 has been greatly underestimated especially the great number of undersea volcanoes which he claims are not properly allowed for. He points to the large number of undersea volcanoes in Antarctica and the Arctic as factors in ice melting.

He has written several books aimed at the public including: "Climate Change Delusion and the Great Electricity Rip-off". He is concerned that by moving away from fossil fuels electricity will be become unaffordable for many, supplies will be erratic and unreliable, with dire consequences for many industries and people. All, he considers, for no good scientific reason. He is very critical of the trillions of dollars, much of it taxpayer funded, going into this exercise. His books are full of solid technical climate and other data with many graphs and logical arguments.

Regarding awards he won the Clarke Medal of the Royal Society of New South Wales, the Leopold von Buch Plakette of the German Geological Society, is a Fellow of the Australian Academy of Technological Sciences, an Honorary Fellow of the Geological Society of London. In 1995 he was awarded the Centenary Medal of the Australian Humanist Society. He has been chair of several professional bodies including the Australian Geoscience Council. He was on the Earth Sciences Committee of the Australian Research Council for many years and has won the Eureka Prize for promotion of science.

He has lived in Broken Hill for many years and takes pride in being solidly based by common sense and an outback outlook. He has aggressively attacked his critics and those of opposing views many times publicly.

In his honour a new mineral is named after him, "Plimerite". It is said to be: " brittle and insoluble in alcohol." In addition a rainforest spider, "Austrogella plimeri" has been named after him because of his: "provocative contributions to the issue of climate change." Plimer has stated that he hopes the spider is poisonous. He is arguably Australia's best known geologist.

Judith Curry:

Judith Curry was born in 1953 in the United States. She gained a Bachelor of Science degree in geography from Northern Illinois University in 1973. She then gained a PhD in geophysical sciences from the University of Chicago in 1982.

She then did research at the University of Wisconsin-Madison, then Penn State University, and was professor of Atmospheric and Ocean Sciences at the University of Colorado.

She then became professor and chair of the School of Earth and Atmospheric Sciences at the Georgia Institute of Technology from 2002 to 2013, and is now professor emeritus there. She retired from that institution in 2017. She has served on a number of high committees including NASA Advisory Council Earth Science Committee, the National Oceanic and Atmospheric Organization (NOAA) Climate Working Group from 2006 to 2009, the National Academies Climate Research Group from 2003 to 2006.

During her career she was especially interested in any connections between hurricane intensity and global warming. No definite link was found, as has been already discussed. Her group did demonstrate a connection between hurricane and tornado development. She published over 130 peer reviewed papers. She was a co-author for a textbook: "Thermodynamics of Atmospheres and Oceans (1999)" and also edited "Encyclopaedia of Atmospheric Sciences."

In 2005 she and P.J. Webster et al co-authored a paper: "Changes in Tropical Cyclone Number, Duration, and Intensity in a Warming Environment." The paper concluded that the number and duration of cyclones has decreased from 1995, but the intensity of them had increased, mixed findings. It was severely criticised by sceptics and she became interested in their view. Over the next few years she became increasingly convinced that scientists deviating from the alarmist mainstream view of global warming faced severe criticism and isolation. She said the "climate community had adopted a fortress mentality defending insiders and refusing access to outsiders." She considered that there were great uncertainties about climate data and that severe action was not scientifically justified. She also became concerned about the processes in the IPCC for dealing with scientific evidence, with selection of findings and influence of non-scientists on final reports being of concern. She has said that the IPCC needs "thoroughgoing reform". She was a reviewer for the third IPCC report and felt data, including on cloud formation, was not properly handled.

Concerning sea level in 2018 she stated in a report "Sea Level and Climate Change" that the evidence showed only a slow creep up over the last 150 years with no recent increase due to higher CO2 levels as has already been discussed. Some U.S. local governments contemplate lawsuits for compensation against fossil fuel companies anticipating sea level rise and her evidence could make that more difficult.

Between 2014 and 2019 she testified before Republican-led Congress committees stating that the dangers of global warming caused by human use of fossil fuels are: "overstated and difficult to predict." And: "Man-made climate change is not an existential threat in the 21st century….the perception of near term apocalypse has narrowed the policy options." She criticised the U.N. climate action plan and President Obama's plan aimed at reducing human CO2 emissions.

The Trump administration, which did not believe in human caused global warming, offered her a senior position at the Nation Oceanic and Atmospheric Administration (NOAA) in 2020 which she declined.

She retired in 2017 from Georgia State University saying there was: "A poisonous nature of the scientific discussion around global warming." She then shifted to running the Climate Forecast Applications Network, a consulting company with government, insurance and energy company clients.

She has won several awards including the Houghton Research Award from the American Meteorological Society and the Presidential Young Investigator Award from the National Science Foundation.

In her own words she was an "Insider" and has become an "Outsider" as being consistent with her conclusions on the evidence.

Guus Berkhouf:

Guus Berkhouf was born in 1940 in the Netherlands and studied electrical engineering at Delft University, obtaining a PhD in 1970. He initially worked for Royal Dutch Shell oil company, but returned to the university as professor of geography and also acoustic imaging in 1979. He later left the university and became involved in seismic research for oil companies, and also advisory bodies to the government on noise issues.

He has argued strongly against climate alarmism and founded CLINTEL, the Climate Intelligence Foundation. This aims to bring together international scientists who consider there is no climate emergency due to human use of fossil fuels and that any climate changes are part of natural variation. In an open letter the group has made a declaration the principal points of which are:

- Natural as well as anthropogenic forces factors cause warming.

- Warming is far slower than predicted by the IPCC showing the models used are inadequate for the purpose.

- Climate policy relies on inadequate models for making global policy decisions. And they fail to deal with benefits of higher CO2 levels.

- CO2 is a plant food not a pollutant. It is essential for all life on Earth. More CO2 is beneficial for nature, greening the planet and good for agriculture.

- Global warming has not increased natural disasters. Statistical evidence shows no increase in frequency or intensity of hurricanes, floods, droughts, and other natural disasters.

- There is no climate emergency. Net zero for 2050 policies are strongly opposed. If better methods emerge they can be considered. Global policy should aim for prosperity for all by the provision of reliable and affordable energy. In prosperous societies people are well educated, birth rates are low and people care about their environment.

CLINTEL reports that 1107 world-wide scientists have so far signed the open letter declaration. They can be seen on their website: https://clintel.org . This shows that 134 Australians have signed and 20 New Zealanders. It is also shows they are not all scientists, indeed a criticism of it is that only 10% are "climate scientists" alarmists say.

The other side of that coin arguably is dispute over the often stated comment that: "95% of the World's climate scientists agree there is a is a climate emergency due to anthropogenic excess carbon dioxide pollution." Disputers say that the "World's climate scientists" quoted have been heavily preselected, and those disagreeing excluded. There is clearly intense polarity in these debates.

There are many other scientists who hold this view that the Earth's climate is very complex, there are great uncertainties about scientific data and especially computer modelling concerning it, and no justification for severe measures, or any measures many would consider, limiting human use of fossil fuels. These include:

Patrick Moore, who was a founder of Greenpeace.

Michael Castillo.

Dr John F. Clauser (Nobel Prize for Physics 2022).

Dr Ivan Giaever (Nobel Prize for Physics 1973. Resigned American Physical Society 2011 over "Global Warming").

Dr Roberrt B. Laughlin (Nobel Prize for Physics 1998).

David Legates.

Anthony Lupo.

Bjorn Lomberg

Sterling Burnett.

Vivay Jaya Raj

Freeman Dyson.

Nigel Lawson.

Frederick Sietz

References and Further Reading for Chapter 9: Carbon Dioxide Heating of the Atmosphere:

www.en.wikipedia.org/wiki/John_Tyndall

www.en.wikipedia.org/wiki/Svante_Arrhenius

"The Temperature of the Lower Atmosphere of the Earth." Hulburt, E. O. www.journals.aps.org/pr/abstract/10.1103/PhysRev.38.1876

www.optica.org/en-us/history/biographies/bios/edward_o_hulburt/

"The Influence of the 15 micron carbon dioxide band on the atmospheric infrared cooling rate – the carbon dioxide theory of climate change." Johns Hopkins Uni 551.521.61 www.geosci.uchicago.edu/-archer/warmiong_papers/plass.1956.radiation.pdf (Paper by Gilbert Plass et al.)

www.en.wikipedia.org/wiki/Gilbert_Plass

www.nature.com/articles/437331a (Keeling)

www.courses.seas.harvard.edu/climate/eli/courses/EPS28r/Sources/Keeling-CO2-curve/Keeling%20Curve%20-%20

https://en.wikipedia.org/wiki/Charles_David_Keeling

www.en.wikipedia.org/wiki/Jerry_D._Mahlman

Mann, M.E., Bradley R.S. and Hughes, M.K.: "Northern Hemisphere Temperatures during the past millennium: inferences, uncertainties and limitations." Geophys. Res. Letters 26 (6) pp 759 – 762 (1999).

www.en.wikipedia.org/wiki/hockey_stick_graph

www.en.wikipedia.org/wiki/Hubert_Lamb

www.en.wikipedia.org/wiki/Gavin_Schmidt

www.en.wikipedia.org/wiki/Susan_Solomon

www.en.wikipedia.org/wiki/Michael_E._Mann

www.en.wikipedia.org/wiki/Raymond_S._Bradley

www.en.wikipedia.org/wiki/Malcolm_K._Hughes

www.climatesciencespace.eu/mauro-facchini

www.climate.copernicus.eu/copernicus-globally-seven-hottest-years-were-last-seven

https://en.wikipedia.org>wiki>Jim_Salinger

https://orcid.org/0000-0002-5782-1411 (Jim Salinger)

https://en.wikipedia.org>wiki>Fred_Hoyle

https://www.famousscientists.org>fred-hoyle

www.en.wikipedia.org/wiki/Joseph_D%627Aleo

https://en.wikipedia.org>wiki>Sallie_Balilunas

https://en.wikipedia.org>wiki>Willi_Soon

Soon, W. Balilunas, S. "Proxy Climatic and environmental changes of the past 1000 years." Climate Research 23:89-110. 31 Jan 2003

www.en.wikipedia.org/wiki/Robert_M_Carter

https://www.desmog.com (Nicola Scafetta)

https://en.wikipedia.org>wiki>Nicloa_Scafetta

Michael Duffy; Michael Manton; Ray Evans; Vincent Gray (8 May 2006) "Nine Lies about Global Warming" Counterpoint. ABC Radio National.

www.en.wikipedia.org/wiki/Vincent_R_Gray

www.en.wikipedia.org>wiki>Roy_Spencer

https://www,desmog.com>royspencer

www.en.wikipedia.org/wiki/Richard_Lindzen

https://www.cato.org>people>richard-lindzen

www.en.wikipedia.org/wiki/William_M_Gray

William M Gray. "On the hijacking of the American Meteorological Society." Meteo Lycée Classique de Diekirch. June 2011.

Gray, William M. BBC News: " Viewpoint: Get off the warming bandwagon." 16 November 2007.

https://atmos.uw.edu>wallace>biographicalnotes

https://en.wikipedia.org>wiki>Fred_Singer

"Hot Talk, Cold Science." S. Fred Singer; David R Legates; Anthony R Lupo. Independent Institute 2021.

https://en.wikipedia.org>wiki>>Ian_Plimer

"Climate change delusion and the great electricity rip-off." Ian Plimer. Connor Court Publishing Ltd 2017.

Curry, Judith; Webster P J et al: "Changes in Tropical Cyclone Number and Intensity in a Warming Environment" Science Vol 309 Issue 5742 16 Sep 2005

https://en.wikipedia>wiki>Judith_Curry

https://www.nature.com>news>judith_curry

https://annual.ametsoc.org>conferences-and-symposia>Kevintrenberth

https://www.theguardian.com>environment>jul>a-p

https://en.wikipedia.org>wiki>Kevin_E._Trenberth

Chapter 10

Methane

Properties and Characteristics:

Methane is made up of molecules containing one carbon atom with four hydrogen atoms attached. It is a gas on Earth. As a liquid it has a boiling point of minus 161°C at sea level pressure, 1013 hectopascals. At minus 182°C it becomes a solid. It is almost colourless but does absorb a tiny bit of red light close to the red edge of the visible light wavelengths at 0.7 microns. This gives it a slightly blue colour and explains why Neptune and Uranus appear blue as they have a lot of methane in their atmospheres. It is much lighter than air.

Methane molecules also absorb infrared heat radiation in a narrow band at 7.9 microns wavelength as explained in Figure 11 in Chapter 1, and some other very narrow bands at 1.5, 2.3 and 3.2 microns. All other infrared heat radiation passing from Earth to space in the wavelengths 4 to 90 microns goes straight through methane which has no effect on it.

Methane Formation:

There are a number of ways Methane is formed on Earth:

1)

Most methane is formed by primitive bacteria-like organisms called **Archaea** which live very widely in nature. They have a unique type of respiration which is **anaerobic,** that is does not use oxygen for fuel and energy. These organisms use carbon dioxide and combine it with hydrogen from water to make methane and more water. Chemically the reaction is:

$CO_2 + 4H_2$ goes to CH_4 (Methane) + $2H_2O$ (two water molecules)

By doing this the primitive organisms, which have just one simple cell with no cell nucleus, use the energy made by this reaction to live.

These organisms are prolific in soils, cattle and other ruminant animals which eat plants, in swamps, rice paddies, landfills, the guts of termites, the deep seafloor and floor of lakes. The methane produced by animals in land based areas escapes into the atmosphere. That produced on the sea floor gets fixed in the water as hydrates and then used by other bacteria like organisms which eat it for fuel. These are called

methanotrophs, which means methane eating. Hence most sea floor methane probably does not escape into the atmosphere, but this is poorly understood and it is possible that large amounts do.

2)

Methane is also formed from the breakup of dead organic matter from plants and animals deep in sedimentary rocks and layers. The high temperatures and pressures there cause methane to form from the hydrocarbons in the proteins, fats and carbohydrates of the dead organisms. This methane is then trapped in layers deep in the rock, and can be released by drilling. It is then "Natural Gas" which is a very common fossil fuel.

3)

Methane can be formed by chemical processes deep in rock without using dead organic matter.

Human activities forming methane include landfills, rice paddies and other crops, pastoral ruminant animals such as cattle and sheep, fossil fuel industries especially coal mining, oil and gas industries, and nuclear power generation. And, as methane is formed in lakes and rivers from both living and decomposing plants and animals it also forms in man-made dams, including hydropower ones in large amounts. Much of this is vented to the atmosphere. This methane production from hydropower dams is related to the amount of electricity production. The study of this by Bridget Deemer and others referenced below suggested about 32 million tonnes of methane is emitted from man-made dams and reservoirs world-wide.

Methane was discovered first in 1776 by Alessandro Volta, an Italian physicist. He found it in gas emerging from marshes. In 1778 he had purified it and shown it was flammable.

When used as fuel methane – natural gas – produces about twice as much heat as coal for the same output of CO_2. It is also more conveniently transported than coal by pipelines and ships. To be transported by ships it needs to be cooled and kept at less than minus 162°C in liquified form. One litre of this makes 600 litres of gas for use.

There are many large specially adapted ships and terminals for this purpose. Australia is very involved in this process.

At normal temperatures, 21° C, methane can only be liquified by huge pressures 320,000 hectopascals, or 320 times sea level atmospheric pressure. It is therefore vital to keep it very cold in the transporting ships!

Methane, natural gas, is transported in carbon steel or special plastic pipelines at atmospheric pressure, and at normal temperatures, very widely, including internationally, and by smaller pipes to homes and other users. Russia is a large producer and seller by this method relied on by Europe especially for energy. There are 3.5 million kilometres of pipelines in 120 countries.

Propane (C_3H_8) and Butane (C_4H_{10}) gases are by-products of methane and petroleum processing. They turn to liquids at far lower pressures and higher temperatures than required for methane and so are commonly used as convenient efficient fuels for homes, vehicles and industry stored in the familiar LPG steel cylinders at normal temperatures.

Methane in the atmosphere rises quickly through the troposphere to the stratosphere because it is so light. There it is quickly broken down to water vapour and formaldehyde, CH2O, by solar radiation. The formaldehyde then quickly reacts with **hydroxyl radicals** which are unstable molecules of oxygen and hydrogen, OH. This reaction produces carbon dioxide and more water vapour. The reaction also can produce ozone, O3.

Hydroxyl radicals are very unstable molecules of a hydrogen atom and an oxygen atom. They are formed partly by breakdown of ozone (O3) aided by solar radiation, then a reaction with water vapour molecules, and partly by other atmospheric reactions. Each hydroxyl radical molecule lasts for less than a second but it reacts quickly with other molecules like methane destroying them. There are fewer than one of them every trillion (million million, 10^{12}) atmospheric molecules but they cause the destruction of massive quantities, about 500 million tonnes of methane each year and many other gases like carbon monoxide, for example. They have only been studied since 1980.

The end result is that huge quantities of methane are quickly broken down in the stratosphere to carbon dioxide, water and some ozone.

Methane Measurement Issues:

Methane at present is in the Earth's atmosphere with a concentration of 1.8 parts per million, about one methane molecule to 500,000 other atmospheric molecules. It is thought, from ice core studies that the level has increased from about 0.7 parts per million in ancient interglacial warm periods and about 0.4 parts per million in ancient ice age glacial periods. But there has to be some uncertainty because modern instrument methods are being compared with these methods for ancient samples which are very different. In addition the very low concentration of methane from the ancient atmosphere mean that large amounts of ancient ice have to be examined, about a cubic meter. This is hard to get from deep ice cores, so surface apparently old ice is used instead. A further problem is that the methods depend on Carbon ^{14} radioisotope dating to get the age of the methane, and Carbon ^{14} is made in the atmosphere from solar radiation effects. So it is far from straightforward.

Similarly working out the age of methane in old lakes, swamps and the atmosphere is difficult because of its very low concentration. Some methods rely on Carbon ^{14} radioisotope dating with similar problems to those discussed above.

So-called "Bottom up" and "Top Down" measurements of methane production from human driven sources are part of the Kyoto Protocol requirements.

"Bottom up" consists of countries recording the level of the various human activities known to produce Methane annually, and then estimating the methane produced form standardized tables or protocols. The method has many assumptions and depends on accurate records, obviously virtually impossible in many countries, and open to manipulation.

"Top Down" depends on measuring methane emissions by tower based instruments, aircraft instruments, vehicles based instruments and satellites. There are now a number of satellites apparently measuring it but the methods are largely very indirect and make many assumptions. The data is processed using computer modelling with its problems and limitations, so there is no precision about these findings.

Admitted difficulties include:

- Quantifying actual amounts.

- Annualizing intermittent sources – that is converting a source of a short time to data for a year.

- Determining methane over water, the oceans due to poor reflection of light from the surface.

- Identifying intermittent emitters,

- Identifying those avoiding detection.

The satellites exam sunlight being reflected from the Earth below, so-called "Earthshine." The European Space Agency TROPOMI instrument on it's Sentinel-5 Precursor satellite 824 km above the surface looks at this light being reflected from a 2600 km wide "swath" as it passes over. The instrument exams how much infrared radiation there is for the wavelength 2.3 microns which is a narrow band absorbed by methane. Then complicated computer models are used to calculate the apparent methane concentration in the column of atmosphere 2600 km wide being looked at. If the instrument finds less infrared radiation at 2.3 microns wavelength, it implies there is more methane in the air below absorbing that.

TROPOMI is able to see changes if they are larger than 7 km, its "resolution". Newer satellites improve this, including the Italian PRISMA instrument with a 30 km swath and 30 meter resolution, the Canadian GHGSAT with a 12 km swath and 50 m resolution, and the planned MethaneSAT joint U.S./ New Zealand satellite planned for launch in 2022 with a 200 km swath and 400 meter resolution, and great sensitivity able to pick up methane when it is just 2 molecules in every billion of the atmosphere.

These satellites all look at the wavelength 2.3 microns for methane. The instruments also look at other wavelengths in the infrared spectrum for other gases of interest as already discussed including CO2. The method has a number of problems including needing an absence of clouds, other aerosols, dust effects on the light reflected from the surface, calibration and computer processing of data issues. But they are clearly massive and exciting advances providing fascinating new information.

Methods measuring methane directly from air samples – vehicles, aircraft, towers – use a spectrometer instrument similar to those discussed already for carbon dioxide measurement. They generate a laser light beam of a wavelength absorbed by methane and a second beam as a control, and measure the difference between the two beams internally in the instrument, then compute the result. These are very accurate, well less than 1% error. Again the wavelength 2.3 microns in the infrared spectrum is used.

Variations of the method have been used for methane dissolved in water, lakes, the sea, swamps by applying a vacuum over the liquid and using the spectrometer instrument on the gas collected. But the method is new and there are many issues concerning accuracy including calibration.

Older methods include gas chromatography often combined with flame ionization, a time consuming laboratory centred non-mobile method with it's own problems and inaccuracies.

A basic principle in trying to determine ancient methane levels from ancient ice core samples is that very old methane arising from reactions in rock trapped in the ice should not contain any Carbon ^{14} isotope, as it will have all changed to Carbon ^{12} over that time. Whereas Methane which has resulted from dead plants and animals, organic material, contains methane which has carbon atoms originally in the atmosphere and processed by photosynthesis into plants, and animal food; this will have some Carbon^{14} as that isotope is made in the atmosphere by the action of solar radiation, as already discussed. So measuring Carbon^{14} concentrations in this ice bound ancient air can help estimate how much of it came from dead organic material. And it also helps date the time of origin of the methane by carbon dating.

Thus measuring methane concentrations in ancient ice trapped atmosphere can both estimate the level of methane in the atmosphere then and how old the air is. The main instrument used is a Mass Spectrometer already discussed for it use for CO2 levels and age.

The whole process is extraordinarily complex and meticulous with great potential for error and many assumptions, as discussed in the paper by Petrenko et al referenced below. Because methane is so scarce in the atmosphere large amounts of ancient ice have to be melted and analysed. Great care has to be taken to avoid contamination with modern air containing new methane. And to avoid new methane being made in the samples by natural radiation, and many other problems.

For all these reasons and more methane atmospheric concentration measurements are far from precise. It follows that dogmatic statements about levels possible rising, and exhortations to take radical action, such as avoiding fossil fuels or pastoral animals, cannot be well founded in conclusive science.

Natural Sources of Methane Compared with Human Sources

The International Energy Agency (IEA) stated in 2020 that 60% of methane emissions into the atmosphere were due to human activity and 40% natural sources, especially wetlands, termites, volcanoes, wildfires, ocean sediments. It seems that this conclusion is based on the above measuring methods with their problems as discussed. Given the very wide spectrum of natural causes and the difficulties with measurement it is hard to see the basis for this statement.

Similar figures, 60% human, 40% natural are used by other organizations such as the Climate and Clean Air Coalition and the United Nations Environment Program. These conclusions were reached by "international teams of scientists using state-of-the-art composition and climate models….".

Therefore there has to be great uncertainty about how much is natural in origin and how much is due to human activity. That large parts of the world, especially Africa, Asia and South America, are very poorly scientifically developed adds to this conclusion.

Two examples of studies working out a "budget" of sources and consumers of methane globally are given here:

The amounts are in millions of metric tonnes per year (Teragrams, 10^{12} grams).

Source: Fung et al (1991)   Bousquet et al (2006)

Natural:

- Wetlands   115   147

- Ocean to Atmosphere   10   15

Human Activity:

- Energy   75   110
- Landfills   40   55
- Livestock   80   not detailed
- Rice paddies   100   31
- Biomass burning (wood fires) 55   50
- Other human sources   no reading   90

Sinks, Consumers:

- Soils   10   21
- Tropospheric with OH   450   448
- Stratospheric loss   no reading   37

Budget:

Total Sources   500   525

Total Sinks   460   506

There are obviously big variations in these study results underlining difficulties and uncertainties in these estimations. It is clear that the troposphere and stratosphere readily consume huge quantities of methane by natural processes resulting in the budget being virtually balanced. This adds to the view that major efforts to reduce methane would be futile.

Missing from these figures is the massive amount of methane produced by **wild ruminant animals.** For hundreds of millions of years huge herds of wild buffalo, wild cattle, sheep, goats, deer, camels, wildebeest have grazed and produced massive quantities of methane from Archaea bacteria in their guts. Their numbers were obviously enormously more before human farming developed and occupied a lot of the land they used.

The paper by Crutzen et al referenced below estimates 2 to 6 million tonnes of methane from wild vertebrate animals such as these annually into the atmosphere but they admit the data is very poor, there have been few studies especially in undeveloped areas. By comparison they estimate 74 million tonnes from farm animals each year. Humans, as non-ruminants, only produce 200,000 metric tons by "passing wind" each year!

But it is obviously very likely that previously when there were much greater number of wild animals their production of methane would have been far higher.

Kangaroos also are ruminants with Archaea in their gut and produce methane. It used to thought that they made less by "passing wind" than farm animals but new research by Adam Munn PhD from Wollongong shows that they make about the same as other animals, although less than cattle, for their weight. Dr Munn found this by shutting kangaroos in a sealed room and measuring the "wind".

Then there are **all the invertebrate animals** – insects, termites, earthworms and many others. Very few of these have been studied. Dr Crutzen in their paper estimates termites produce about 28 million tonnes of methane each year into the atmosphere. But it is known that many insects, beetles, worms use archaea bacteria in the gut to digest food and emit methane. If they live deep in the soil it is thought some of the methane is used by Methanotrophic bacteria eating it for food. But much of it escapes, especially from surface creatures, into the atmosphere clearly. Estimates vary. One by Zimmerman et al referenced below estimates 150 million tonnes of methane into the atmosphere each year from these animals. And that too has been happening for hundreds of millions of years.

But the major criticisms of methane being regarded as a cause of global warming are:

1)

It is a very rare gas in the atmosphere compared to others, 1.8 per million, 0.00018%, one in every 500,000 molecules. Even if a methane molecule does get hotter it will dissipate its extra heat quickly to surrounding air molecules and have no effect practically speaking on the overall atmospheric temperature.

It is often said in alarmist comments that methane is between 34 and 104 times stronger as a greenhouse gas than CO_2, for periods of 10 or 20 years in the atmosphere respectively. More commonly 28 times is quoted as a 100 year figure. But that is comparing *equal masses* of the two gases, that is one ton of CO_2 compared with one ton of methane.

But that is far from the reality in the atmosphere. For each methane molecule there are at least 200 carbon dioxide ones. And usually 20,000 water vapour molecules. It is not rational to assert than the tiny amount of methane will dominate water vapour effects.

2)

Figure 11 is repeated here to help explain the second major criticism.

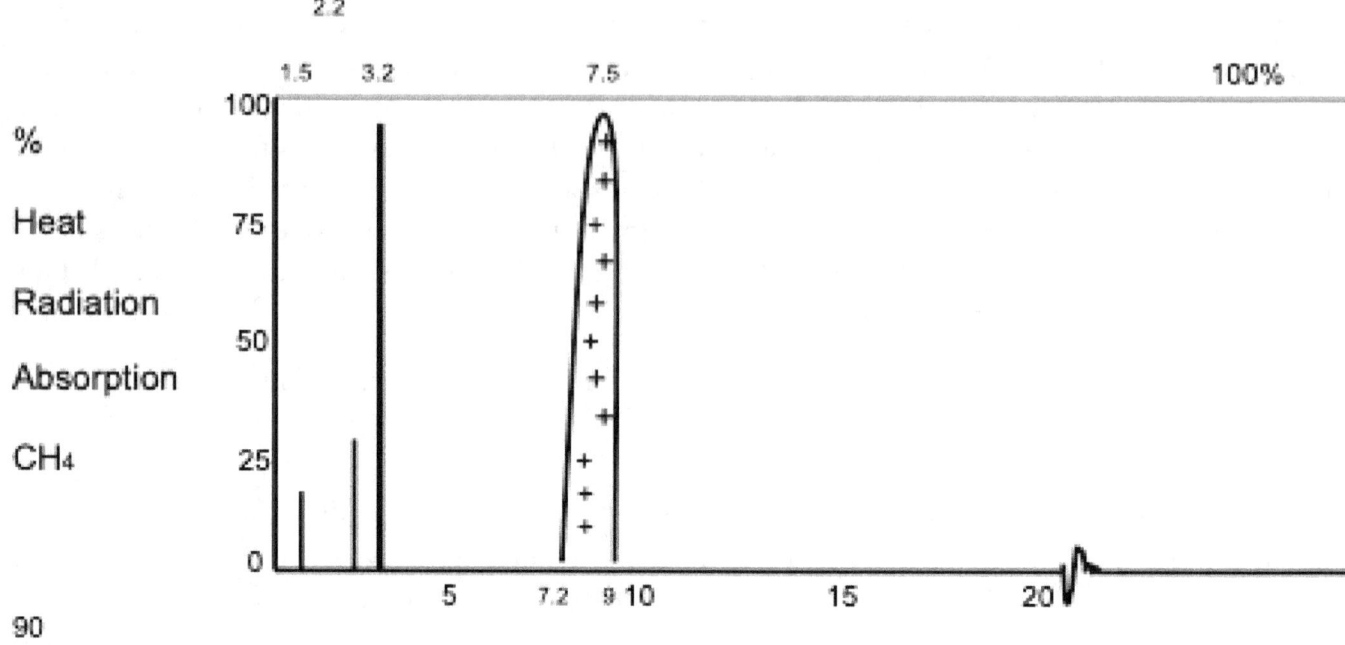

Figure 11: Radiation Absorption by Methane Molecules CH₄

Concerning Methane's effect on the Earth's heat radiation leaving its surface there is a band of intense absorption from 7.2 microns to 9 microns, peaking at 100% absorption at 7.5 microns. There are no other absorption bands. Looking back at Figure 7 which shows the heat radiation wavelengths of the Earth's radiation it can be seen that methane would have only a small overall effect, perhaps about 1%, blocking the radiation. When the tiny amount of methane in the atmosphere, 1.7 molecules per million, is considered the effect becomes very tiny.

As explained methane molecules only absorb a small portion of the infrared heat radiation escaping from Earth into space maybe 1%, mainly in the wavelength band 7.2 to 9 microns. All other wavelengths pass freely. Further the amount of absorption in this band is not as severe or total as it is for carbon dioxide concerning it's band of absorption from 14 to 16 microns, and nothing remotely like the very wide absorption band for water vapour molecules. The tiny bands at 1.5, 2.3 and 3.2 microns are outside the infrared spectrum.

This makes it impossible to see how methane can be seriously regarded as a major, or indeed any, cause of atmospheric heating, global warming. On the basis of this basic science.

And, indeed, there is no practical evidence it is causing warming. It is one thing to construct a computer model, even a very advanced one, taking into account all known factors, then feed in increasing levels of methane and watch the program produce predictions of rising temperatures. It is another to observe what is really happening in the very complex, poorly understood Earth environment.

It follows that proposals by alarmists to severely cut human methane sources involving massive changes to agriculture, fossil fuel industry especially at great expense both financially and on quality of life for many people are certain to fail to reduce methane because of the lack of knowledge of it's origins, the very likely huge natural production, the massive natural destruction of methane in the atmosphere anyway and also certain to be ineffective in reducing global temperature for the reasons stated regarding the tiny concentration of methane in the atmosphere and it's small infrared absorption wavelength band.

References and Further Reading on Chapter 10: Methane:

https://en.wikipedia.org>wiki>methane

https://www.epa.gov>ghgemissions>overview-green-

Petrenko, V. et al. Nature 548, 443 (2017)

Graven, H et al. Detection of Fossil and Biogenic Methane at Regional Scales using Atmospheric Radiocarbon. Dept Physics Imperial College London UK. Earths Future Research Article 10.1029/2018EF 001064. www.agupubs.onlinelibrary.wiley.com (Nuclear power methane).

Deemer, Bridget R. et al. Greenhouse Gas Emissions from Reservoir Water Surfaces: A new Global Synthesis. Bioscience Vol 66 Issue 11 1 November 2016.

www.sciencedirect.com>science>article>methane.

Dean, J F et al. Water Research Vol 115 15 May 2017 pp 236-244

https://watson.brown.edu>news>researchbriefs>methane

Atherton, E. et al. Mobile measurements of methane emissions from natural gas developments in northeastern British Columbia, Canada. Atmos. Chem. Phys. 17, 12405-12420 2017.

Hartmann, J. et al. A fast and sensitive method for the continuous in situ determination of dissolved methane and its C^{13} ratio in surface waters. Limnology and Oceanography Methods. 27 April 2018. https://doi.org/10.1002/lom3.10244

Petrenko, V. et al. A novel method for obtaining very large air samples from ablating glacial ice for analysis for methane radiocarbon. Cambridge University Press 8 Sep 2017.

https://www.epa.gov.>ghgemissions>overview-greenh

Sources of methane emissions. IPA 25/7/20.

https://public.wmo.int>media>news.global-methane

www.en.wikipedia.org>wiki>Atmospheric_methane

https://sgp.fas.org>crs>misc (Satellite methods for methane measurement).

Pandey, S. et al. Earth, Atmospheric and Planetary Sciences 116(52) 26376-26381 (Satellite methane instruments – TROPOMI)

Irakulis-Loitxate, I. et al. ScienceAdvances 7(27):eabf4507 30 Jun 2021 doi:10,1126/sciadv.abf4507 (satellite methane measuring spectrometers results.)

https://group.met.com>media>energy-insight>naturalgas

https://en.wikipedia.org>wiki>Pipeline_transport

https://soilsolution.org>2017/07>ruminants-and-methane

https://www.horsetalk.co.nz (kangaroo methane)

Crutzen, P. et al. "Methane Production by Domestic Animals, Wild Ruminants, other Herbivorous Fauna, and Humans." Tellus B:Chemical and Physical Meteorology. 38:3-4, 271-284 DOI: 10.3402/tellusb.v38i3-4.15135

Zimmerman et al. "Termites: a potential large source of atmospheric methane." Science 218, pp 563-565, 1982

Li, M. et al. "Tropospheric OH and stratospheric OH and Cl concentrations determined from CH_4, CH_3Cl, and SF_6 measurements." Climate and Atmospheric Science. 29 (2018) 11 Sep 2018

Chapter 11

The Earth's Surface Temperature

The Earth's surface temperature has varied greatly over the 4.5 billion years since it formed. During the Archaean period more than 3 billion years ago the surface temperature and that of the sea is likely to have been between 50 and 75°C as continuous hot rain fell, the opaque atmosphere with 25% at least carbon dioxide, a similar amount of methane, the rest mostly nitrogen, no oxygen. Volcanoes belched continuously, large meteors frequently crashed into the planet as the Solar System settled and the only life were primitive bacteria and the beginnings of the blue-green algae which would transform the planet about 2 billion years ago by causing the "Great Oxygen Event" discussed above in the section on photosynthesis development.

The era before the Archaean, appropriately called the "Hadean" after Hades, or Hell, was even worse. The Earth was just a truly hot, partly molten rock.

After the Archaean the Proterozoic era followed with the formation of oxygen using life, mostly single celled bacteria type forms. This lasted for about 2 billion years until about 560 million years ago. The Earth did gradually cool but was still about 30°C averaged over the planet.

For the next era, the Paleozoic, from about 570 million years ago to 240 million years ago, and the more recent eras, the Mesozoic from then until 65 million years ago and the Cenozoic from then until the present, there is a better idea of surface temperatures thanks to modern paleontological methods especially Oxygen[18] isotope studies, studies of the ratio of Magnesium and Calcium in the shells of ancient shellfish called **Forams**, and studies of **Alkenones** which are found in ancient phytoplankton remains.

Carbon dioxide levels were over 6000 parts per million 570 million years ago and have steadily dropped since, further during major ice ages. They are currently at very low levels comparatively. There is a relationship between CO2 changes and temperature changes in that CO2 *follows* temperature changes by a lag of several hundred years as already discussed.

Data from the Vostok ice cores covering the last 400,000 years, a period covering 3 other warmer times of about 10,000 years each with cold periods each of 100,000 years in between, shows that these other warm periods were apparently warmer than now, with much higher sea levels. CO2 levels dropped to about 180 – 200 ppm during the cold times, then rose to about 280 ppm in the warmer periods, following the temperature change by about 800 years as said. The explanation given for this lag is that the ocean contains 50 times as much CO2 as the atmosphere, and it takes time to warm or cool, and thus release or take more

CO2. This period is within the present Pleistocene ice age which has been going for 2.5 million years as discussed already.

The graph in Figure 31 below shows the average global temperature for the past 570 million years when complex life started to rapidly evolve.

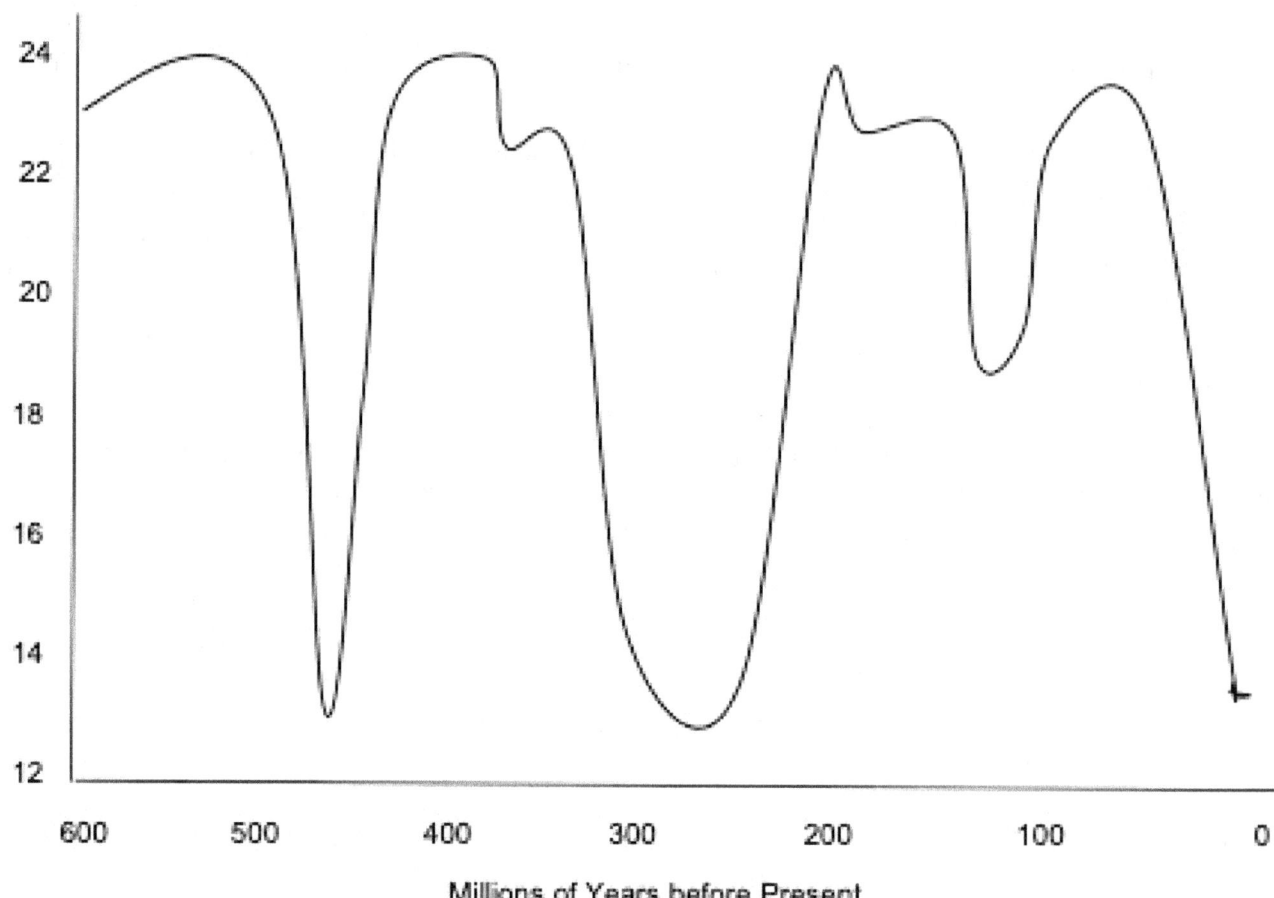

Figure 31: showing the global average temperature over the last 570 million years. There have been many millions of years when it has been much warmer averaging about 22°C, then shorter periods when it has dropped about the present level, 14°C. These colder period are during major ice ages including our own cold period during the current **Pleistocene Ice Age** which commenced 2.5 million years ago, as discussed before in the section on solar radiation. The last million years of the current ice age have featured about 100,000 year episodes of severe cold with extensive ice covering polar and temperate regions, but not tropical ones,

with brief 10,000 year warmer periods between. We are in such a period now called the **Holocene** which started about 11,000 years ago.

As commented before CO2 is predicted to eventually disappear from the atmosphere and oceans and eventually all be incorporated in carbonate rocks – a few hundred million years away yet! Life as we know it would then be impossible. That is the explanation for the long term trend of decreasing CO2 levels.

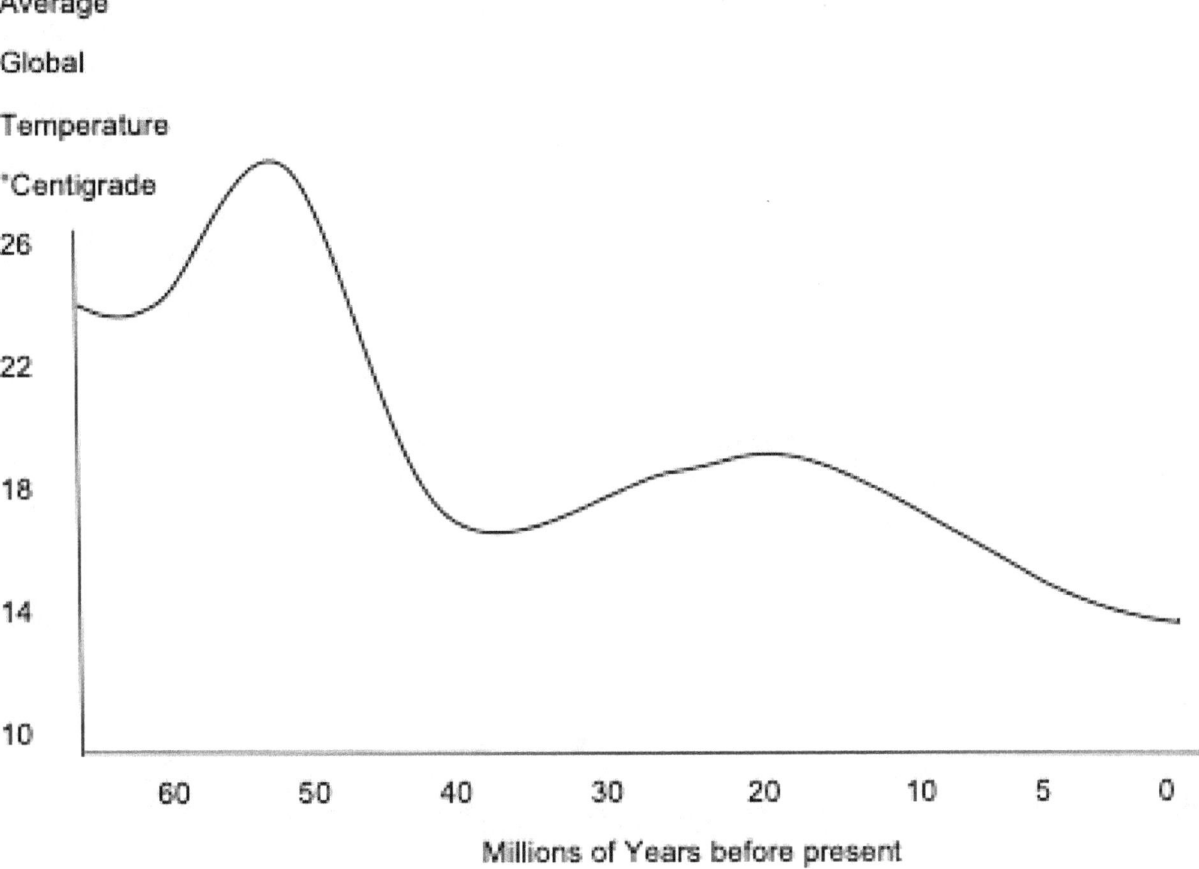

Figure 32: showing the average global temperature for the last 65 million years. The peak at about 28°C was 50 million years ago and is called the **Eocene Thermal Maximum.** The dip at 40 million years coincides with the Antarctic continent shifting to the South Pole by tectonic plate movement and becoming ice covered, thus cooling the planet. There was then a warmer period until more cooling started 15 million years ago leading to our **Pleistocene** current ice age starting 2.5 million years ago.

The last time the Earth was as cold as at present was 260 million years ago during the **Karoo Ice Age.**

Concerning CO2 levels during this time 65 million years ago they were about 1000 parts per million, then slowly dropped to about 250 parts per million 2 million years ago following the long term trend mentioned above.

CO2 levels were initially 250 parts per million, rose slightly to 265 ppm 10,000 years ago. They then rose gradually to 280 ppm 900 years ago *at a time when the temperature was falling*, the opposite of an alarmist prediction. This is called the **Holocene Conundrum** by alarmists.

Figure 33 below shows the average global temperature for the last 11,000 years, the **Holocene** period, the current warmer period in the Pleistocene ice age. For the first 5000 years the temperature was about 1°C higher than now, then dropped slowly to the current 14°C. The dip a few hundred years ago is the Little Ice Age which stopped about 1850, and was about 0.5°C cooler than now.

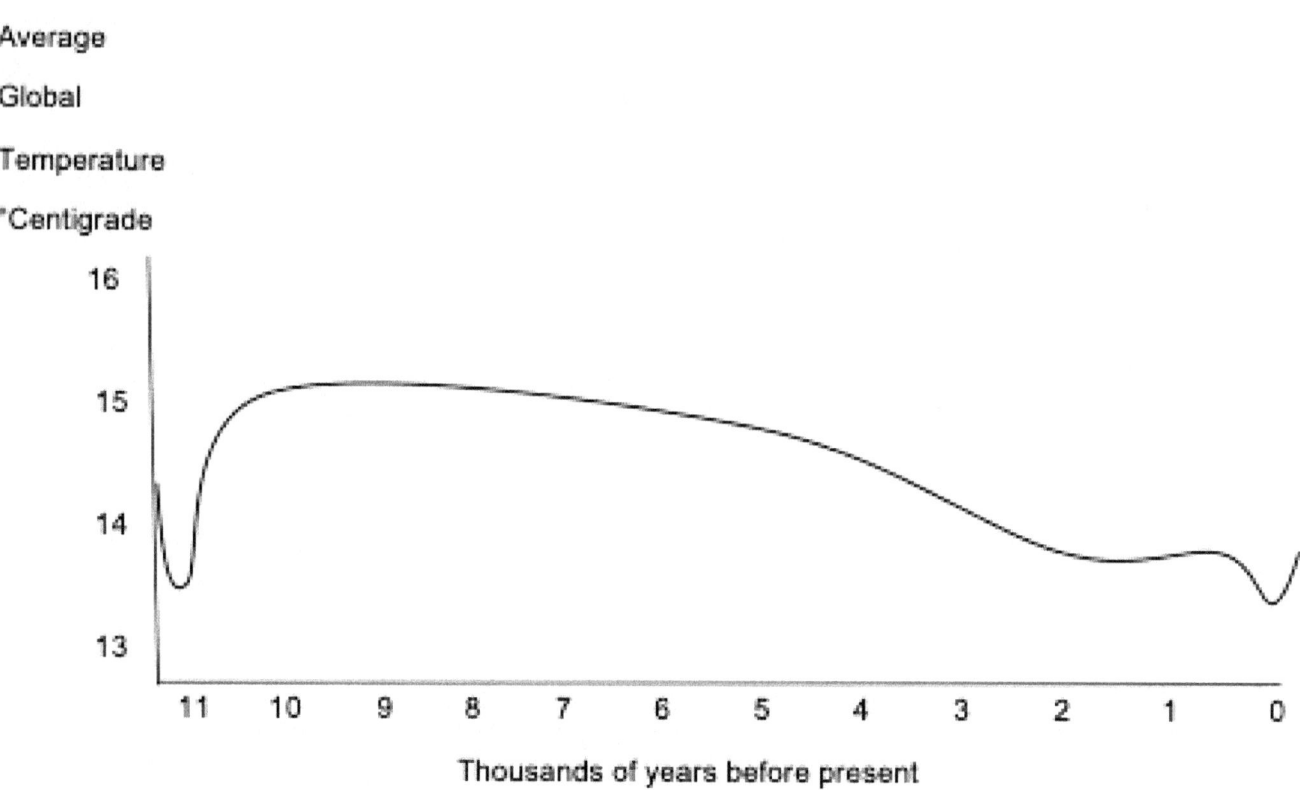

Figure 33: shows the average global temperature for the last 11,000 years which is the **Holocene** period.

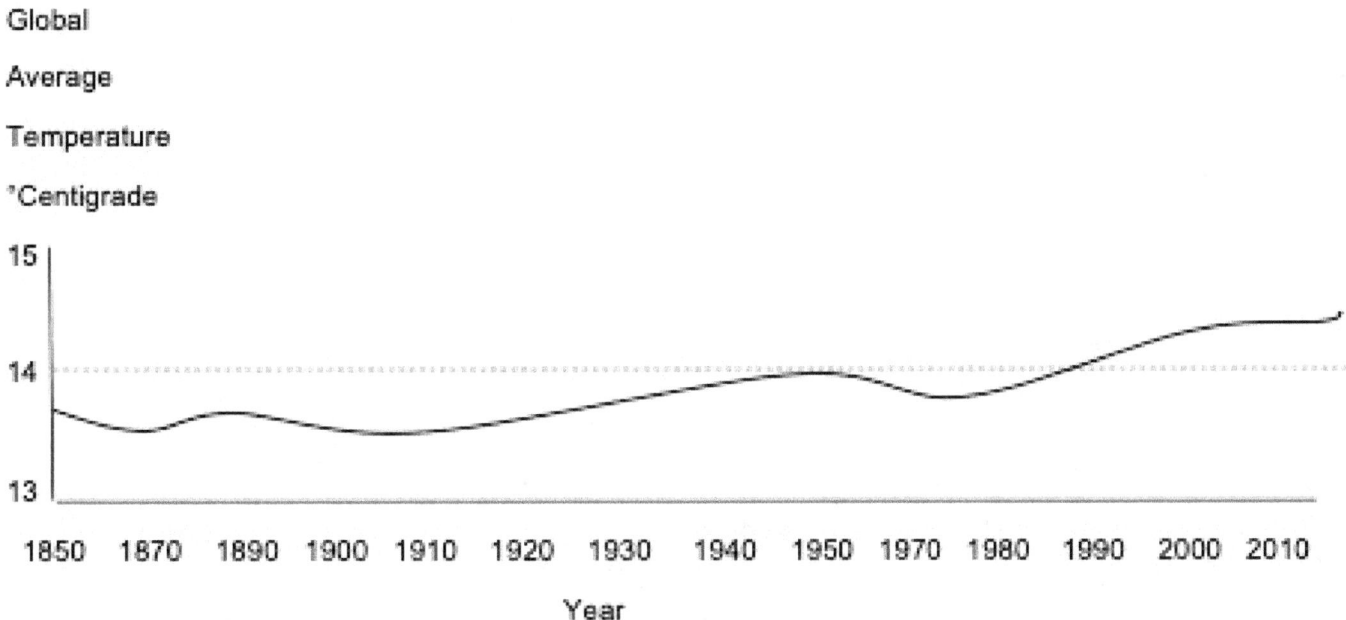

Figure 34: Showing Global average temperature since 1850 after data from the Hadley Centre of the UK Met Office and University of East Anglia Climatic Research Centre. Obviously there are a mix of measurement methods and data processing here. Only surface measuring devices like thermometers were examined. But it does show emergence from the Little Ice Age, and the gradual warming from 1910 to 1950, which in the thirties occurred at a faster rate than the warming after 1980. The pause in warming from 1998 is shown also. The drop in temperature from 1950 to 1976, which caused speculation that the ice age might return, is also shown.

The original graph from which this was drawn (referenced below) expressed the temperature change in deviations from the 1961 – 1990 mean, but it has been simplified here by assuming 14°C as the average.

An increase of about 1°C has occurred since 1870 after the Little Ice Age, and 0.6°C from 1980 after the cool period which started in 1950.

From 1950 to 2010 carbon dioxide production from human made sources increased from 5 to 31 billion tonnes, as already discussed.

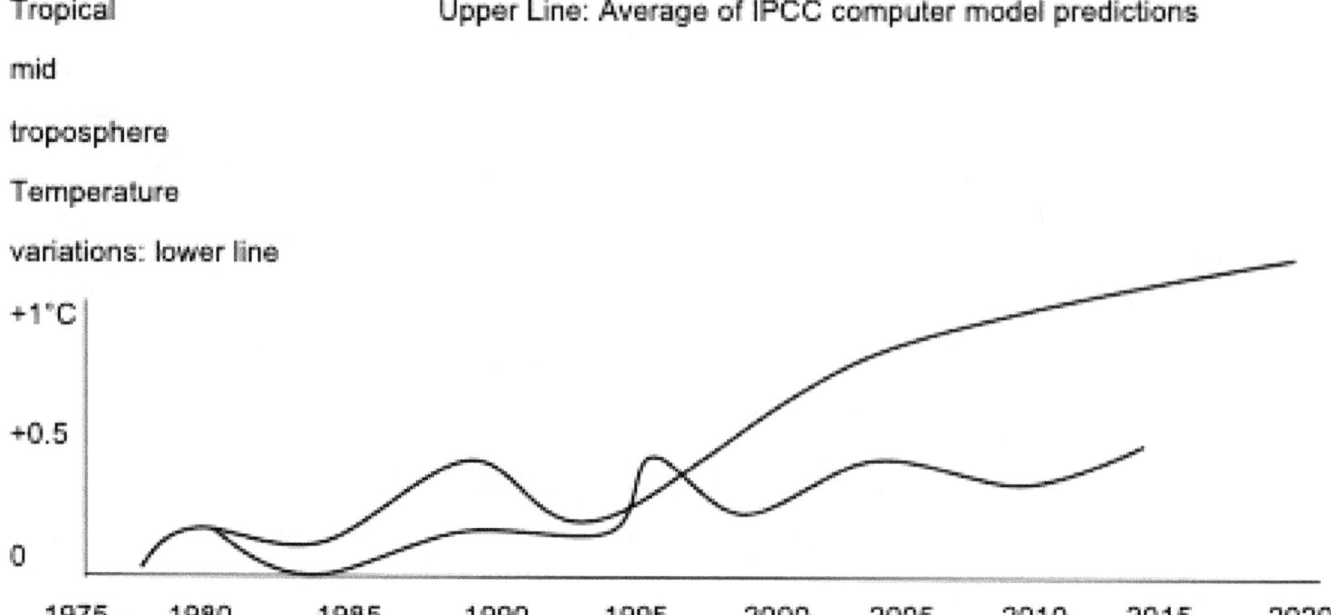

Figure 35: The lower line shows the temperature variation as computed from a combination of 4 balloon measurement and 3 satellite sets – essentially based on observations. Whereas the upper line shows the combined result of IPCC computer modelling predictions. The IPCC computer models predicted a 1.1°C increase by 2015 but the observations showed only a 0.5°C increase. Even that conclusion is not certain due to the variations shown with a warmer year in 1981, cooler times around 1985, warmer in 1996 and 2005 when there was EL Nino, a cooler period around 2010.

The temperature change shown on the vertical scale on the left side of the graph deals with atmospheric temperatures in the tropics half way up the troposphere, which is about 18 km thick there, about 9 km up. This method focuses on *trends* not the actual temperature.

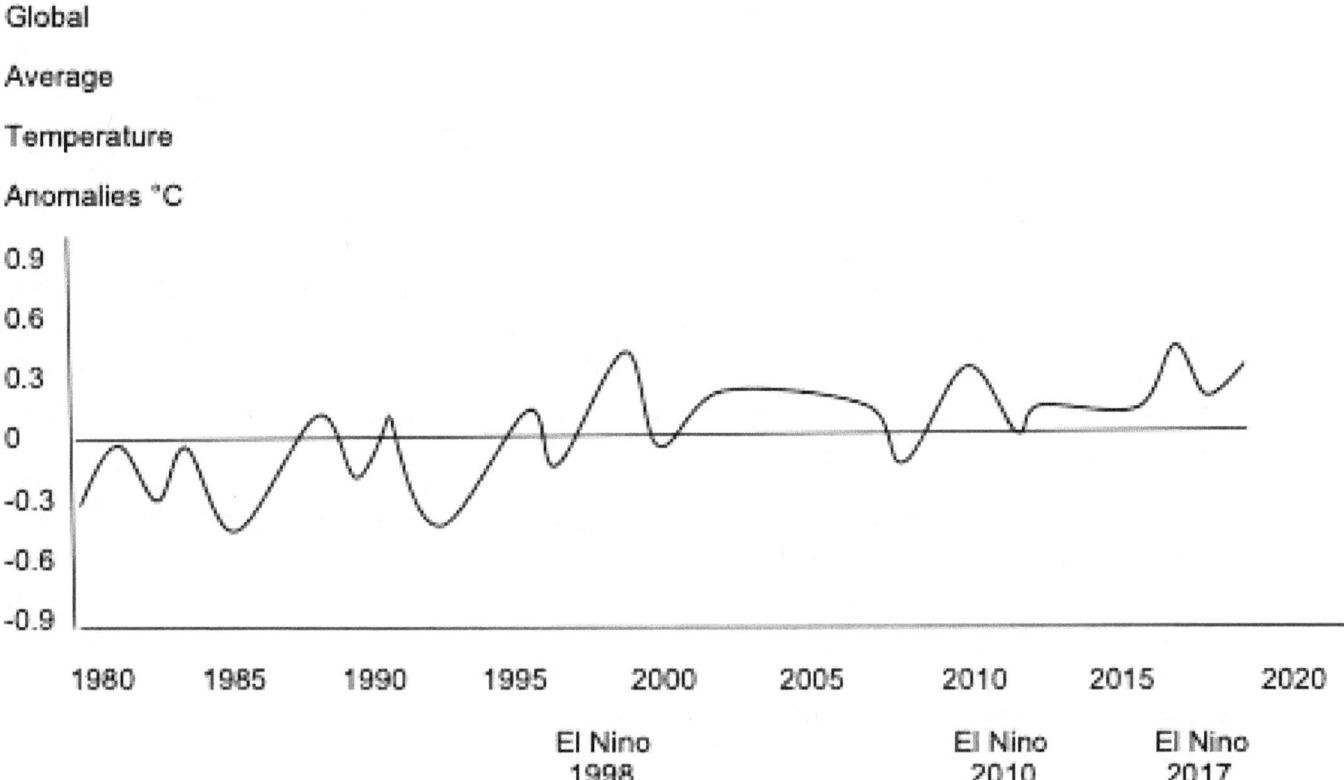

Figure 36: This shows temperature variations from the average temperature for the period 1981 to 2010. The temperature was examined by multiple satellite observations. It was measured for the lower atmosphere close to the surface. Again for each place around the Earth the measurement was taken by satellite the difference from the 1981 – 2010 average temperature for that place was calculated from the satellite readings, then all these differences, referred to scientifically as *anomalies,* for all the places combined to give the world-wide "global" result graphed here. So again this graph looks at *trends up or down* for temperature, not the actual temperature itself.

This is the period of overall warming which has occurred since the cooler period from 1950 to 1976, and which has caused the alarmist speculation. It is clear there has been a lot of variation up and down over this period, with upward spikes in 1998, 2010, and 2017. These are considered due to El Nino ocean cycle events then.

The long "pause" in global temperature change from 1998 to 2017 is shown. The general overall increase is about 0.5°C. But it is clear there are times when the difference was much less, for example in 1980 it was zero compared to the 1981 – 2010 average, and in 2018 it was +0.2°C, and so many would argue that real change is hard to be certain about and is slight if there.

The original graph from which this was drawn is referenced below. The work was done by the Earth System Science Centre at the University of Alabama-Huntsville by Dr Spencer and colleagues in 2019.

Some Actual Surface Temperatures for Australian Locations:

The next series of graphs deal with daily maximum temperatures for twelve different Australian places. The places have been chosen as being away from large metropolitan cities where "urban factors" such as extensive paved and roofed areas, industry, traffic can have a big impact on temperature as already explained. Rather the natural temperature is sought. The information is from the superb Australian Bureau of Meteorology website and the negotiation of the site is described in the references.

The data starts with the maximum temperature for each day of each month. Then the middle reading, or *mean* for each month is recorded. Then all these monthly means for the 12 months in the year are *averaged*, that is they are added up and the total divided by 12 to get the *average* for the year. Technically this is the *average mean monthly daily maximum for the year.* But it can be thought of simply as the average daily maximum for the year accurately enough.

The graphs are of these yearly averages at the beginning of each decade, that is 1880, 1890, etc. not every single year. This is to show long term trends but keep the graph simple and uncluttered. The warmest and coolest years for the whole period of the places observations are described in the text. Usually these are not the first year of each decade and therefore do not appear on the graph.

The horizontal line corresponds to the *mean* for all the annual readings for that place. This includes every single year recorded, not just the first year of the decade.

These temperatures taken at these places are simply that: a simple record of the measured temperature. In contrast to satellite data which involves highly complex processes as discussed before, is indirect with inevitable computer calculation, modelling and assumptions with inevitable errors – and "adjustments".

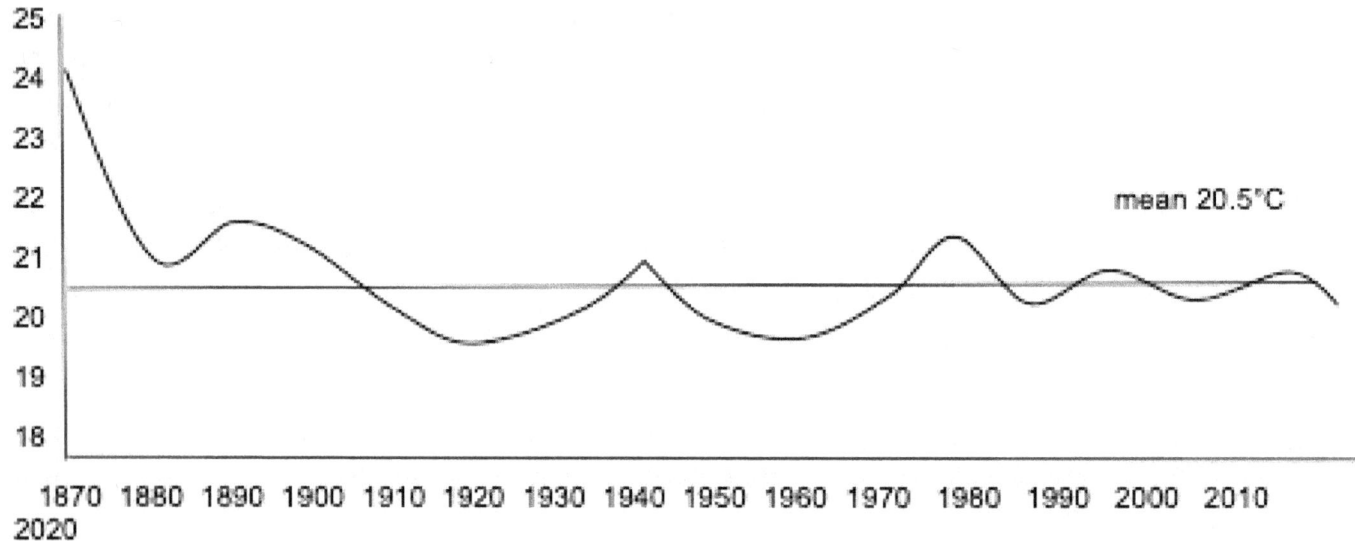

Figure 37: **Actual Surface Temperature Measurements for the Moruya Heads Pilot Station South Coast, New South Wales, Australia from 1876 to 2022.**

The Moruya Heads Pilot Station has a continuous temperature record from 1876 and it is kept and available online at the Australian Bureau of Meteorology website referenced below. The temperatures graphed are average annual temperatures concerning daily maximums for each year as described above.

The highest annual average was the first year, 1876 when it was 24.2°. The mean temperature for the whole period is 20.5°C and the horizontal line is drawn for that. The last recording is for 2021 when it was 20.1°C, 0.4°C below the long term mean. Clearly global warming has not come to Moruya !

The cooler period 1950 to 1976 is shown. The lowest year here was in 1961 when it was 19.4°C. The warmest year recently was 1980, an El Nino year, when it was 21.2°C.

The Moruya River was a busy commercial port in the 19th century, hence the long term record. The town and region have no large cities or industrial areas to affect readings from the natural temperature.

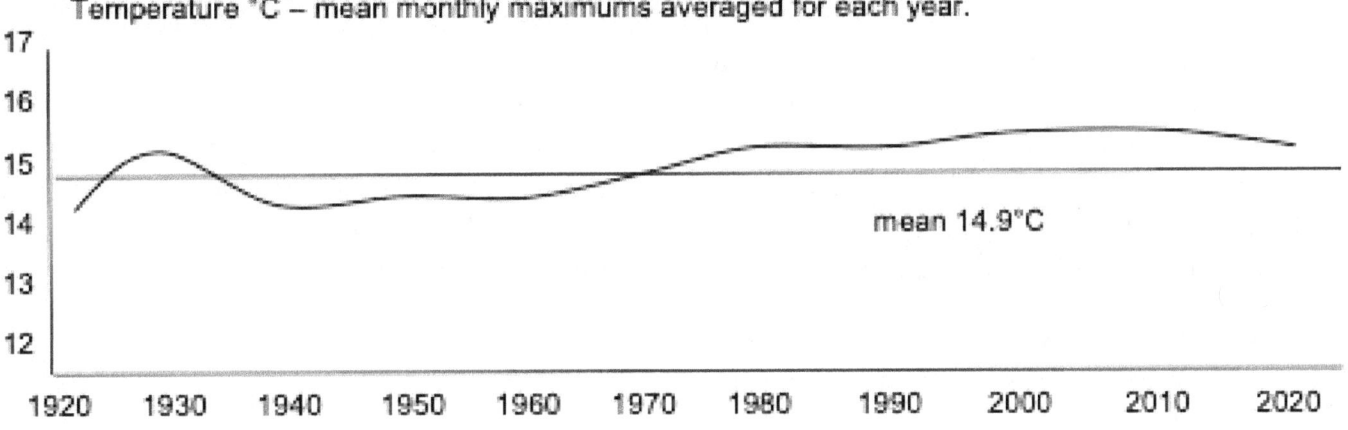

Figure 38: **Actual Surface Temperature Observations for the Cape Bruny Lighthouse, near the Port of Hobart, Tasmania from 1923 to 2021.**

The Cape Bruny lighthouse was opened in 1871 to aid navigation to Hobart and the temperature record is available on the BOM website from 1923. Again the annual average daily maximums are graphed here. The mean for the period is 14.9°C where the horizontal line is.

The warmer period in the 1930's observed world-wide is shown here too, when a reading for 1934 was 15.1°C. The cooler period leading up to 1976 is shown, then slight warming from 1980. The warmest reading was 16.3°C in 2016, but in 1988 it was 15.9°C, and in 1972 15.8°C. For the last year calculated, 2020, it was 15.4°C.

Again, global warming has not arrived at Cape Bruny.

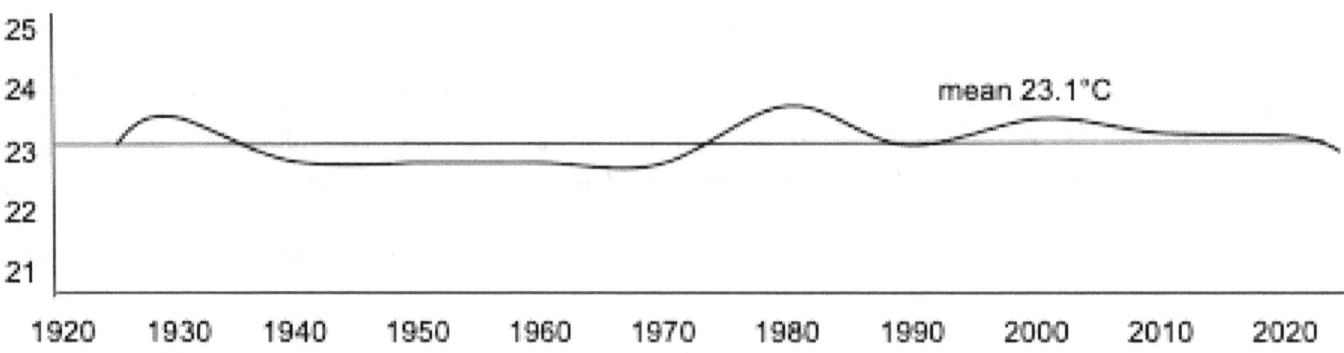

Figure 39: **Streaky Bay, South Australian coast west of Adelaide.**

Streaky Bay is another old port from the early days in the 19th century, now a small coastal town with no nearby city or industrial area. It used to be a wool loading port and was named by Captain Flinders for its oil streaked water due to oily seaweed when he visited in 1801. The temperature record started in 1926. The mean annual daily maximum is 23.1°C corresponding to the horizontal line. The maximum was 25°C in 1961, the lowest 21.7°C in 1957. Again in 1980 there was a warm reading of 23.7°C corresponding with El Nino then. The last available annual average daily maximum reading is 23.0°C in 2021.

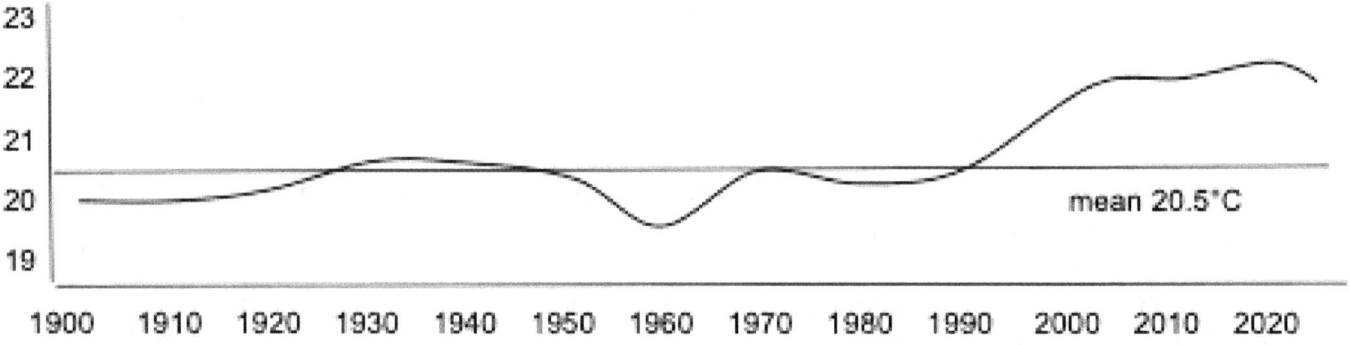

Figure 40: **Cape Naturaliste Lighthouse.**

Cape Naturaliste is on the south western part of Western Australia south of Perth. The lighthouse was built in 1904 and the temperature record started then. Again the graph plots the annual averaged daily maximums. The warmest is 22.8°C in 2011, the coolest 19.2°C in 1960, the mean since records began in 1904 20.5°C . Again the cooler period world-wide from 1950 to 1976 is seen. Temperatures remained similar to previously until 2000 when they increased by 1 to 1.5°C as shown. In 2021, the last calculated year, it was 22.2°C.

Interpretation of this is complicated by the major Indian Ocean Dipole cycle which makes a big difference to temperatures and rainfall in this part of Australia, and indeed the whole country. When this is in a negative phase this part of the Indian Ocean off Western Australia is warmer and there is more rain. There was a negative phase in 1992 and a strong one in 2010. But there have been several positive phases during this time also from 1980, including in 2006. During positive phases the ocean there is cooler and the rainfall less. So it is not safe to attribute this increase in temperature solely to the Indian Ocean Dipole. It does not fit the "global warming" profile either as the warming started too late and has remained the same since 2000.

Temperature °C – mean monthly maximums averaged for year.

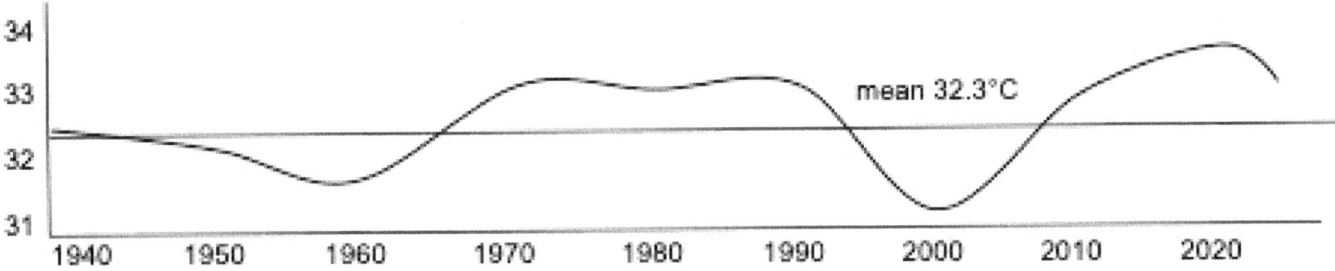

Figure 41: **Broome Airport.**

Broome is a major tourist town on the northern tropical coast of Western Australia and is a historic pearl fishery also. The record started in 1940 and the mean is 32.3°C. The maximum year was 33.7°C in 2019, the lowest year 2000 when the average was 31°C. The latest year calculated was 2021 when the average was 32.9°C.

The pattern is similar in a way to Cape Naturaliste shown in Figure 40 with a warmer period this century, but cooler around 2000. Again the Indian Ocean Dipole is a likely factor.

Temperature °C – mean monthly maximums averaged for each year.

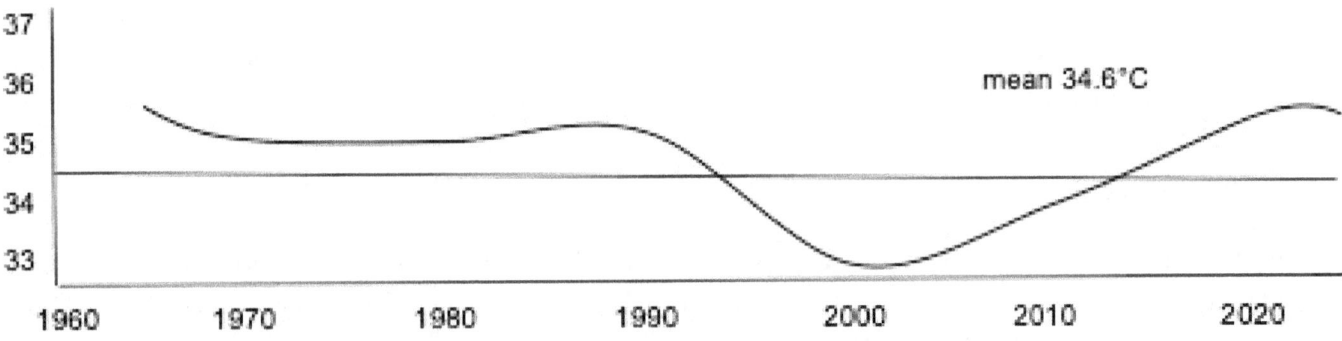

Figure 42: **Victoria River Downs.**

Victoria River Downs is a huge outback cattle station in the Northern Territory of Australia south of Darwin. It has an area of 8900 square kilometres and has been operating for since 1883.

The temperature record commenced in 1965. The mean is 34.6°C corresponding to the horizontal line. The warmest year had an average for daily maximum of 36.4°C in 2019 when El Nino was strong. The coolest was in 2000 at 33.0°C. The most recent year calculated in 2021 with an average of 35.5°C, similar to when the record started in 1965 when it was 35.6°C. Again the pattern is not that of "global warming".

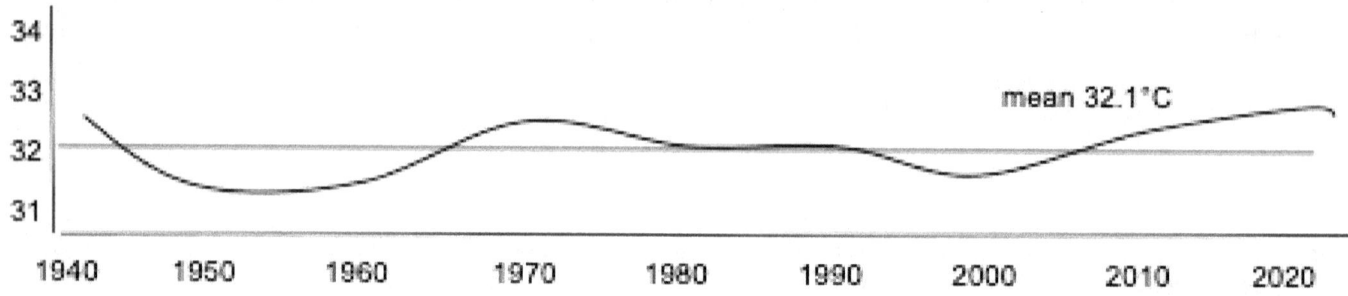

Figure 43: **Darwin Airport.**

This record starts in 1942 when it was 32.6°C. As seen it meanders along in a tight band between the lowest, 31.0°C in 1943 and 33.2°C in 2016. The mean is 32.1°C for the period.

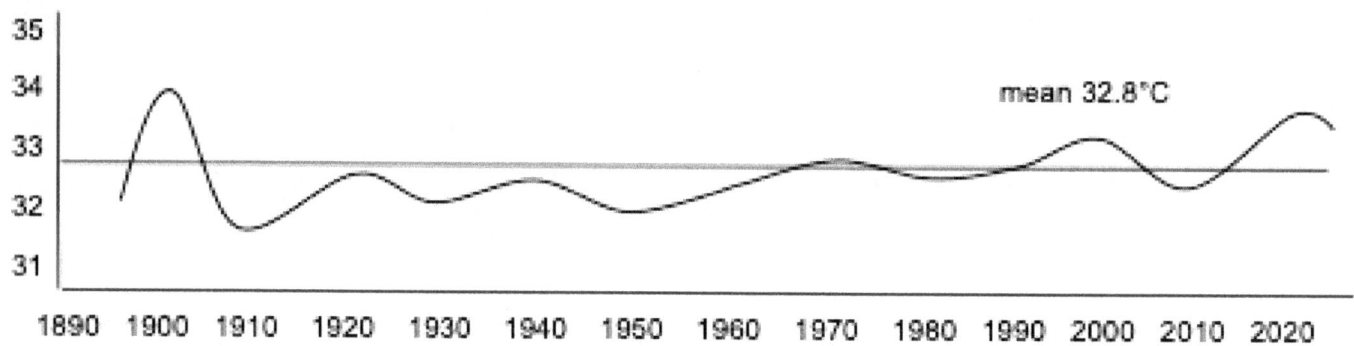

Figure 44: **Palmerville.**

Palmerville is another large cattle station of 134,000 hectares in far north Queensland north of Cairns close to the Dividing Range. The temperature records starts in 1896 when it was 32°C. The mean is 32.8°C. There was a warm year in 1990 when the mean monthly maximum averaged for the year was 34°C. The highest was for 1988 at 34.1°C and the lowest year 1914 when the yearly average was 31.5°C. In 2020 the average was 34.0°C and the last calculation is for 2021 when it was 33.5°C. The mean for the whole period is 32.8°C.

Again no pattern suggesting global warming.

Temperature °C – mean monthly maximums averaged for each year.

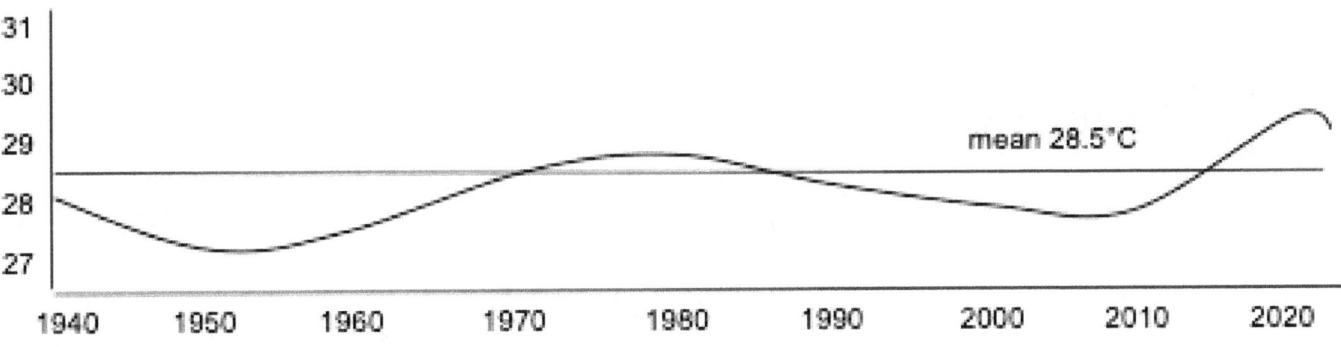

Figure 45: **Rockhampton.**

Rockhampton is a moderate sized city close to the North Queensland coast. The temperature record starts in 1940 when the average for the year of mean monthly maximums was 28.1°C. The highest was for the 2019 year at 30.2°C coinciding with a strong El Nino. The lowest was 27.2°C for the 1950 year. The last yearly average is for 2021 and was 29.3°C. The mean is 28.5°C.

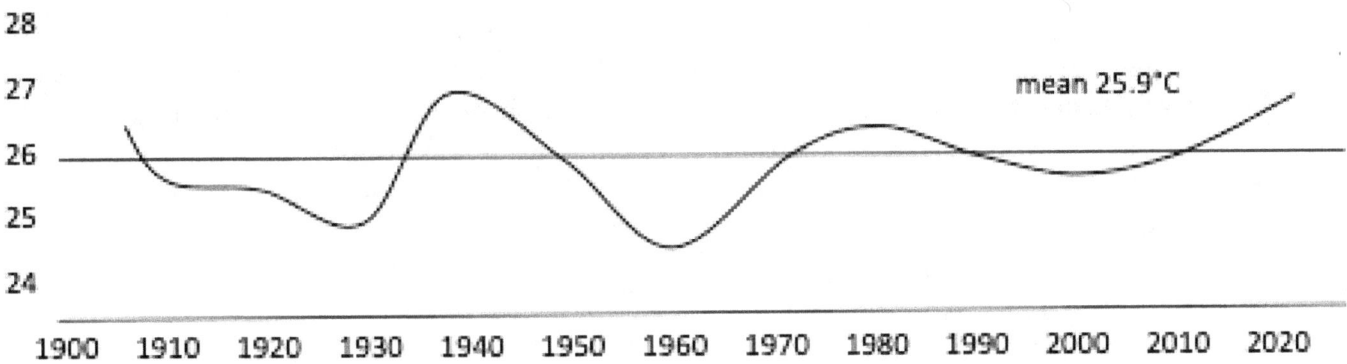

Figure 46: **Sandy Cape Lighthouse.**

Sandy Cape Lighthouse is at the northern tip of Fraser Island off the mid Queensland Coast north of Brisbane. The temperature record here has been going since 1907 and the mean yearly average of mean monthly maximums is 25.9°C corresponding to the horizontal line. The highest year was 1942 with 27.5°C and the coolest 1928 with 24.4°C. The last year calculated was 2019 when the average was 27°C. This record is clearly a strong argument against global warming.

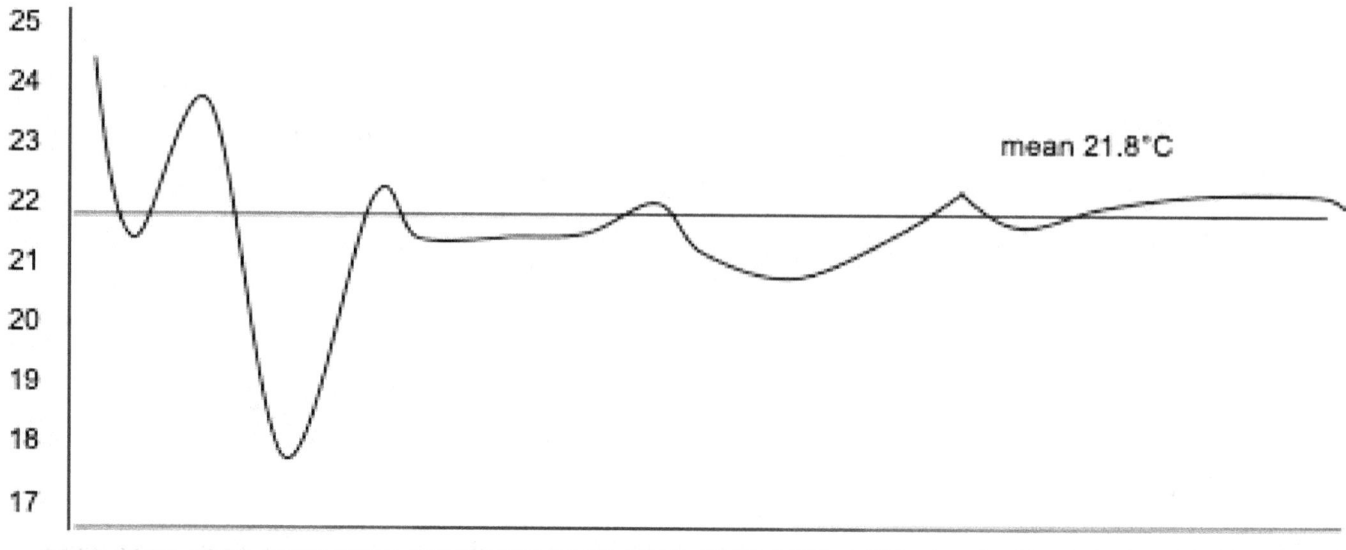

Figure 47: **Newcastle Nobbys Head Signal Station.**

This opened in 1858 and is at the entrance of Newcastle Harbour a major port in New South Wales. This is a very historic continuous record of great value from 1862. The highest yearly average was in 1877 at 25.2°C and the lowest that century also in 1890 at 17.8°C. There might have been some instrument issues then perhaps. Since 1900 the yearly average of monthly maximum means has been very flat with the last year calculated in 2021 at 21.9°C close to the mean for the whole period which is 21.8°C. No global warming here!

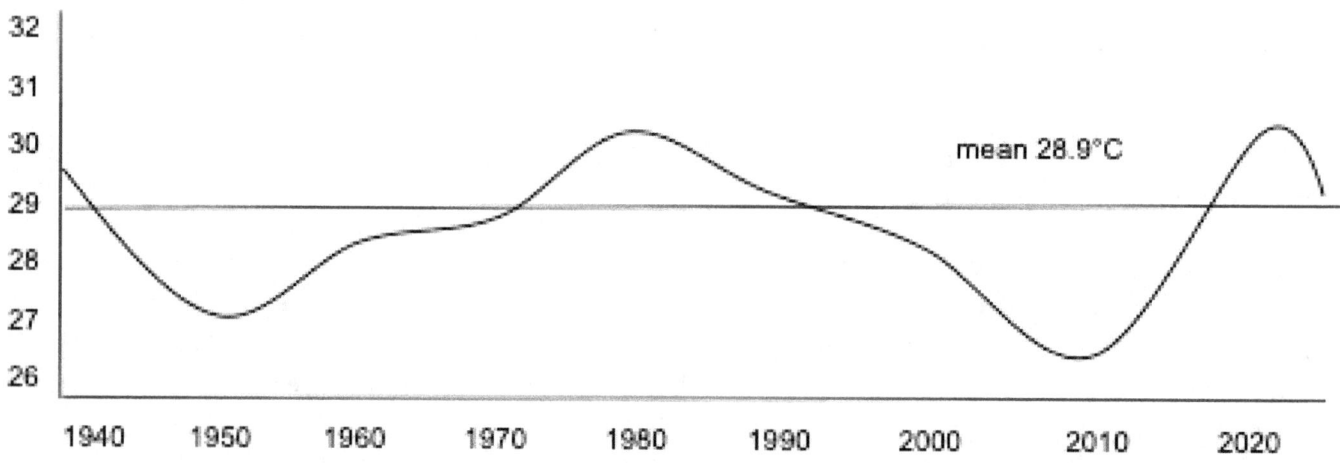

Figure 48: **Alice Springs Airport.**

Alice Springs is right in the arid centre of Australia. This record started in 1942 and the mean yearly average is 28.9°C. The highest year was 31.1 in 2019, an El Nino year, and the lowest 26.6°C in 2010. The last year calculated was 2021 when the average was 29.1°C, close to the mean.

Again the pattern is far from that promoted for global warming, increases since 1980. Rather it is variable in a different way.

Please note the temperature scale is every one degree not every two as in other graphs. This makes the variations appear larger.

The next five graphs are from five of the places detailed above: Moruya Heads Pilot Station; Bruny Lighthouse near Hobart, Tasmania; Cape Naturaliste in Western Australia; Darwin Airport; and Newcastle Nobby's Head Lighthouse. The graphs are looking at the summer and winter temperatures, January, the hottest month, and July, the coldest month, from 1980 to 2021. The mean of daily maximum temperature is calculated for the month, either July or January, for each year. The upper horizontal line is at the mean for January for the whole period the station has been open, not just the 40 years from 1980 to 2020. The lower horizontal line is at the mean for July. The readings are again all taken from The Australian Bureau of Meteorology (BOM) website referenced below.

Monthly means for every second year are graphed as is the last year calculated. The highest and lowest years during the 40 years are described in the text.

Alarmists say the temperature has been climbing since 1980 along with CO2 levels. These graphs again look at what has happened especially since 1980.

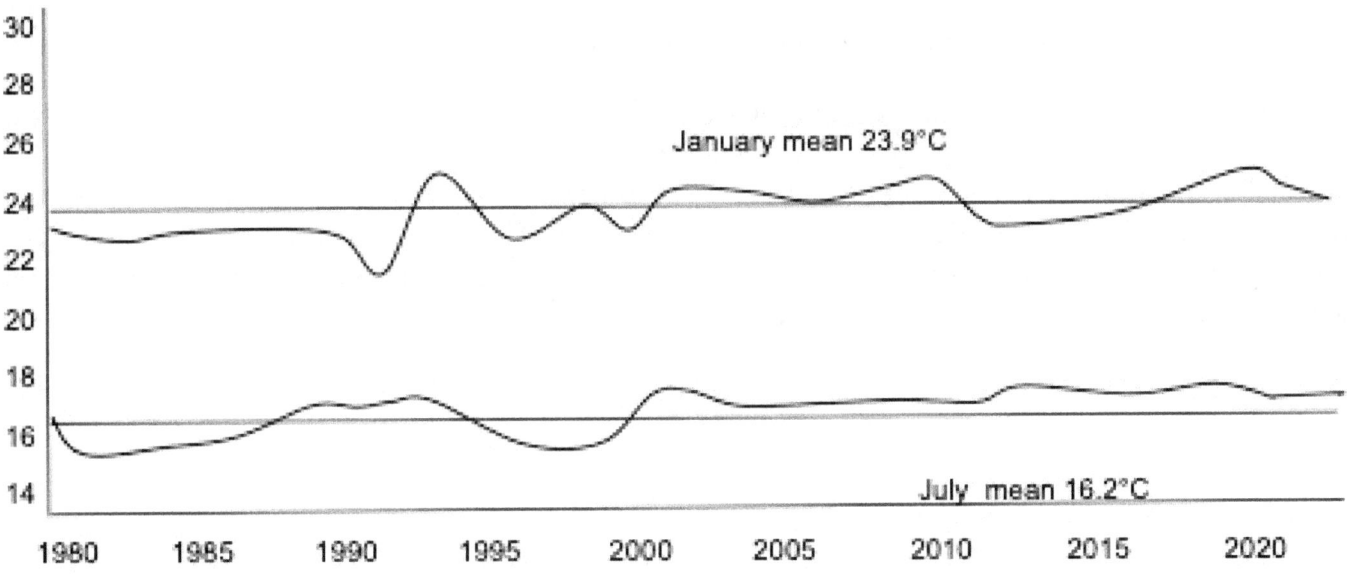

Figure 49: **Moruya Heads Pilot Station, South Coast, New South Wales.**

For the January mean daily maximums the mean for the whole period from1876 to 2021 was 23.9°C, where the upper horizontal line is. For the July readings the mean was 16.2°C, where the lower horizontal line is. The upper line is the mean daily maximums for January for each year, the lower blue line is for July. As said every second year is graphed. The last reading for 2021 for January was 24°C, close to the mean for the 40 years. The last reading for July in 2021 was 17.1°C, 0.9°C above the mean.

The highest January average daily maximum was 25.1°C in 1994 and 2001. And the highest average daily maximum for July was 17.9°C in 2014.

Temperature °C – mean daily maximums for the month. Upper line – January. Lower line – July.

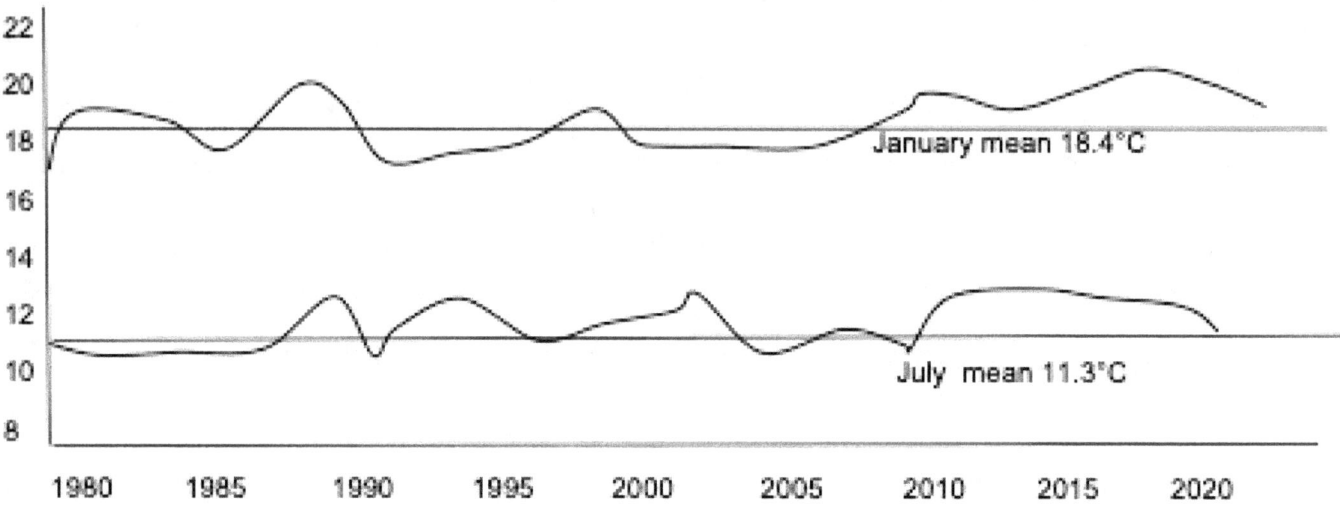

Figure 50: **Cape Bruny Lighthouse, near Hobart, Tasmania.**

The mean for the January readings for the whole period from 1871 is 18.4°C where the upper horizontal line is, and that for the July readings 11.3°C, the lower line. The highest year from 1980 for January was 2018 when the mean daily maximum was 21.2°C. For July the highest temperature year was 2002 when the mean was 12.8°C. The last January calculated was 2021 when the mean was 19.5°C, close to the mean for the period. For July the last year calculated was 2020 when the mean was 11.5°C, also close to the mean for the 40 years.

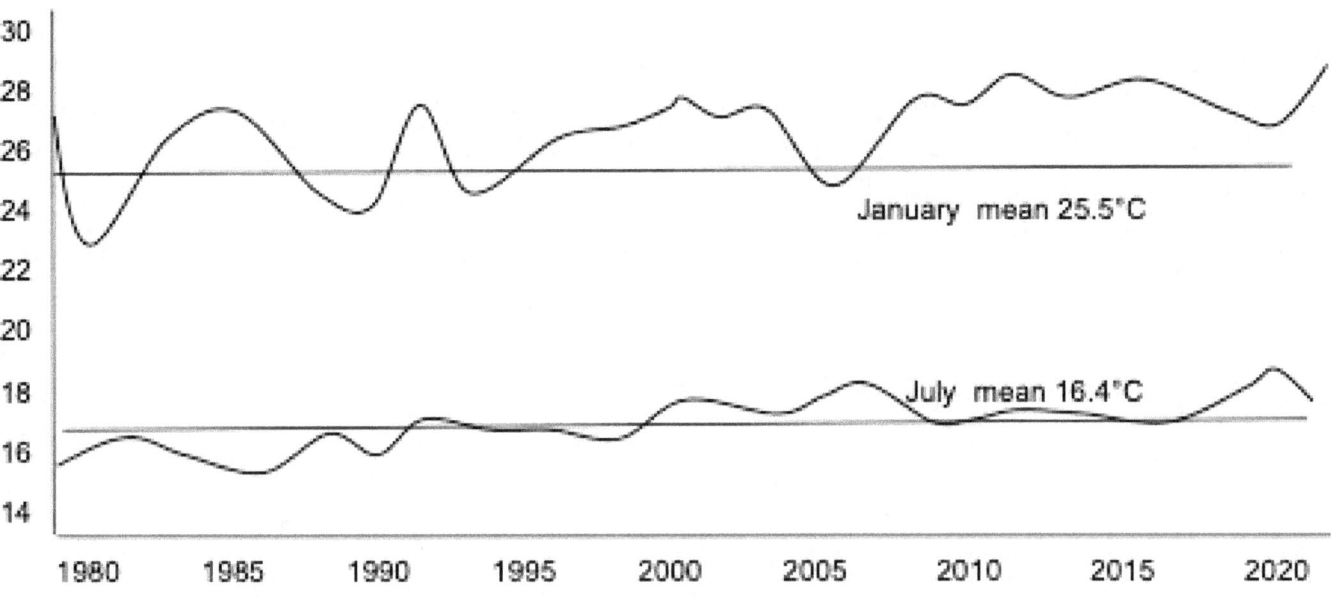

Figure 51: **Cape Naturaliste, Western Australia.**

Again the upper horizontal line shows the long term mean of January readings from 1904 to 2022, and the lower line the long term mean for July readings from 1904 to 2021.

The mean for all Januarys from 1904 is 25.5 and for Julys 16.4°C.

The highest January mean was 29.6°C in 2019, and that for July 18.3°C in 2020. The last reading for January was 29.2°C for 2022, and that for July 17.3°C for 2021, the same as the mean for the whole period from 1904. Obviously the winter record is very stable and for the last 20 years has been close to the long term mean. The summer one is more variable, affected by the Indian Ocean Dipole arguably.

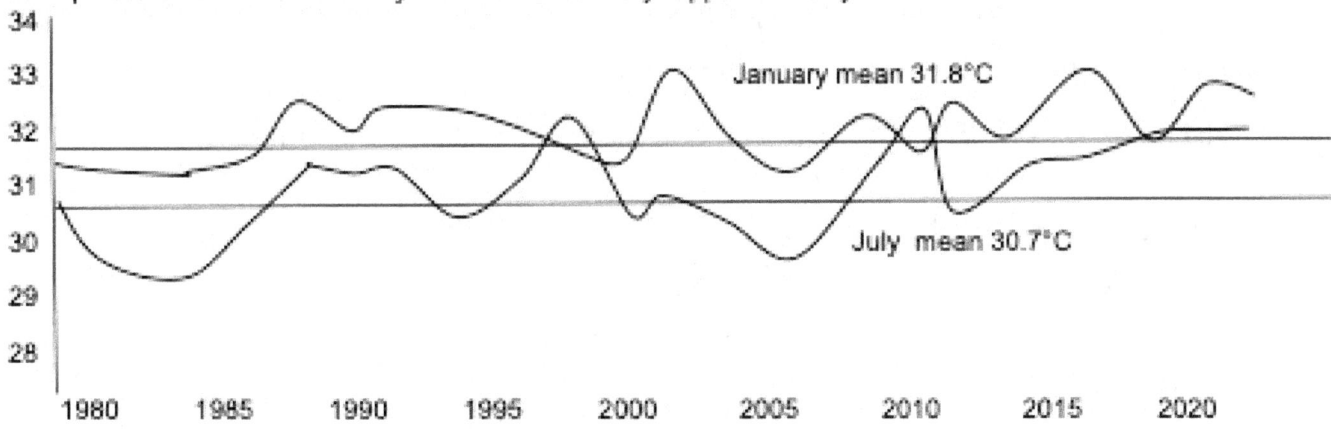

Figure 52: **Darwin Airport.**

The mean for January from 1942 to 2022 is 31.8°C, the upper horizontal line and the lower line shows the mean for July, 30.7°C. The warmest January in the 40 years 1980 on was 33.1°C in 2002, and the warmest July 32.6°C in 2010. The last reading for January is 32.6°C in 2022 and for July 32.3°C in 2021. As expected in a tropical region there is ongoing variation but no long term trend evident during this 40 years, except it was cooler in the early 80's.

Again note the temperature scale is every one degree, not every two as in the other graphs making the variation appear larger.

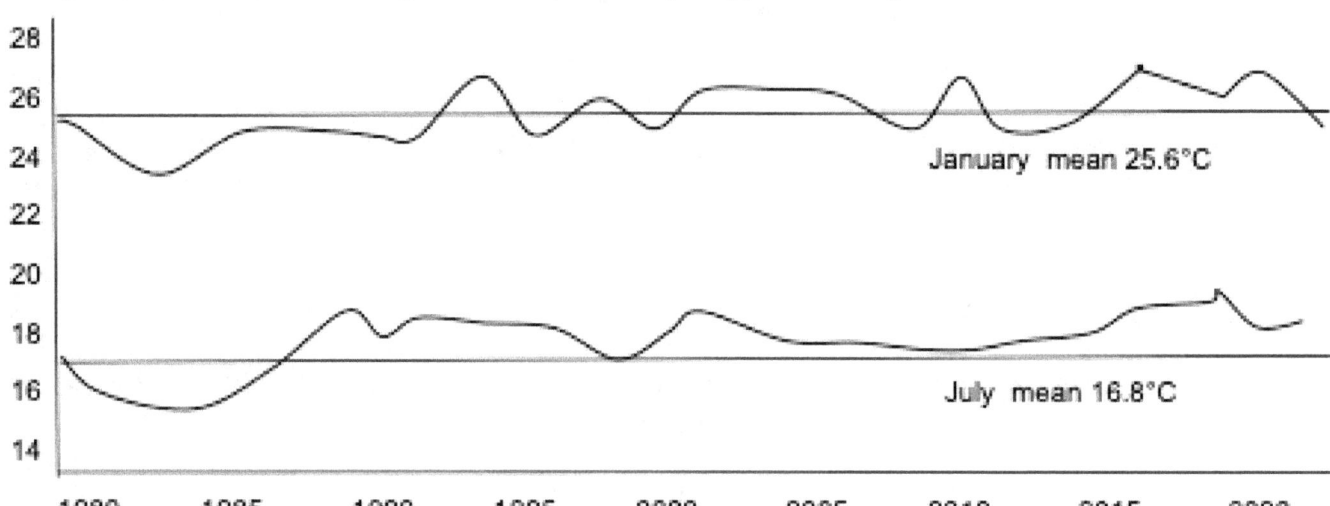

Figure 53: **Newcastle Nobbys Head Signal Station.**

The mean for January for the whole period the station has been open from 1862 to 2022 is 25.6°C, the upper horizontal line. That for July, the lower line is 16.8°C. The warmest January in this 40 year period from 1980 was 27.9°C in 2017. The warmest Julys were 2017 and 2019 with 19.2°C. 2019 was a strong El Nino year.

The last reading for January was 24.9°C in 2022, below the mean. That for July was 18.0°C in 2021.

Some Surface Temperature Records in Canada:

Again places outside the main metropolitan cities have been chosen to avoid urban factors. The Canadian Government has an excellent website at www.climate.weather.gc.ca which provides temperature data of thousands of Canadian places. Many, perhaps the majority have been discontinued in recent decades but many remain. In the references below there is a description of how to negotiate the website. Three are given here scattered across Canada. Again country places away from metropolitan areas have been chosen.

For each one daily maximum temperatures for either the month of January or July are averaged to give an average daily maximum for that month and year. The graphs focus on the 40 years from 1980 to 2000 which alarmists are concerned about. The graphs are drawn every two years for that period and every 5 years for earlier records than 1980. Again these are simple direct temperatures not "adjusted" as are satellite and other processed readings.

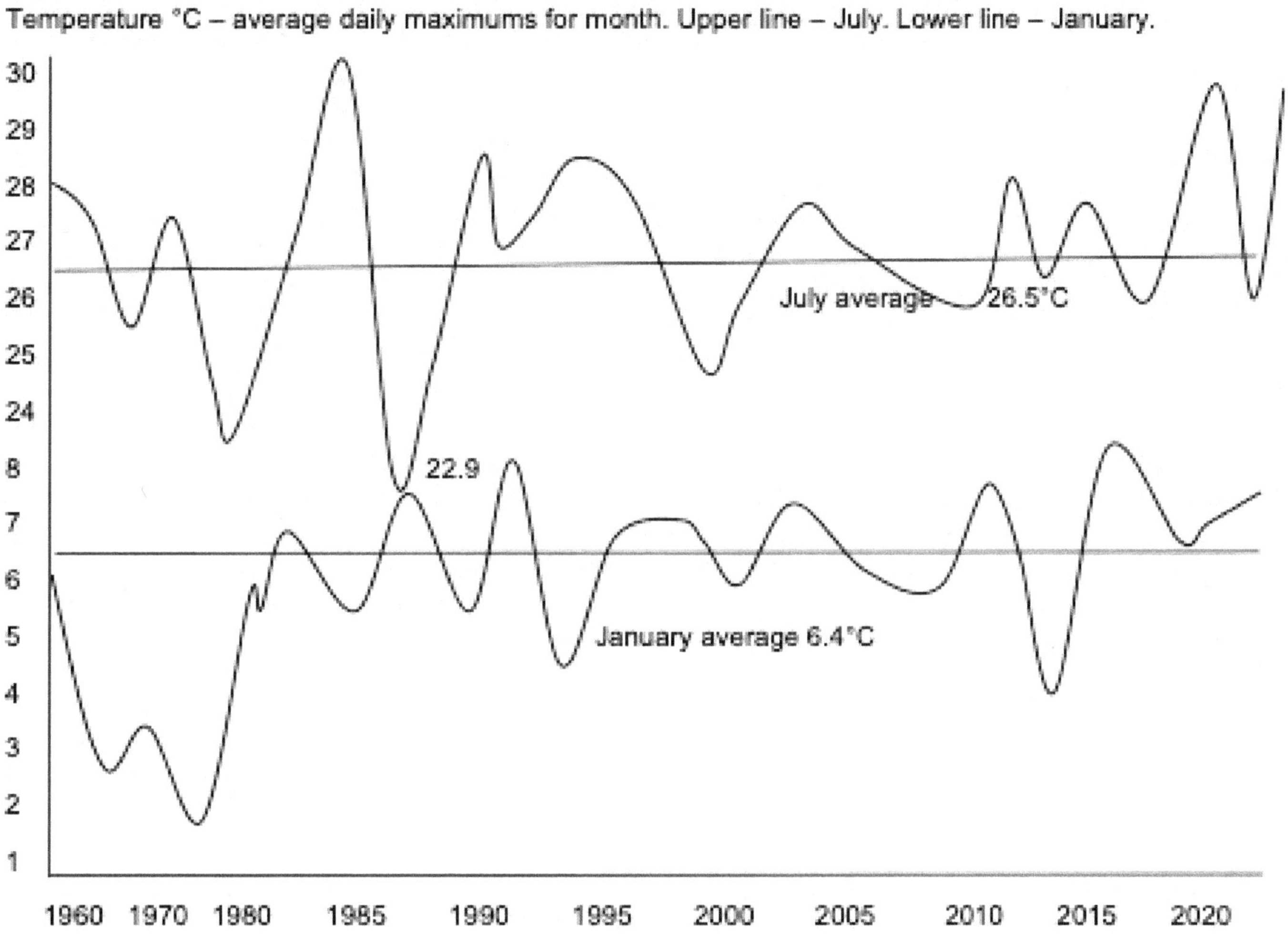

Figure 54:

Alberni Robertson Creek, Vancouver Island, British Columbia.

This is a fish hatchery on a river in the centre of Vancouver Island near Port Alberni. It has kept weather data since 1961. There is a big difference between summer and winter, so the temperature scale "skips" from 24° to 8° as shown. The average for July for the whole period from 1980 is 26.5°C and that for January 6.4°C as shown. For July the highest monthly average was in 1985 when it was 30°C. In 2018 and in 2022 it was 29.6°C.

For January the warmest average was 8.7°C in 2016.

The temperature scale is marked every single degree.

The world-wide cool period before 1976 is suggested here. There is a lot of variation otherwise, and no steady upward trend since 1980 is demonstrated.

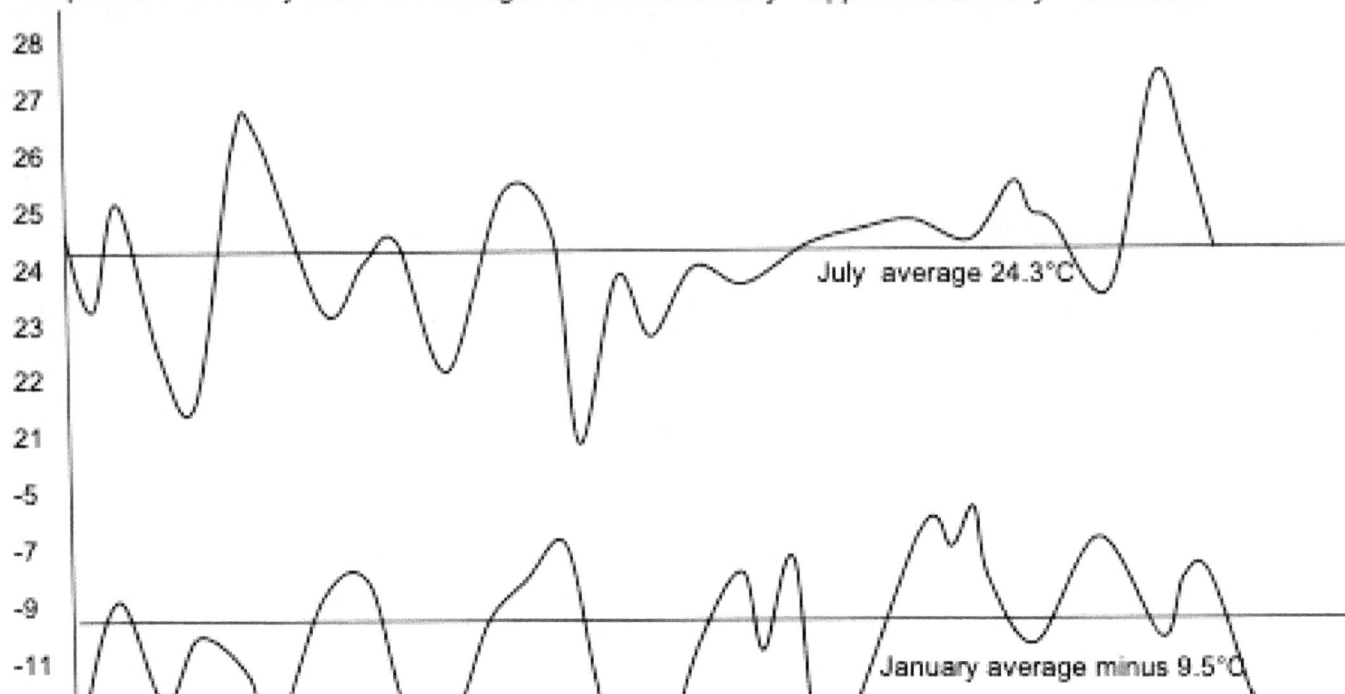

Figure 55: **Bagotville Air Force Base, Quebec.**

This is a famous air force base well known for its renowned air show. It is several hundreds of kilometers north of Quebec City at latitude 48° North. The record goes back to 1945 and again there is a big difference between summer and winter. For July averages the warmest year was 2018 when it was 27.9°C, and for January minus 5.6°C in 2010. The average for the period 1980 to 2021 for July is 24.3°C, and that for January from 1980 to 2022 is minus 9.5°C. The last reading for July was 24.3°C in 2021, and that for January minus 13.4°C in 2022.

The temperature scale is marked every single degree for July and every second for January.

Again there is no consistent upward trend since 1980.

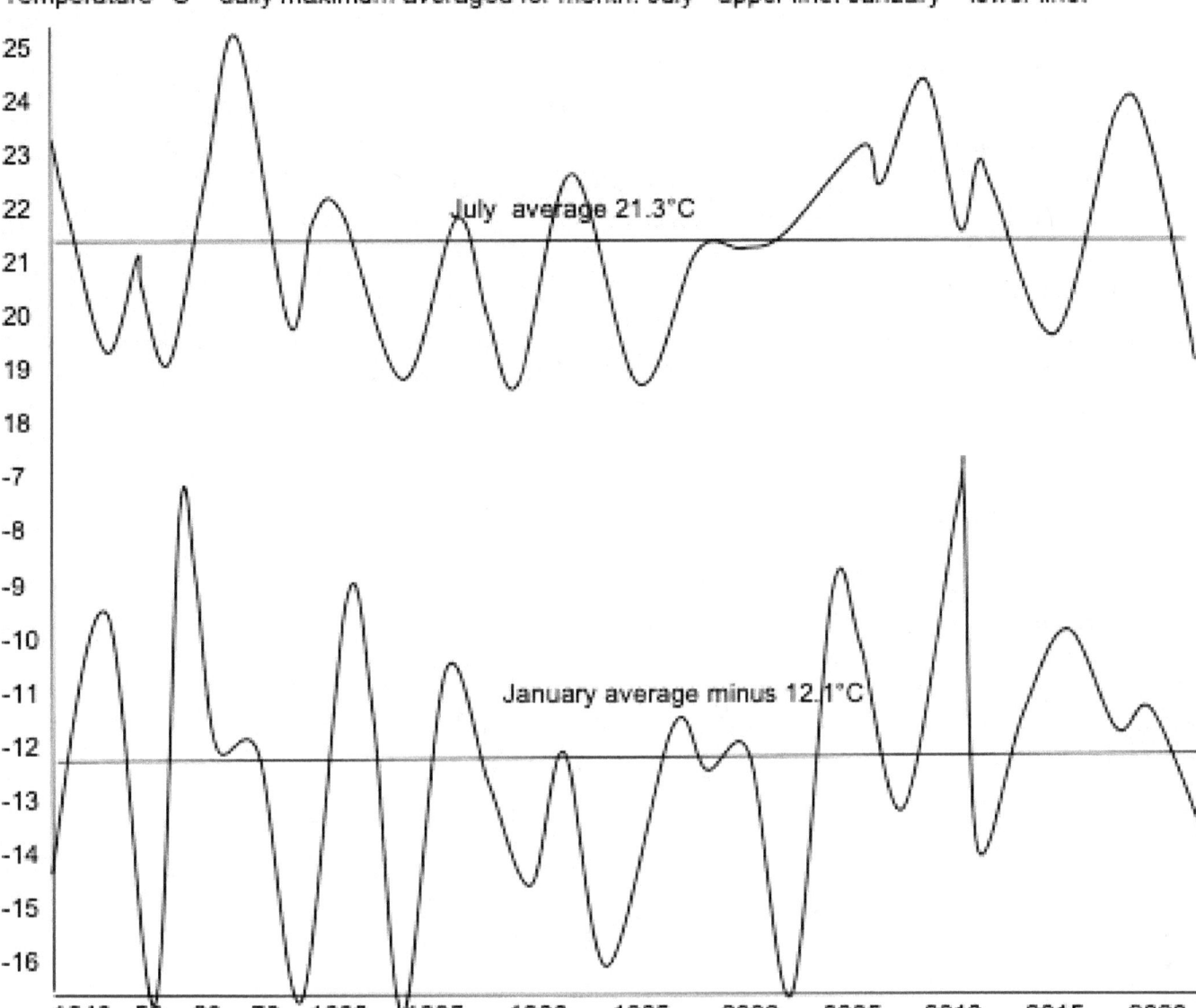

Figure 56: Goose Bay, Newfoundland.

This is another Canadian Air Force Base and international airport with long term weather records from 1941 to 2022. It is on the Atlantic Coast at 53° North latitude. Again there is a big difference between July and January temperatures. The highest average daily maximum for July was in 1975 at 25°C, and that for January minus 7.0 in 2010. The average for the period 1980 to 2022 for July was 21.3°C, that for January minus 12.1°C. The last average daily maximum for July was 18.8°C in 2022, and that for January minus 14.4°C also in 2022.

The temperature scale is marked every single degree.

A big variation from year to year is seen but no clear upward trend.

Some Surface Temperature Records from the U.S.A.

Four places are shown. Again an effort was made to find smaller rural places but this was hampered by many having incomplete records for the period 1980 to 2022 of interest. Many have been closed down in recent years. Also the official government sources, mainly centred around the National Oceanic and Atmospheric Administration (NOAA), do not seem to have available detailed records of different places for different months and years, just long term average data which may have been "adjusted". So the website "wunderground.com" has been used for data as explained in the references below.

The temperature scale on the left side of each graph varies as shown to show as clearly as possible trends.

These U.S. records are in degrees Fahrenheit, °F, as this is used there. 0° Centigrade is 32° Fahrenheit. To convert from Fahrenheit to Centigrade subtract 32, multiply by 5, and divide by 9. Again these are simple direct readings not "adjusted".

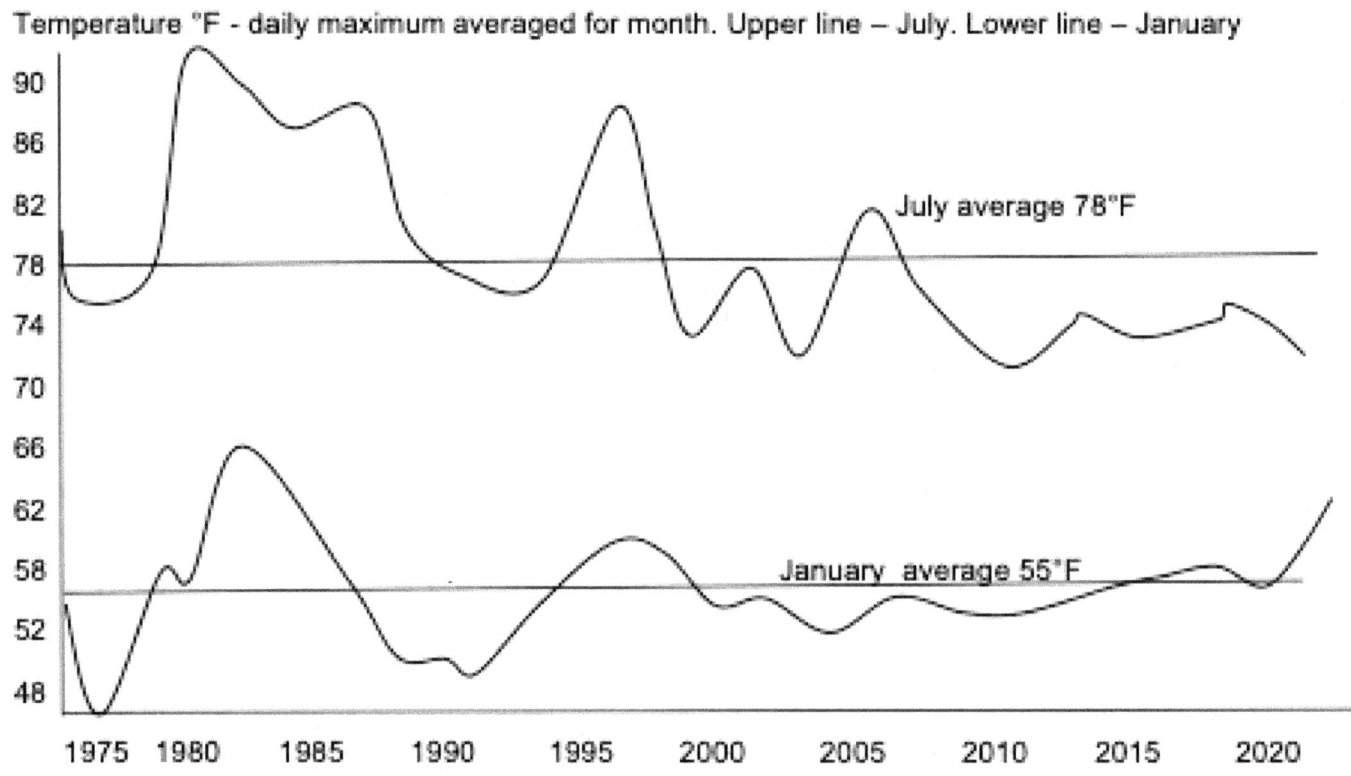

Figure 57: **Santa Rosa, California.**

Santa Rosa is a medium sized city in California north of San Francisco and inland in a rural wine growing area. It has a population of 180,000. The temperature record goes back to 1974.

The monthly average of daily maximums is again shown. The highest July average in this period 1980 – 2022 was 91°F (32°C) in 1982, and the highest for January was 63°F (17°C) in 2022. The average for the period was 78°F (25°C) for July and 55°F (13°C) for January.

The temperature scale is marked every second degree.

It is clear there is no upward trend from 1980 to 2022 for Santa Rosa, indeed there is a downward trend for July, and January is steady.

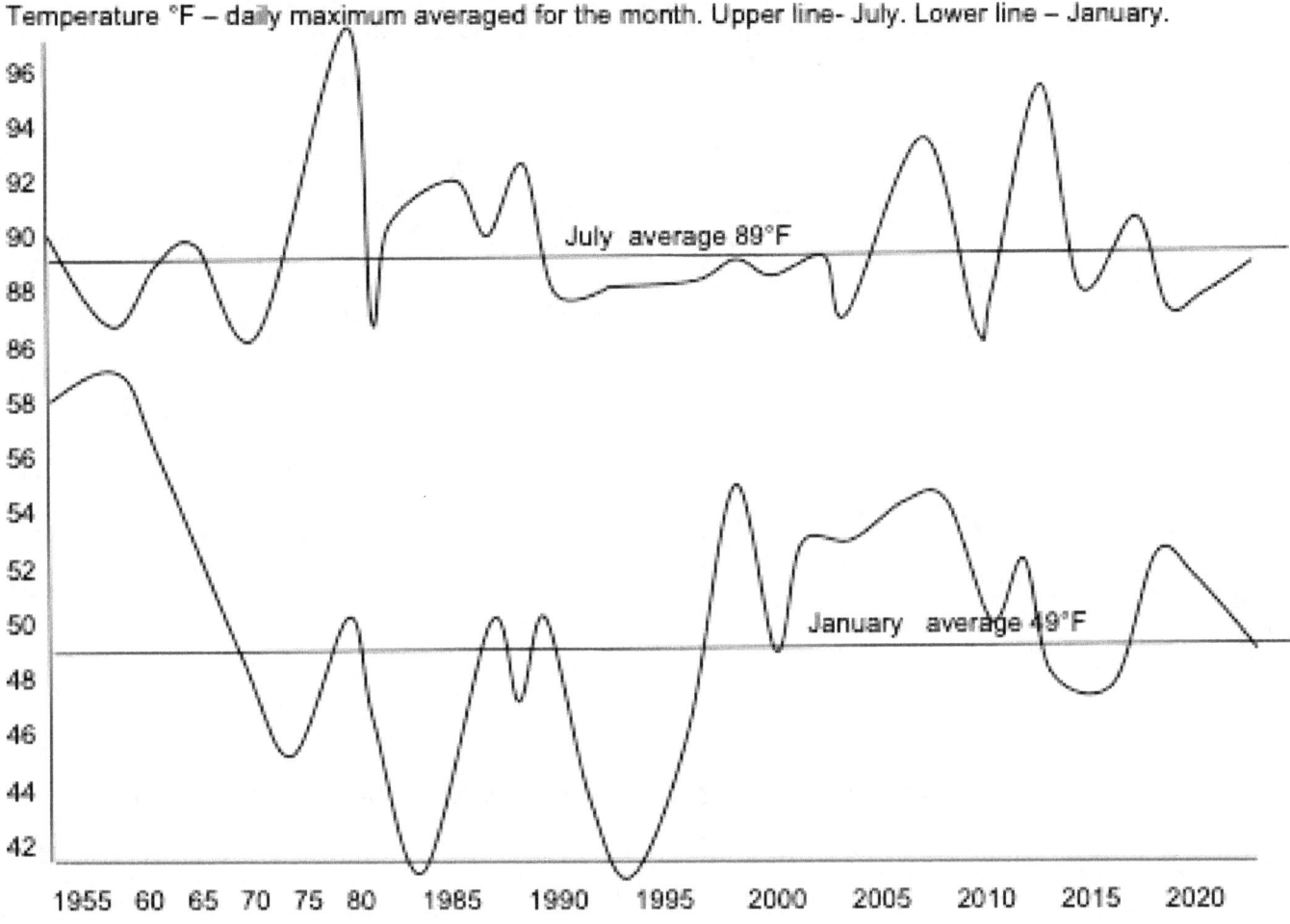

Figure 58: **Wichita, Dwight D Eisenhour Airport, Kansas.**

Wichita is a moderate sized city in Kansas, population 390,000. Because there is a major airport and USAF base there the temperature record goes back to 1955. For the period 1980 to 2022 the highest July daily maximum monthly average was in 1980 and was 97°F (36°C). That for January was 59°F (15°C) in 1958. The average for this period for July was 89°F (32°C) and for January 49°F (9°C).

The last record monthly average of daily maximums for July was in 2022 at 88°F (31°C) and for January also 2022 at 49°F (9°C), both at or lower than the average for the 40 years.

The temperature scale is marked every second degree. Again there is no upward trend suggesting long term warming.

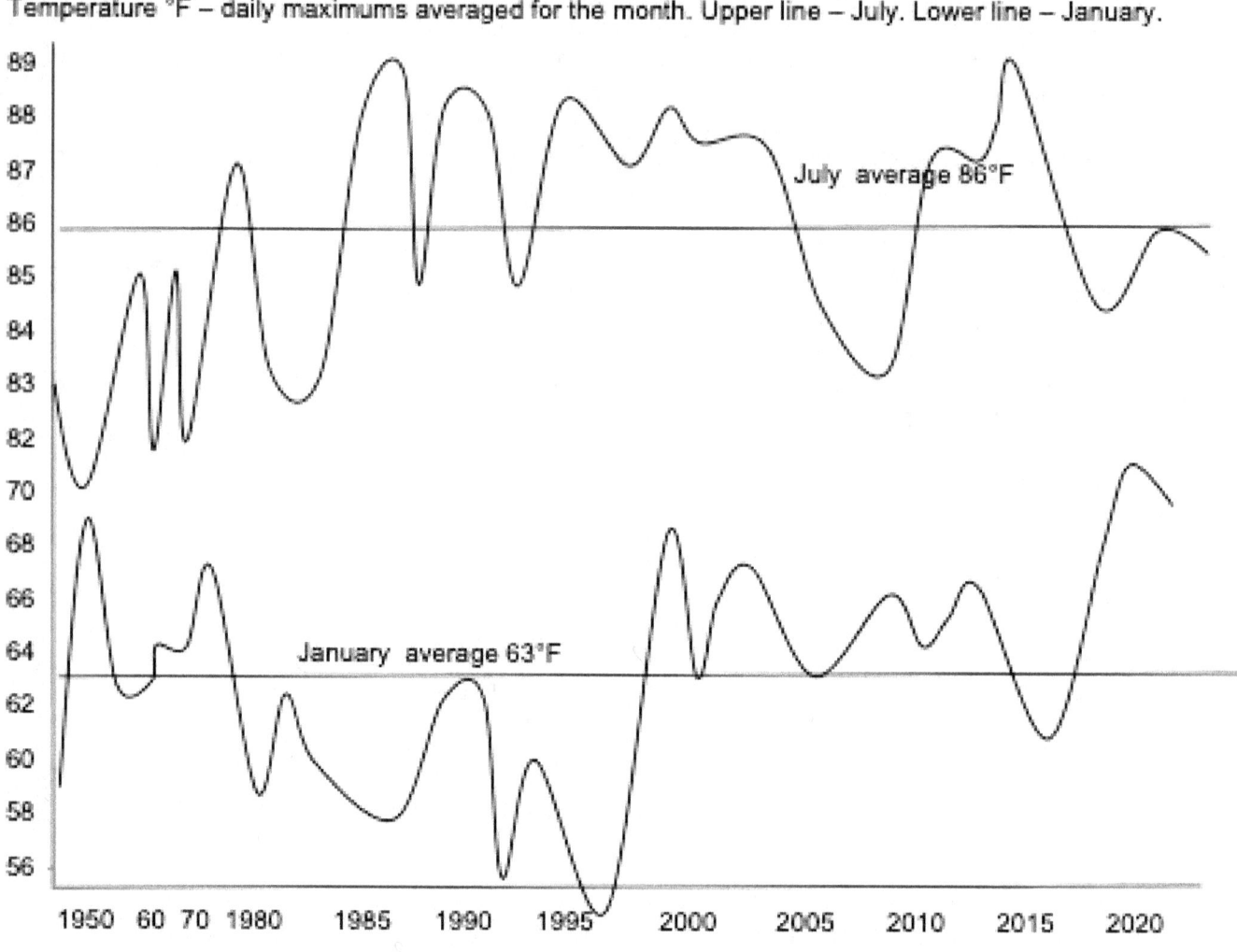

Figure 59: **Charleston International Airport.**

Charleston is a moderate sized city in South Carolina, population 134,000. It was founded in 1670 and has many fabulous old buildings and streets. The temperature record at the airport starts in 1945. The highest monthly average of daily maximums for July was 89°F (32°C) in 1986, and that for January was 71°F (21°C) in 2020. For the period 1980 to 2022 the average in July was 86°F (30°C) and for January was 63°F (17°C).

The world-wide cooler period from 1950 to 1975 is suggested by the summer graph, but not the winter one. The summer one varies around the average from 1985 and the last temperature average, 85°F in 2022 is below that. For winter, after cooler periods from 1980 to 2000 it has been close to average with the last 2 years warmer, 69°F (21°C) for 2022.

The temperature scale is marked every single degree for July and every second for January.

Again there is no pattern here to suggest a steady warming since 1980.

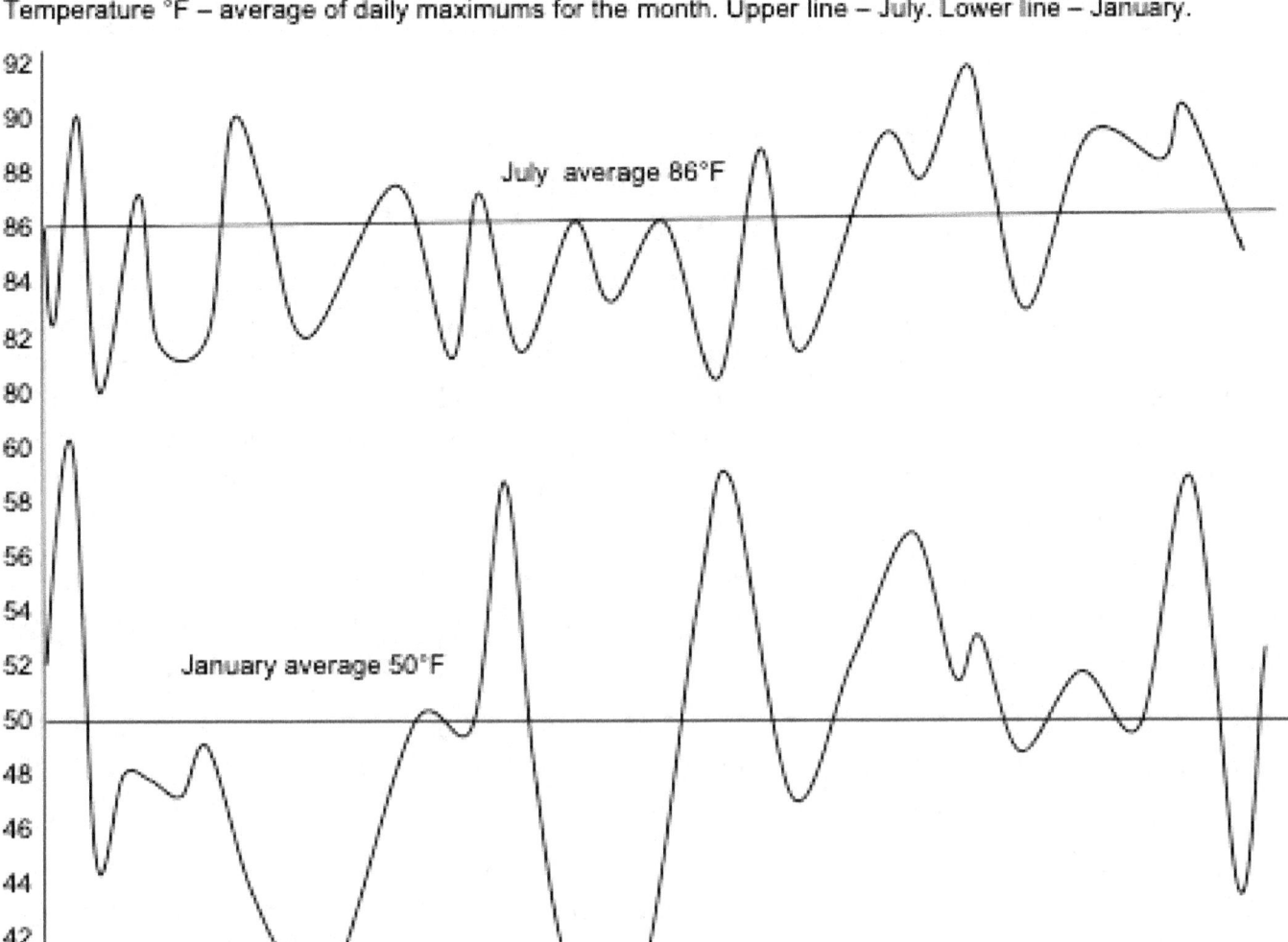

Figure 60: **New York City – La Guardia Airport.**

La Guardia Airport is close to the centre of New York City. In general an effort has been made to study smaller, rural, non-metropolitan areas to avoid "urban factors" which tend to increase temperatures in big cities due to large paved areas and roofs absorbing the Sun's heat, industrial pollution, human energy production, and also to try to get the most "natural" Earth surface temperature away from human influence if possible. But an exception has been made here because the Goddard Institute of Space Studies is located at Columbia University on Broadway in New York City; and that is the headquarters of James Hansen, Gavin Schmidt and others at the forefront of alarmist promotions.

The temperature record starts in 1949. The warmest average of daily maximums for July was in 2010 at 92°F (33°C) and for January was 60°F (15°C) in 1950. For the period 1980 to 2022 the July average was 86°F (30°C) and that for January 50°F (10°C). The last July average was 85°F (29°C) in 2022, below the long term average, and the last for January in 2022 was 53°F (12°C). In January 2021 the average was 44°F (7°C) as shown on the lower graph.

The temperature scale is marked every second degree here. This record, from the heartland of the alarmists, does not show any upward trend, just variations around the average.

Some Surface Temperature Records from the United Kingdom.

Again an effort was made to obtain daily historical temperature records from the official government websites especially the U.K. Meteorological Office but only long term adjusted data seems to be available. So the website weatherspark.com has been used. This website does give historical temperature records back to 1952 for some places, mainly airports. The data appears "unadjusted". The method used to get the monthly averages for each year especially from 1980 to 2022 is given in the references. It involved getting daily maximum temperatures for every second year for July and January, and calculating the monthly average for those months for each of those years.

The graph below applies to Cambridge. The data comes from the nearby Stansted Airport for 2021 back to 1980. From 1953 to 1975 the data comes from Mildenhall RAF base near Cambridge. This base is now largely used by USAF bombers.

It seems that although this data comes from Stansted Airport it is arising from records taken in Cambridge, or, for the years 1953 to 1975 as said above, from Mildenhall RAF Base. There are small differences in the temperature records for these places resulting in this conclusion.

It is clearly disappointing that simple unadjusted temperature records for specific locations are hard to get concerning the U.K. . Many weather stations have closed. The official sites feature adjusted data for multiple locations clumped together.

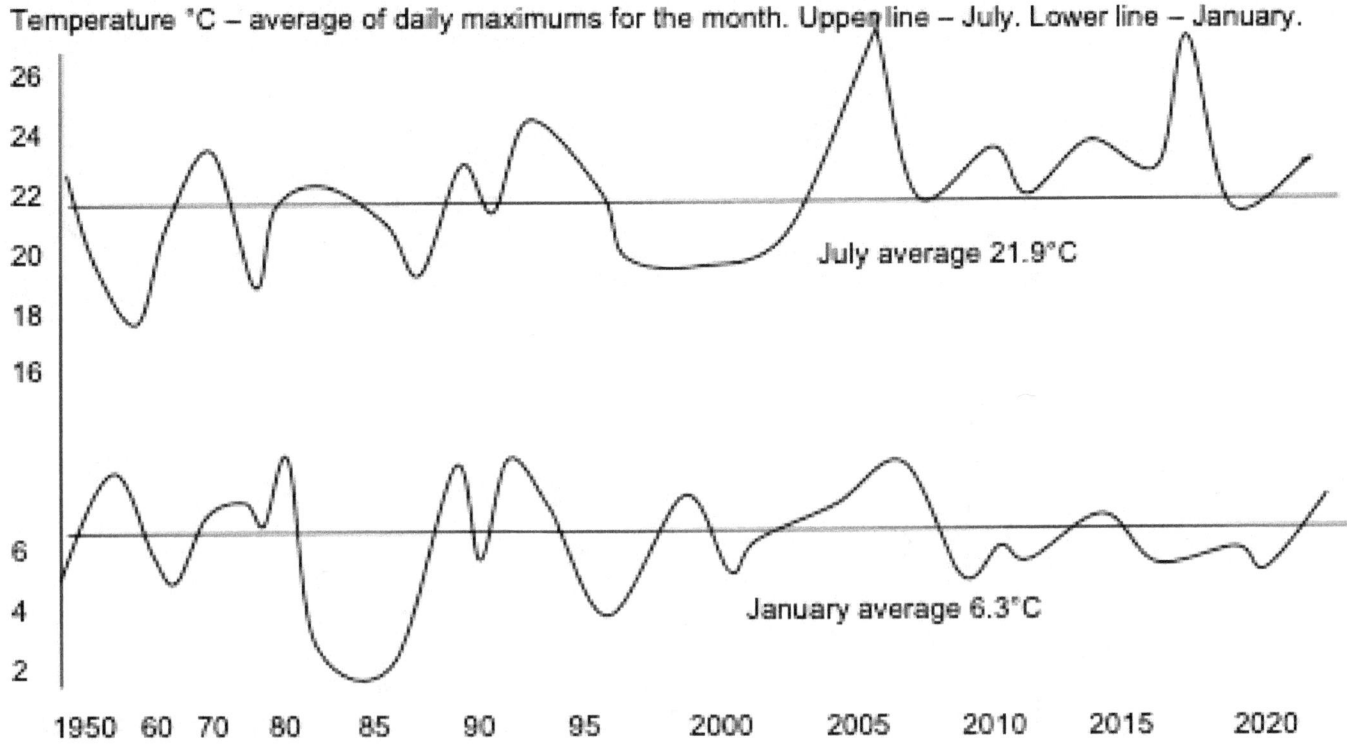

Figure 61: **Cambridge, U.K. (nearby RAF Mildenhall for 1953 to 1975).**

The average for July daily maximums from 1980 to 2021 was 21.9°C and that for January 6.3°C. The warmest July was in 2006 with an average of 26.8°C daily maximum, and the warmest January also was 2006 at 9.5°C. The last temperature in July was 22.1°C in 2021, below the average. The last for January was 7.8°C also in 2021.

There is no long term warming shown here, just variations around the averages.

The temperature scale is marked very second degree here.

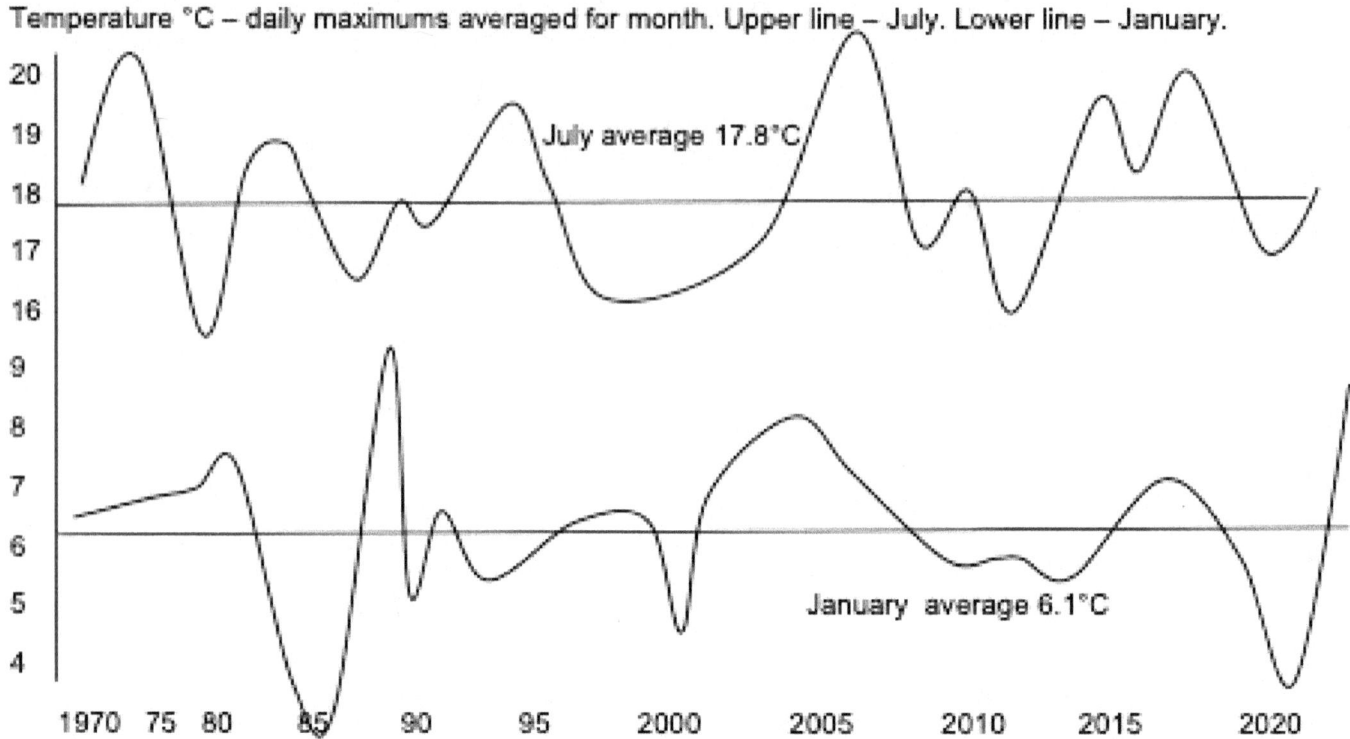

Figure 62: **Aberdeen Airport.**

Aberdeen is a port city of 480,000 population on the northeast coast of Scotland 57° North latitude. The airport where these records were taken is located on the northern city outskirts. The available record, again on weatherspark.com website, starts in 1972. The monthly average for daily maximums for July for the period 1980 to 2021 is 17.8°C, and that for January 6.1°C. The highest July average was in 2006 at 20.5°C, and the highest for January 9.6°C in 1988. The last reading for July was 18°C in 2021, and for January 8.7°C in 2021.

It is clear there is no long term warming trend, just variations around the averages.

The temperature scale is marked every single degree here.

Some Temperature Records for Europe:

Official government meteorological websites were again searched for simple direct temperature records for places but only long term averaged "adjusted" data could be found. So again the website weatherspark.com has been used, the same method to negotiate the site as discussed in the references below. Weatherspark.com largely uses historical data from airports and military air force bases. Again country non-metropolitan areas have been sought.

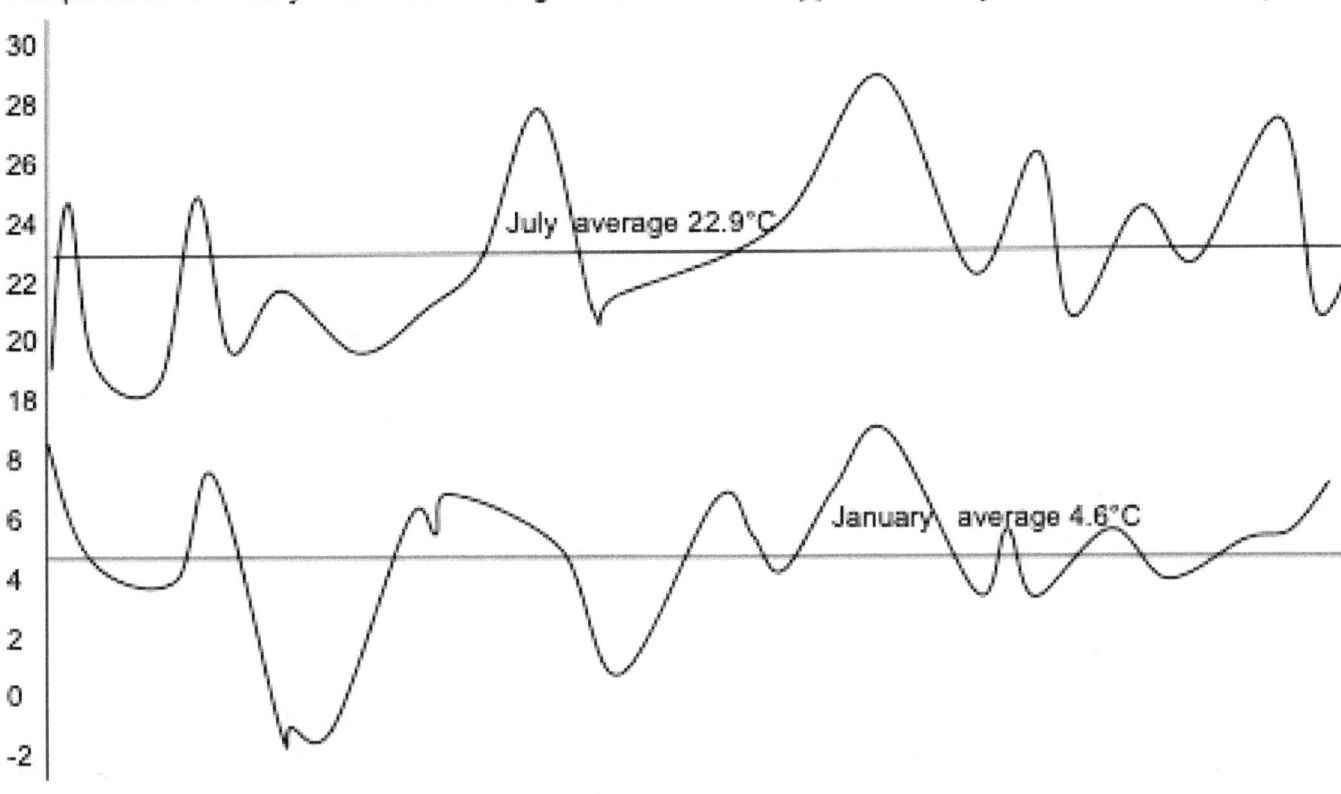

Figure 63: **Deelen Air Base, The Netherlands.**

Deelen is a tiny hamlet in Gelderland, The Netherlands. It is rural east of Amsterdam close to the German border, latitude 52° North. There is a large Royal Netherlands Air Force base there. The temperature record on weatherspark.com starts in 1974. For July there is a gap from 1998 to 2002. The monthly average of daily maximums for July for the period 1980 to 2021 was 22.9°C, and that for January 4.6°C. The highest July average was 29.1°C in 2006, and that for January 7.7°C in 1982. The last average for July was 22.3°C in 2021, and the January average for 2022 was 6.8°C.

There was a short cooler period in 1984, otherwise the graphs reveal variations around the average with no long term warming trend from 1980. A feature of the daily record on temperatures here is big swings of up to 20°C from one day to the next, reflecting its more inland location away from the moderating influence of the sea.

The temperature scale is every two degrees.

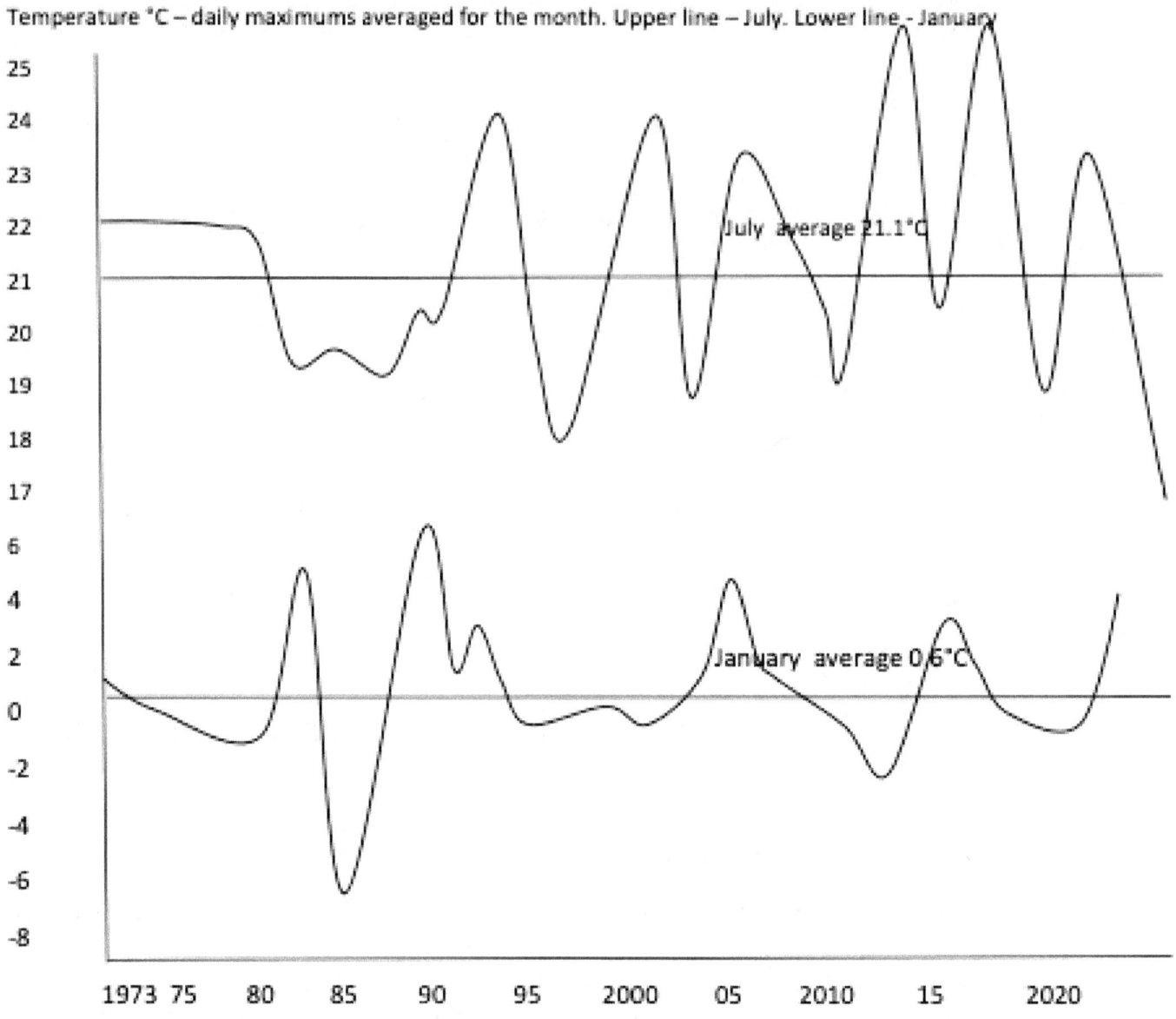

Figure 64: **Rygge Airport, Norway.**

This is a small town of 13,000 population 60 km south of Oslo on the sea of Oslofjord latitude 59° North. It was the site of a Royal Norwegian Airforce base then a commercial airport. Again weatherspark.com was used for the temperature record which starts in 1973. The monthly average for daily maximums for July for the period 1980 to 2022 was 21.1°C, the highest July average year was 2018 at 25.9°C, and the lowest July average year 2022 at 17.5°C, the last year in this record.

For monthly average for daily maximums for January for the period 1980 to 2022 the average was 0.6°C, the highest January average monthly maximum was 6.4°C in 1989 and the lowest minus 7.3°C in 1987. The last January in the record is 2022 when the average daily maximum was 4.1°C.

Every second year is graphed from 1973 to 2022. Again there is no long term trend, just variations around the average.

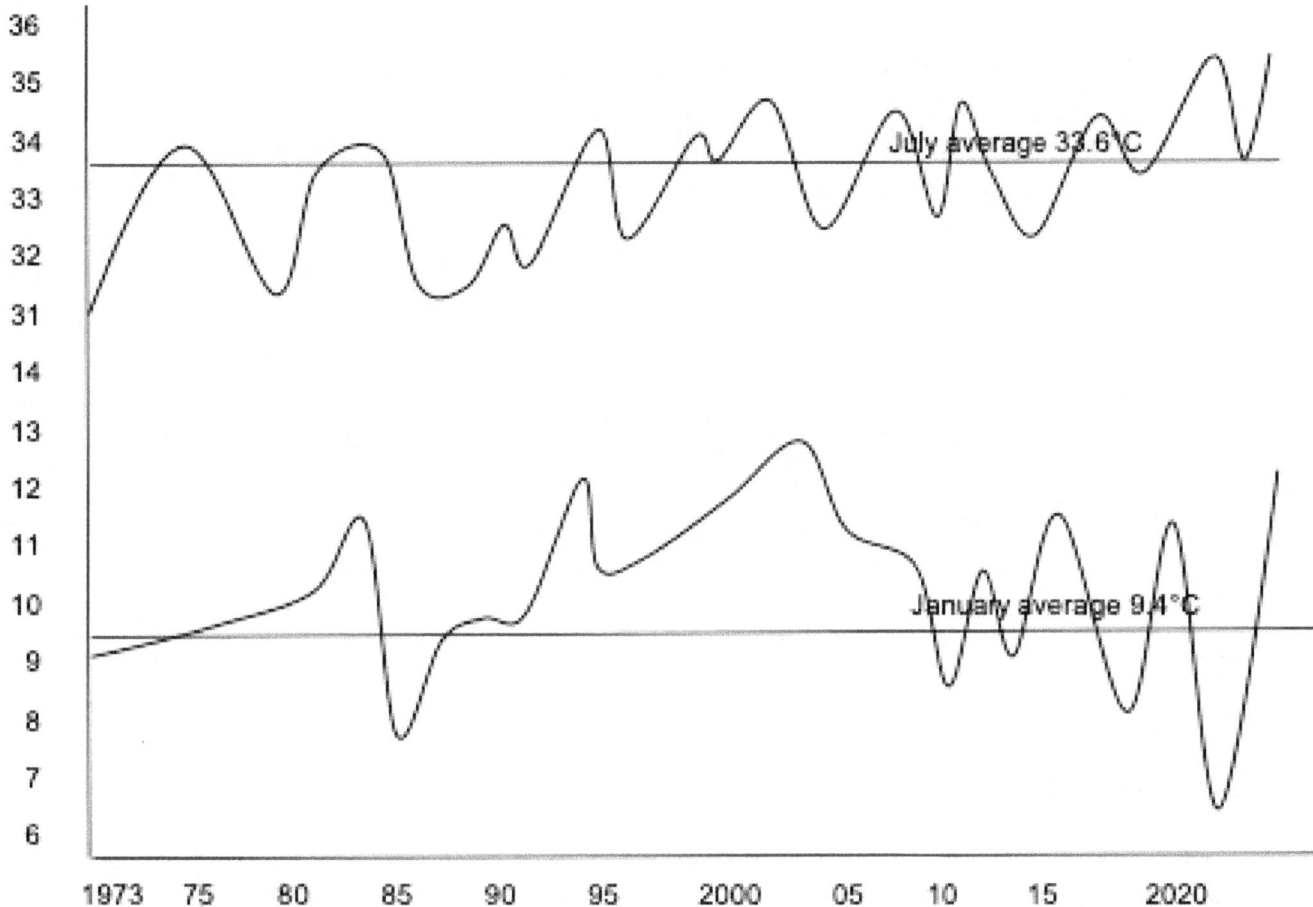

Figure 65: **Albacete, Spain.**

Albacete is an inland city of 170,000 population located in high rich farmland at 700 meters height in southeast Spain, latitude 38° North. The temperature record on weatherspark.com starts in 1973, readings are taken at the international airport there.

For July the average of monthly averages of daily maximum temperatures for the period 1980 to 2022 was 33.6°C, the highest was 35.5°C in 2022, the coolest 30.9°C in 1973. The last July recorded was 2022 when the monthly average was 35.5°C.

For January the average for the period 1980 to 2022 was 9.4°C, the highest 13.1°C in 2004, the coolest 6.3°C in 2021. The last January was 2022 when the average was 12.2°C.

A case could be made for a trend upwards - but minimal - concerning the July results but not for the January temperatures.

Lastly, Two Pacific Records:

Figure 64 below shows the long term temperature at Vunisea, a village on Kadavu Island, Fiji. It is south of the most populated island Viti Levu at latitude 19° South. The airport for Kadavu Island is there at about 2 meters height above sea level.

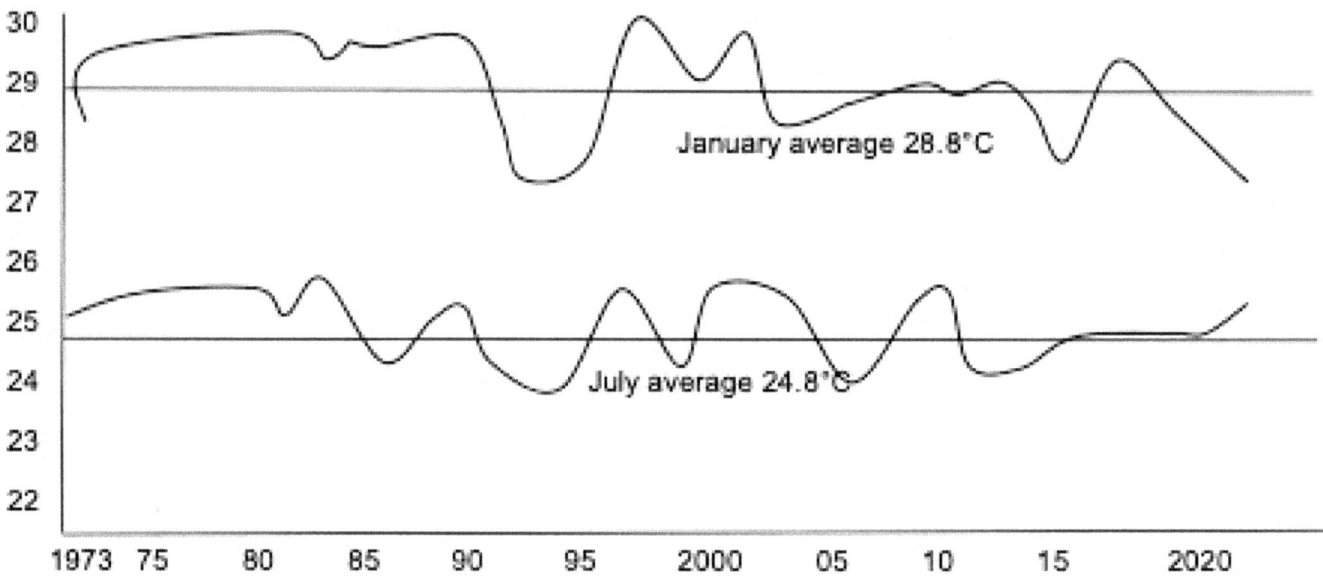

Figure 66: Vunisea, Fiji.

The temperature record for the airport on weatherspark.com starts in 1973. The monthly average of daily maximums for January was warmest in 1997 at 30°C and coolest in 2021, also 1993, at 27.1°C. The average from 1980 to 2021 was 28.8°C. The last reading in 2021 was 27.1°C.

For July the average for the period 1980 to 2021 was 24.8°C, the warmest 25.9°C in 1984 and the coolest 23.9°C in 1994. The last reading in 2021 was 25.5°C.

The graphs again variation about the average. For January (summer, the wet season) there has been apparent cooling over the 48 years of observations. For January (winter, the dry season) temperature has been steady.

Figure 65 below shows the record at Christchurch, the largest city in the South Island of New Zealand with a population of 380,000. It is located on the east coast close to the Pacific Ocean at latitude 43° South. The temperature record on weatherspark.com starts in 1952 and is taken at the international airport close to the inland boundary of the city, height above sea level 37 meters.

For the monthly average of daily maximums for January (summer) for the period 1980 to 2022 the average was 21.8°C. The warmest month was in 2019, a strong El Nino year, when the average was 24.4°C, the coolest in 1987 at 18.7°C. The last reading was 21.8°C, the same as the long term 1980 – 2022 average.

For July (winter) the average for the period 1980 to 2022 was 10.9°C, the warmest 13.2°C in 2018, the coolest 8.0°C in 1952. The last average was in 2022 when it was 10.8°C, just below the average.

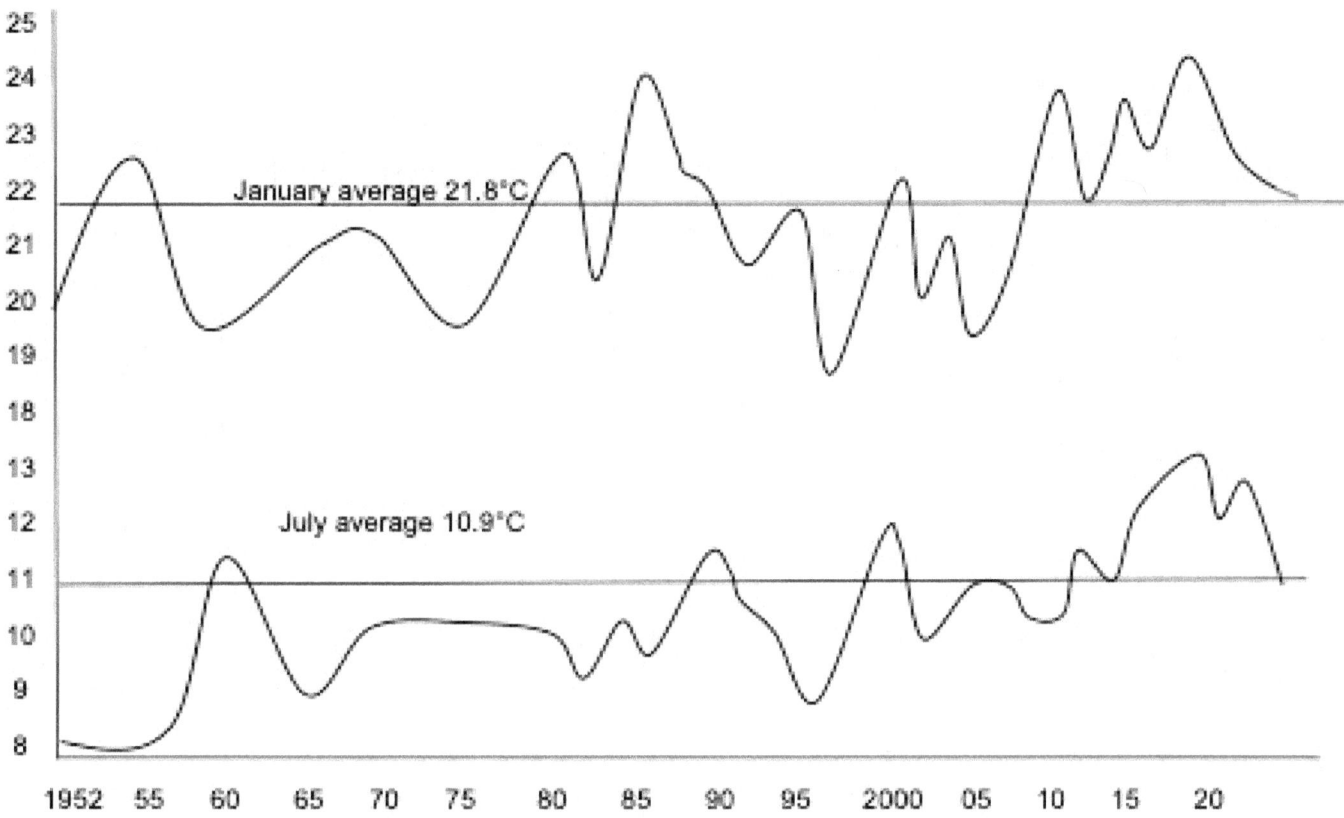

Figure 67: **Christchurch, New Zealand.**

The graphs show the world wide cooler period from 1950 to 1975 discussed before. Following that there is variation around the averages except for a transient warmer period in 2018-19 during the strong El Nino then, then cooler in 2021 and 2022 close to the long term average again.

Again there is no steady long term warming evident over the 42 years from 1980.

The overwhelming conclusion from this series of long term simple non-adjusted temperature records from a wide range of places globally is that there is no evidence of a rising temperature trend. Just features consistent with natural variation.

The Daily Temperature Cycle:

Everyone everywhere expects that there will be a temperature variation over the 24 hour day and this has been already mentioned in Chapter 3. Two places from those discussed before are detailed here as examples of how big the temperature variation over the 24 hours is. The first is Christchurch, New Zealand shown in Figure 68 for January 1985, midsummer.

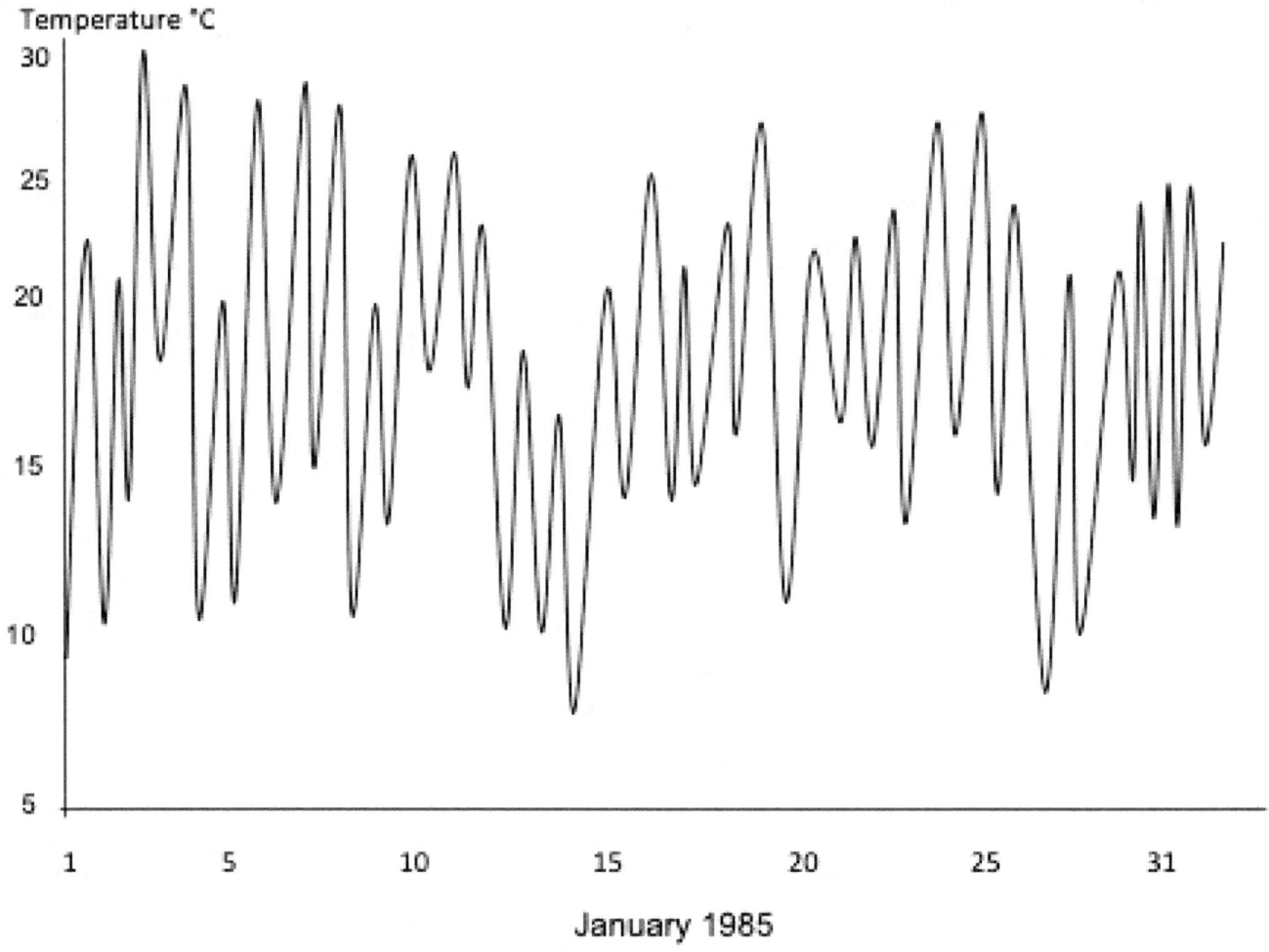

Figure 68: **Christchurch, New Zealand showing daily maximums and minimums for January 1985, summer.**

The graph shows big swings over each 24 hours from night to day ranging up to 17°C.

The second is Albacete, Spain shown below in Figure 69. The daily variation for these midsummer months in these years ranges up to 20°C for Albacete and 17°C for Christchurch. The variation is usually greater for inland places away from the moderating effect of the ocean especially large inland deserts such as the Sahara and central Australia.

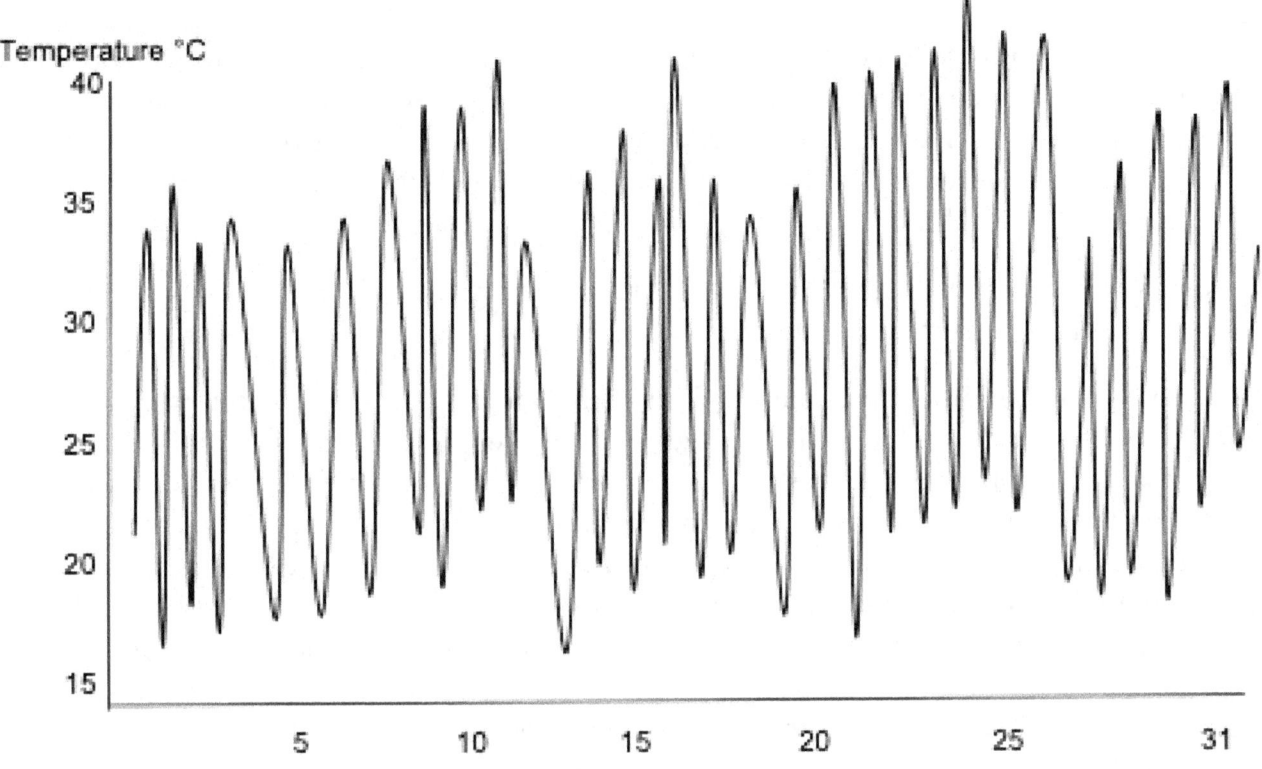

Figure 69: **Albacete, Spain showing daily variations for July 1984.**

There are two major points to make about this world-wide phenomenon which all regard as commonplace and take completely for granted.

The first is that it demonstrates how *dynamic* the Earth's situation is concerning heat energy arriving and leaving. As soon as the Sun sets the temperature drops very quickly. Massive amounts of heat energy rapidly flow from the Earth's surface to surrounding space which is at minus 270°C. Everything on the night side of the planet experiences this.

Then the Sun rises and huge amounts of heat energy flow to the surface, the temperature of everything on the daylight side rises markedly. These mechanisms have been discussed in some detail in Chapter 3.

But the point to make here is the massive dynamism of this system, infinitely greater, more complex and more rugged than anything humans do. It is far from being controlled by one factor, as, for example the temperature of a house can be controlled by a thermostat regulating heat inflows. Especially it is not controlled by one trace gas, carbon dioxide. Atmospheric Carbon dioxide levels rise during the night due to photosynthesis stopping, as has been explained, but the temperature falls. And during the day CO2 levels fall as photosynthesis restarts, yet the temperature rises. This may seem to be an unduly simple, naïve argument but behind it is the realization of the huge forces at work all the time concerning our planet and it's temperature.

These daily variations in surface temperature are clearly far greater than any postulated long term rise due to CO2 by alarmists.

Similar huge temperature variations involving massive heat energy transfers occur with thunderstorms and other weather phenomena as discussed in Chapter 4. Frequently the temperature for large areas quickly drops 10°C with these events as is commonly experienced. The enormous heat energy loss from the planet due to the 16 million thunderstorms which occur each year, 38×10^{15} kilowatt hours, 300 times human creation and usage of 1.25×10^{14} kilowatt hours annually, is another example of this.

As are cyclonic storms, discussed also in Chapter 4. A single one transfers about 600 billion kilowatt hours of heat energy away from the surface to the stratosphere and space.

The second major point is that the impossibility of determining the Earth's **Average Temperature** with any precise accuracy is illustrated. This has been recognized by many prominent scientists in this field including James Hansen, a leader of the alarmists, as has been mentioned already. When one considers the great variation that occurs in surface temperature in *just one place* through the 24 hour daily cycle from one minute to the next the difficulties in determining the accurate average temperature *for that one place* are obvious enough. Figures 68 and 69 just deal with the places where the temperature sensors are at Albacete and Christchurch airports. And then only two time moments each day are considered, the warmest and the coldest. The temperature just a few meters away from those locations is almost certainly different. It is clearly completely impossible to determine the temperature everywhere because of the infinite complexities involved, and therefore impossible to calculate the true average.

Yet some claim to be able to do this and make confident predictions based on their findings. But all that they are doing is making spot measurements of a minute number of places – compared to the huge number on the planet – at brief moments in time – compared to all the time passing – then announcing the results. These limitations are true even for satellite measurements of millions of places and times on Earth even if the errors in processing results, alluded to many times already, did not exist.

Scientists of quality are aware of these limitations and prudent concerning the meaning of results concerning the Earth's surface temperature; but, as is common in human existence, there are others who are not.

Readers can obtain the simple unadjusted long term temperature records for their own locations using the methods described in references below.

References and Further Reading for Chapter 11: The Earth's Surface Temperature:

https://en.wikipedia.org>wiki>>geologic_temperature

www.science.org>content>article>ancient-global-temperature

Scotese, C.R. "Analysis of the temperature oscillations in geological eras" 2002

Hansen, et al. The Open Atmosphere Science Journal. 2. Pp 217-231. 2008. https://pubs.glss.nasa.gov/docs/2008/20058_Hansen_ha00410c.pdf (graph last 65 million years).

Javier et al. "Nature unbound III: Holocene climate variability (Part A). Climate Etc April 30, 2017. https//judithcurry.com/2017/04/30/nature-unbound-iii-holocene-climate-variability-part-a/ .

https://crudata.uea.ac.uk/-timo/diag/tempdiag.htm (global average temperature from 1850 to 2010 from the Hadley Centre).

Christy, J.R. et al "Satellite Bulk Tropospheric temperatures as a metric for Climate Sensitivity." Asia-Pacific Journal of Atmospheric Sciences 53 (4) pp 511- 518. (comparing IPCC computer model predictions versus balloon and satellite observations Fig 35)

Spencer, R. "UAH Global Temperature Update for October" www.drroyspencer.com/2019/11/uah-global-temperature-update-for-october-2018-0-46-deg-c/. (Fig. 36 graph 1980 – 2020).

https://en.wikipedia.org>wiki>indian_ocean_dipole

https://www.bom.gov.au>enso>history>IOD-what

https://www.bom.gov.au go to "climate and past weather", then "climate data online" at bottom screen, then select "data about – temperature", then "daily", then "maximum temperature", then "I would like – monthly mean maximum temperature" in the box alongside map. Then scroll over the map and zoom in to the place, they have little diamonds. Double click the diamond, box appears with the name of the place and "mean maximum temperature". Double click that and the data table appears.

www.climate.weather.gc.ca go to "historical climate data"; then to "historical data"; then to "search by province or territory"; then to "search" – choose a station; then "daily data" and at the bottom "average daily maximum for month".

www.wunderground.com go to "more", then "historical weather", then "place", then "date", click "monthly", go to the summary at the page bottom, average maximum temperature for the month is there. To change places in the US it is easiest to go to "change" next to the name of the place; a google map appears to scroll over and select a place.

www.weatherspark.com : For U.K. and European historical temperatures: google search: "temperature Cambridge 1980." Go to weatherspark.com , click "history", for location box put "Cambridge UK", scroll down to the daily graphs. The top border of these has years to select. Clicking on the July part of the graph enlarges that section. Arrows at the side enable the month to be changed. The graph may be printed enlarged, the temperature scale marked in single degrees and the daily maximum temperature read. These are then added up for the month and divided by the number of days in July – 31 – to get the average for that years July. The process is repeated for each year of interest.

Chapter 12

Conclusions

The Earth's surface temperature is entirely due to it's position in the Solar System 150 million kilometers from the Sun. If it was a little closer or a little further away *liquid water* would not exist on the surface and the planet would be either hot and arid or a frozen ice ball.

The Earth is unique in the Solar System in having surface liquid water. It is also large enough to hold the water to the surface by *gravity*, and its *rotation* results in a daily cycle which allows moderate heating to occur in daytime, and moderate cooling at night.

The Earth's size and gravity results in retention above the surface of the *atmosphere*. Because large parts of the Earth away from the Poles have temperatures above the freezing point of water and below it's boiling point, due as said to it's distance from the Sun, *water vapour* exists in the atmosphere in balance with the surface liquid water which covers 70% of the surface as the *Oceans*.

A very large amount of heat energy arrives continuously from the Sun. The Earth must lose this same amount to remain at stable temperatures. It does this largely by *water phase changes* - evaporation of surface water, then condensation of water vapour to liquid and freezing, sublimation (vapour directly to ice) in the atmosphere as explained in chapter 4. These processes with *convection* result in huge transfers of heat energy from the surface to the high atmosphere where it is radiated away into space.

Water is the ***only substance*** on Earth capable of handling the huge heat energy loss necessary to keep the temperature stable and compatible with life. If there was no water on Earth in daytime it would heat up to a temperature where the heat loss equalled that arriving from the Sun, much hotter than now. This is because none of the other atmospheric gases undergo phase changes in the atmosphere to aid heat energy transfer from the surface to space. And at night the temperature would plummet far more because none of the other gases absorb heat radiation from the Earth as water vapour does, heat energy would be free to *radiate* directly from the surface to space.

But the Earth is in the "sweet spot" where liquid water covers most of the surface so that does not apply.

Liquid water and it's atmospheric component, including all the water phases of vapour, liquid and ice, has a massive ***moderating*** effect on the surface temperature. If the surface liquid water gets warmer it evaporates more, losing heat energy and cooling. The vapour containing this heat energy then rises in the atmosphere by convection. It then cools because it expands with less atmospheric pressure higher up, and the vapour condenses back to liquid or sublimates directly to ice, losing massive amounts of heat energy high in the atmosphere where it radiates to space freely because there is no water vapour at that level, the stratosphere and above, to stop it. Thus the surface liquid water, covering 70% of the planet, is a huge **negative feedback mechanism** moderating the surface temperature.

The liquid water and ice in the higher troposphere forms **clouds.** The warmer the surface water gets the more clouds there are. They cover about 30% of the Earth's surface. They are powerful **reflectors** of arriving Sun energy back to space, preventing up to 80% of it reaching the surface. They also are a massive negative feedback mechanism moderating the surface temperature. If it gets warmer cloud increases.

And if it gets cooler, there is less surface evaporation and decreased cloud cover resulting in the surface liquid water getting warmer and more Sun energy making it to the surface, thus moderating the cooling trend.

This ability of water to be liquid on our planet's surface and have these effects is due to **only to it's molecular structure, H2O.** This structure determines the freezing and evaporation temperatures of water, which in turn determine the behaviour of water in the atmosphere. And the structure also determines the **fingerprint absorption** of heat energy radiation of water molecules which is so extensive across the wavelengths of **infrared** heat radiation, from 4 to 90 microns as explained, leaving the Earth's surface. Without the resulting delay in the transfer of heat energy from the surface to space the Earth would be frozen. Liquid water would not exist on the surface, the planet would be lifeless.

But water molecules do have these **intrinsic properties**. Nothing from outside the water molecules influences these qualities. And they control the nature of our planet, including it's ability to host life and it's surface temperature. Nothing else can override the water effect providing Earth remains in the "sweet spot" in relation to the Sun.

The environment of our planet's surface including temperature is **predetermined** by the amount of heat energy arriving from the Sun permitting liquid water to exist here.

There are other intrinsic properties of water molecules which are also determine aspects of the Earth's surface. Not least of these is that ice is lighter than liquid water; water freezing to ice **expands** as we all see when we leave wine in the freezer for too long! That is different to most substances where the solid is denser, heavier than the liquid, therefore the solids sink to the bottom as freezing occurs. If that were so for water the oceans would freeze from the bottom up. The ice would be positioned where the Sun's heat energy could not reach it as it is all absorbed in a few meters of surface water. The ice would be permanent there and much of the ocean away from the tropics would be solid ice to the bottom.

But ice is lighter than water, it floats on the surface where it is warmed and melted by the Sun. And the ocean is liquid all the way to the bottom because of this intrinsic property of water. We take this completely for granted.

Another intrinsic property of water shaping our environment is the change in water density at different temperatures which occurs with the addition of salts in ocean water, already discussed. Fresh water gets denser and heavier as it cools, warmer water floats to the surface as we all know from swimming in lakes. But fresh water is most dense at 3.5°C so water at that temperature is found at the bottom of fresh water lakes. Cooling below 3.5°C results in lighter water which rises. This causes turbulence as the cooling water first sinks, then rises.

If this were so for salt water in the sea the oceans would be very turbulent. But the addition of salt changes this as water then just gets denser and heavier as it gets colder all the way down to freezing. So the ocean depths are very cold but calm, not turbulent, due to this intrinsic property of water when salt is added.

Our environment at the Earth's surface depends entirely on the planet remaining in the "sweet spot" concerning the heat energy arriving from the Sun. This varies as explained in chapter 3. The Sun's output varies in several cycles and much is still being learnt about that. There are changes in the Earth's distance from the Sun due to it's orbit around the Sun being elliptical, not a perfect circle, and changes in the axis tilt. As a result there are inevitably big changes in the surface temperature here including ice ages and warmer periods.

Other factors in these cycles include local ones on the planet itself especially plate tectonics. If there is no large land mass at a pole the surface is much warmer, and if there is one there, as at present concerning Antarctica, much cooler. This is discussed in Chapter 3 and Chapter 5. Large volcanic eruptions cool the planet sometimes for years, as do large meteors striking it.

What, then, is the place of carbon dioxide concerning surface temperature ? There are some key points:

The first is the overwhelming evidence that changes in atmospheric carbon dioxide levels *follow* changes in the surface temperature. This is discussed in several places in this book but especially in Chapter 8. The evidence includes both very long term, millions of years, and more recent evidence. The inevitable conclusion from this is that carbon dioxide levels simply follow surface temperature, they do not cause it to change.

The second is the huge amount of CO_2 stored in the oceans and the way that interacts with the atmosphere defined by Henry's Law concerning partial pressure of a gas above the surface of a liquid in which it is dissolved. This discussed in Chapter 8. As said the oceans contain 50 times the amount of carbon dioxide as the atmosphere. And the amount held in different places varies considerably with the ocean temperature. If it gets warmer, much less can be held in the water. The CO_2 then rapidly vents to the atmosphere, raising levels there. And if the water is colder more is held, CO_2 leaves the atmosphere and atmospheric levels drop. This mechanism has been shown conclusively to be the reason atmospheric CO_2 levels drop during ice ages, sometimes to 180 parts per million, 0.018%, then rise during warmer periods, commonly 1000 ppm, even up to 3000 ppm 150 million years ago and 6000 ppm, 0.6%, 500 million years ago as already discussed. At present levels are towards the lower end of this long term range because the Earth is cooler as discussed in Chapter 11, especially Figure 31.

Generally, as discussed in Chapter 8, warmer seas in tropical regions emit CO_2 to the atmosphere and colder polar one absorb it. The great mixing which occurs in both the turbulent troposphere and the oceans explains why the atmospheric level is similar in different places globally. Huge amounts of carbon dioxide are involved in this process – between 90 and 390 or more billion tonnes annually as discussed. So any extra CO_2 arriving in the atmosphere disappears into this huge churning machine.

This mechanism is *a sufficient explanation* for the changes in atmospheric carbon dioxide levels observed in recent decades, which are low compared to the very long term as said. No other explanation is necessary. The Earth is emerging from the Little Ice Age which lasted from 1300 to about 1850. And it is such a powerful mechanism that the idea of being able to **override** it by artificially increasing atmospheric CO2 and causing ocean levels of CO2 to increase *despite the ocean maximum level determined by temperature* thus reversing the natural process is not credible. Expecting rivers to flow uphill would be similar. If the ocean gets colder it will take up atmospheric CO2, and it vents it if it gets warmer. The oceans contain the maximum amount of CO2 they can for that temperature; if they contain less they will take up more usually from the atmosphere. If more again arrives it can be precipitated out as solid carbonate salts, usually calcium and magnesium carbonate to make limestone rocks on the sea floor as already discussed. Indeed this is the very long term predicted fate of CO2 – to be all incorporated in limestone rocks with none in the atmosphere or ocean as discussed in Chapter 8. The atmospheric level is governed by the ocean temperature.

The third point concerns the **trace amounts** of CO2 in our atmosphere. 400 parts per million, 0.04%, 1 in every 2500 atmospheric molecules. This means that should an atmospheric CO2 molecule get hotter it will quickly dissipate heat to surrounding molecules – especially water vapour as discussed, which can readily handle more heat energy avoiding a temperature rise, and is far more common in the atmosphere, 2 to 4%, one in 25 to 50 atmospheric molecules. This is fully discussed in Chapter 9. This means that CO2 has no ability to heat the atmosphere generally because it is present there in trace amounts only.

The fourth point concerns the ***fingerprint heat radiation absorption*** of CO2 molecules extensively discussed Chapter 1 and Chapter 9. Atmospheric CO2 molecules only absorb a limited range of the Earth's infrared heat radiation from the surface, about 7%, although some would say 15%. It is very limited compared to water vapour as discussed. The absorption is in the wavelengths 13 to 17 microns. The Earth's heat radiation is from 4 to 90 microns. CO2 molecules let all other wavelengths through without any heating effect apart from the band 13 to 17, and some narrow bands at 4.8, 9.2 and 10 microns as explained. So the ability of CO2 molecules to absorb heat energy passing up through the atmosphere from the Earth's surface is greatly limited compared to water vapour on this consideration alone.

But there is another factor: the absorption in this band of 13 to 17 microns, especially 14 to 16, is **very intense**, 100% as discussed already. This means that increasing the concentration of CO2 in the atmosphere does not have any additional effect on heat energy absorption – it is already flat out at much lower concentrations than currently present. There is debate about the "shoulders" just outside the intense absorption band of 14 to 16 microns as discussed; but these probably only make a 5% difference, not nearly enough to make meaningful change.

So fingerprint heat radiation absorption is another indication that CO2 is far less effective at absorbing the Earth's heat radiation than water vapour. This is a further weakness of atmospheric CO2 compared to water vapour regarding atmospheric temperature changes.

Increasing atmospheric carbon dioxide levels do have one indisputable effect: plant photosynthesis is considerably increased. Plants quickly take advantage of increased levels by growing faster and processing more CO2 faster to carbohydrates and oxygen. This includes ocean blooms of sea algae, where 70% of the planets photosynthesis occurs, as well as land plants. This "greening of the planet" is a very positive consequence of higher levels. The dependence of plants on atmospheric and ocean CO2 is illustrated by the consideration that if atmospheric CO2 falls below 150 parts per million, 0.015%, plants die. And as all life ultimately depends on plants life would cease on Earth should that occur long term. But so far during ice ages the level seems not to have gone below 180 part per million, 0.018%.

While photosynthesis does increase and decrease with changes in carbon dioxide level and therefore is a potential **negative feedback mechanism** helping to control CO2 concentrations it is very weak as shown by the long term historical record where CO2 levels drop to about 180 parts per million **despite** plant life having slowed with cooling, and similarly CO2 levels have risen substantially in warmer times again **despite** photosynthesis have increased. A factor in this is that when plant life increases so does the death and turnover of plants and other animals – with production of CO2 with decay. Each plant grows, takes up CO2, then dies and releases it.

The answer, then, to the question posed about the place of carbon dioxide concerning the Earth's surface temperature after considering these four points has to be that it does not have a role determining the temperature, it is a follower.

So human efforts to reduce the Earth's surface temperature by reducing human sources of carbon dioxide are clearly futile. One does not even have to raise the consideration that humans only contribute about 4% of carbon dioxide entering the atmosphere, the other 96% is from the natural world and is not going to alter with human changes. Or look in detail at the temperature records discussed in Chapter 11 which do not in any case show any convincing change outside of natural variation.

Why, then, has the belief that the Earth's surface temperature is rising dangerously due to human activity producing more CO2, especially from the use of fossil fuels, and must be brought down by ceasing that, become the dominant expressed conviction of our world?

There appear to be several reasons:

First, this belief is held principally in the western liberal democracies where there is freedom of speech and an active free press. So convinced scientists and others are able to promote their views widely. By contrast the governments in the "autocracies" – one party states or dictatorships especially China and Russia – can suppress and mould expressed views to suit their purpose. And it clearly suits them to pay lip service to western climate alarmist views but continue to develop their economies vigorously around fossil fuels. As shown by the ongoing massive construction of coal power stations in China and gas facilities in Russia. By contrast many coal power stations in the western democracies have been demolished and those remaining are often on notice.

The western free press is a major factor. Their primary interest is to survive, hopefully expand their operation and make a bigger profit. A large disaster story will achieve much more than a low key "nothing to see here." story. All news is bad news.

Politicians in the democracies also are a major factor. Again their primary interest is to survive which means gaining votes. Up to the present there have been more votes gained by promoting the climate alarmist cause than not. Virtually all the western democracy politicians are scientifically unskilled to make their own decision on the issue. They are frequently heard saying: "I am not a scientist - - -". But they can see which way the wind blows concerning votes.

By contrast in the autocracies this is not a factor. What matters there is the economic wellbeing of the state including it's military; and ongoing control of the population. They are arguably more likely to assess realistically complex scientific arguments and reach a rational conclusion in their interest compared to vote seeking western politicians.

Second, many alarmist scientists have very effectively promoted the cause in this receptive environment apparently sincerely. Their concern and conviction have won many converts as has the professionalism of their presentations.

Third, large vested interest groups have emerged who stand to substantially gain from change. The most obvious are those involved in the "renewable energy" sector including solar, wind, batteries and hydrogen. They promote the cause.

Another vested interest group is some third world countries exploiting the opportunity to blame the rich western democracies for ruining their environment, including alleged rising sea levels and increased destruction from major storms. These allegations do not stand up to critical analysis as explained already. But these countries are succeeding in claiming huge aid funding from the rich democracies and so are highly motivated to push the climate alarmism argument. The United Nations and it's subsidiary organizations has become very prominent concerning this.

Fourth, there is a more subtle group in the education sector, including that for young children, teenagers and universities. They promote the cause as being apparently kind to the planet and avoiding claimed inevitable environmental disaster if change does not occur. And many of their pupils are motivated to avoid the obviously callous mistakes of previous generations. There is no evidence that alternative views are seriously put to these young people. They are a powerful often intolerant force as shown by pupil strikes and "extinction" protesters.

Last, there is a big group in the democracy populations who believe they are good people who wish to look after the planet; that means moving away from fossil fuels; and that should be pretty simple, they think, because renewables, batteries, hydrogen are all very practical and able to maintain the high western standard of living with few problems. So why not go with it and change ? A proportion believe that nuclear is the answer. This group now includes many large corporations virtuously embracing change.

All this has fuelled the movement. And that has resulted in massive funding promoting it. This in turn has led to increased professionalism promoting the cause in a single-minded way. From a scientific viewpoint there are aspects of this which cause concern.

Scientific information is presented in a way which leads to the observer agreeing with the alarmist cause. Examples are:

Graphs being drawn to imply alarming trends:

For example the "Keeling Curve" of atmospheric CO_2 levels from the Mauna Loa volcano station is commonly drawn with the y axis (the vertical line on the left) starting at the bottom at 320 ppm instead of zero. And the top of this axis is commonly about 420 ppm. Then the x axis (the horizontal line at the bottom), which deals with time, often 1960 to 2020 in this case, is usually short to produce a square shape to the graph. These methods result in a much steeper line of measurements than would be the case if, for example, the y axis went from zero to 500 ppm, and /or the x axis was longer. While people with scientific training quickly spot this manoeuvre those without that, the majority, are more likely to agree with alarmism, the aim of the presentation. Figure 70 below shows this.

Another example is the "Hockey Stick" graph of temperature over the last 1000 years discussed in Chapter 9. The scales at the side and bottom of the graph are selected to show recent warming resulting in a very steep upward curve. And the warming trend of the Middle Ages Warm Period from 900 AD to 1300 AD was completely omitted, as was the Little Ice Age cooler period, to help imply recent warming was exceptional.

Another example is the many published "Anomaly Graphs" of temperature change showing trends in temperature; they always have only a few degrees drawn on the side "y axis" line so that slight changes result in a large and steeper line representing temperature. Figure 71 below shows a representation of the "Hockey Stick" graph as commonly drawn from Mann, Bradley and Hughes 1999 paper. This demonstrates the very small range of temperature change used on the vertical y axis to maximize the effect of the change, whereas it was only 0.6° C from the end of the Little Ice Age in 1850 to 1999. Often the graph is shown with data from the Hadley centre added on to 2013, a further 0.3° C, to again increase the effect. The many controversies, errors and difficulties with this have been referred to already.

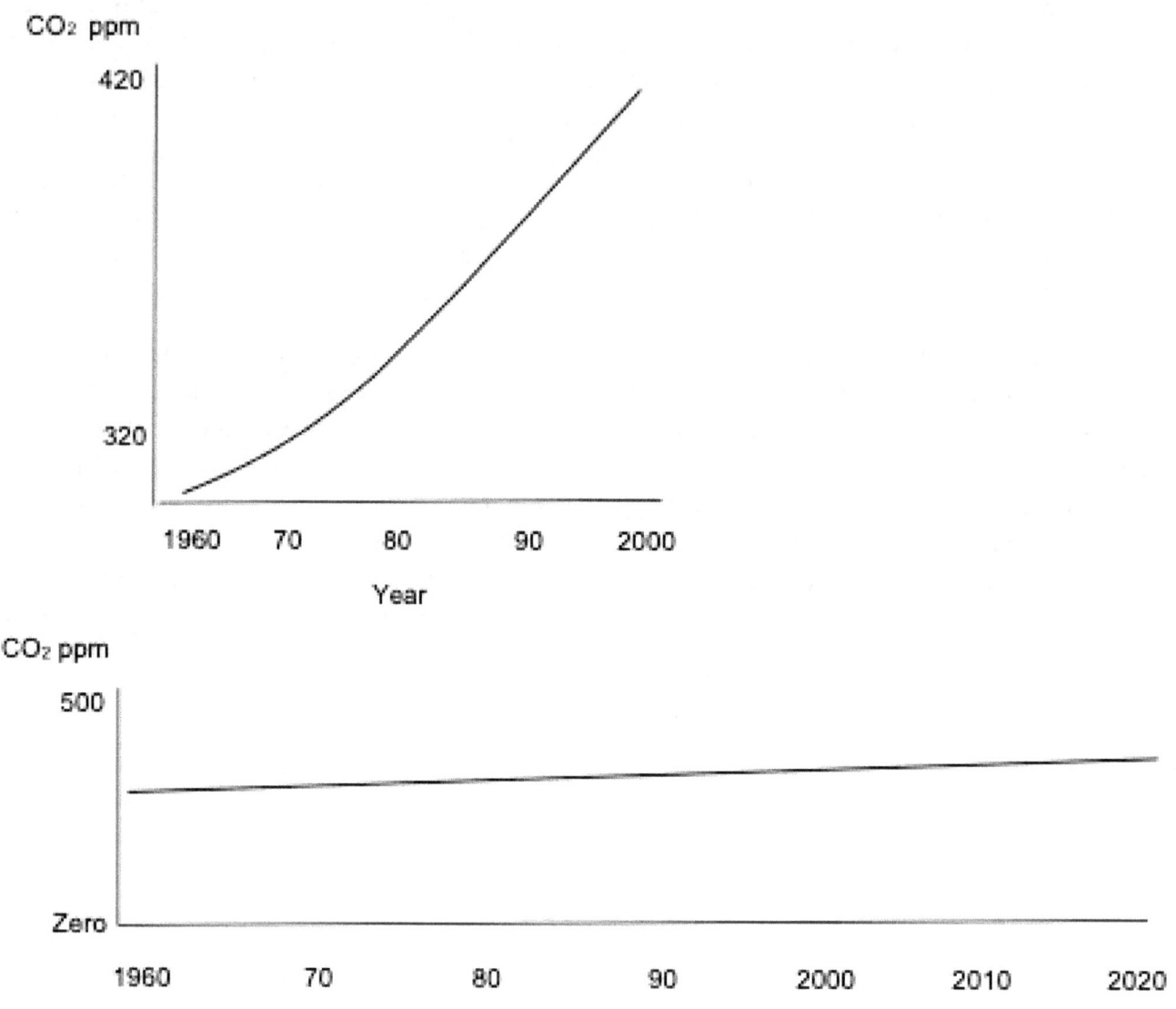

Figure 70: The upper graph shows the Keeling Curve data as commonly drawn to maximize the impression the CO2 rise is severe; while the lower graph shows the same data, essentially a change from 313 ppm in 1958 to 416 in 2022, drawn differently with the y vertical axis from zero to 500 and the time scale longer.

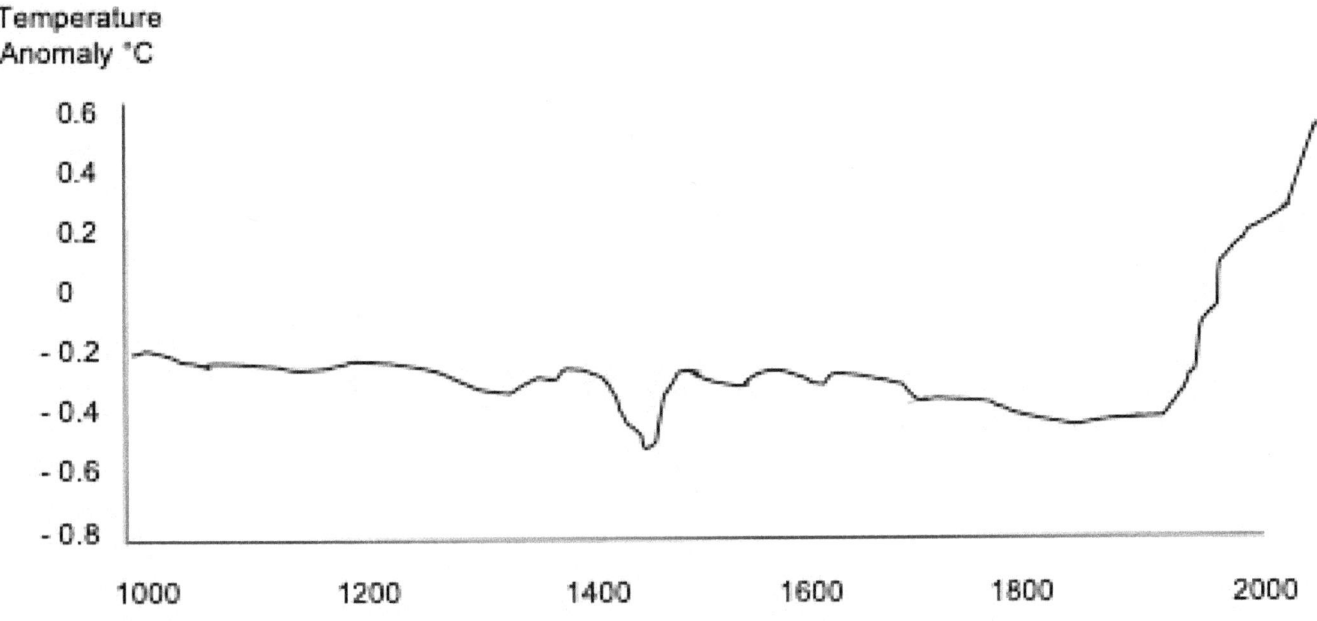

Figure 71: a representation of the "Hockey Stick" graph as commonly shown based on MBH – 1999 with Hadley Centre data added, the last steep upward part, also commonly shown.

As a complete contrast in data presentation Figure 31 is again shown below. This shows the Earth's average surface temperature (according to research data) for the past 570 million years since complex life began:

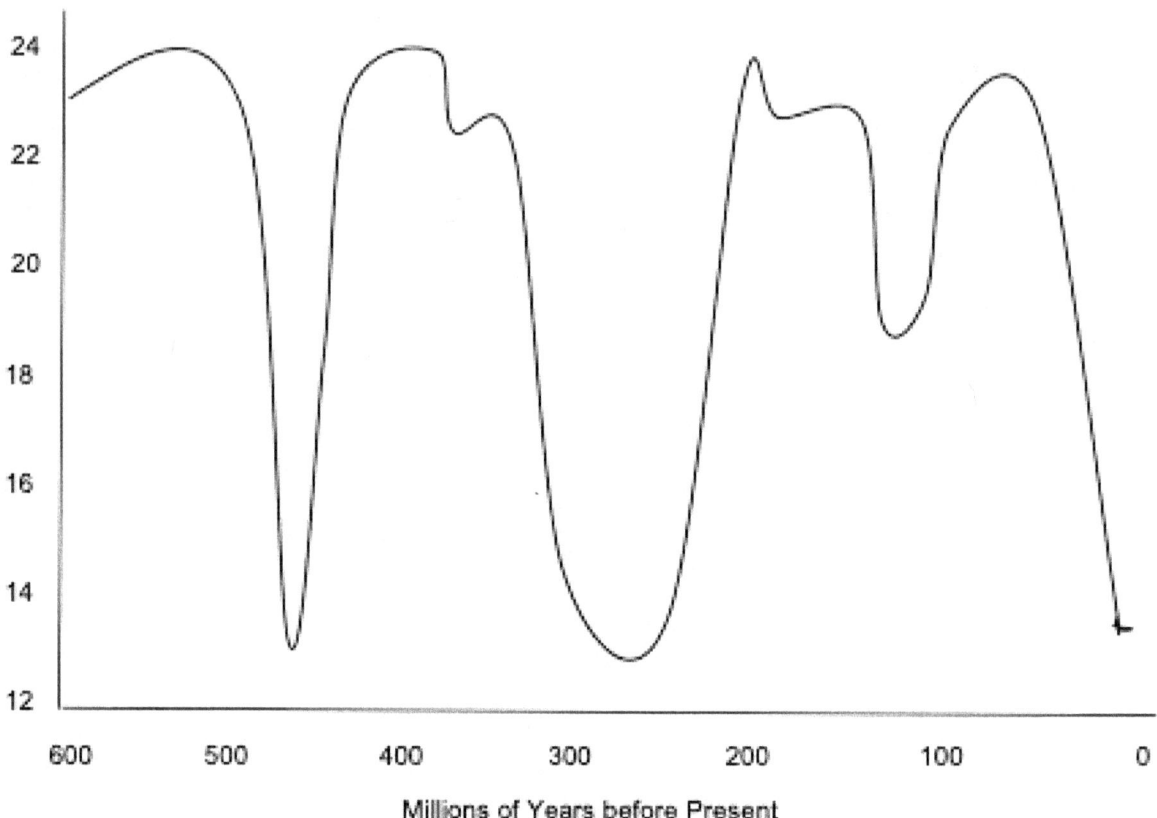

Figure 31: showing the global average temperature over the last 570 million years according to research.

It is clear that apparent tiny recent temperature changes would not be apparent in this graph where the y axis goes from 12 to 24° C, and the x axis from 600 million years ago.

"Adjustments":

Virtually all easily available temperature records, especially on government sites, feature "adjusted" or "normalized" readings. It is quite hard to find simple non-adjusted historical records. And it is also difficult to find records for a particular town or place, just simple readings on the surface. This problem is alluded to in Chapter 11. Exceptions are the government weather sites in Australia and Canada as said, but for other countries examined resort had to be made to airport records to get this information – which conclusively showed no warming trend since 1980 as discussed.

The government sites in Europe, U.S.A and New Zealand have grouped together locations and "adjusted" the temperature readings historically. These "adjusted" readings usually show a warming trend on these sites. Many sites use satellite data regarding surface temperature, including NASA. There is scant mention of the many difficulties, errors, computer modelling and complex calculations in these methods discussed in Chapter 11 and elsewhere. Nor is there mention of the selection of final data to go in a report, often only a small proportion are allowed through, 17% only is mentioned in one study due to these problems. At the very least there should be a "healthy suspicion" about these results and the motives of those presenting them given the contrast with simple surface measurements discussed in chapter 11.

Simplistic but Beautiful Diagrams:

These are frequently used on many internet sites and publications. They are usually wonderfully clear to look at. There is no mention of any uncertainty or doubt about the obvious conclusions, nor are problems, errors, or difficulties with data gathering, especially from satellites, discussed. The immensely complicated and mostly poorly understood natural Earth system is shown to be completely known by the infallible experts behind the diagrams. Far be it for an ordinary mortal to question these!

Inconvenient aspects of data are often not presented:

Examples include graphs showing the fingerprint heat absorption of carbon dioxide. Often these just show the Earth's heat radiation being from 4 to 20 microns wavelength. But, in fact, as discussed, it is from 4 to 90 microns. This means that graphs showing the heat absorption of CO_2 from 13 to 16 microns, as discussed, show it occupying a much larger proportion of the Earth's heat radiation band, up to 25%, than would be the case if the graph showed the true position, about 7%.

Another common example is the poor or absent representation of the massive effects of water vapour in the atmosphere including clouds; and often no discussion that CO_2 is a trace gas, 0.04% of the atmosphere.

Colour use in diagrams:

There are many maps and diagrams on the internet showing temperature changes including global maps. Always red is chosen as representing hotter temperatures and blue as colder. And as time passes the red areas are shown getting more red, often an alarming deep red. But when one looks at the actual temperature change from blue to red it is often fractions of a degree only. And there is no mention of the many problems gathering this data, by satellite for example, which have been discussed above.

Often apparently scientific data as presented is not from the scientists at all:

An example is the International Panel for Climate Change (IPCC) reports. Many scientists have complained that the final versions are produced by bureaucrats influenced by others including politicians; and that critical parts of the conclusions were altered against the scientists' wishes. This has been discussed above including in the section on Judith Curry in Chapter 9. A gross example is described by S. Fred Singer in his book: "Hot Talk, Cold Science" where he described the writer of a children's book on climate change as simply reversing graphs showing CO2 increases *following* temperature increases to falsely show CO2 increases *preceding* temperature increases.

Commonly the massive effect of the oceans on atmospheric carbon dioxide is ignored or downplayed:

The huge amounts of CO2 freely transferred to and from the seas depending on the sea temperature has been discussed in detail. Coupled with this is the assertion that human produced CO2 lingers for ages in the atmosphere, based on extremely difficult mass spectrometer data as discussed, whereas the reality is free and huge transfer to and from the oceans.

Extreme climate events are claimed to be due to human carbon dioxide production:

Activists seize on floods, fires, droughts, ocean storm surges especially in the third world claiming they are historically exceptional and due to the use of fossil fuels in advanced countries, and that compensation should be paid. The United Nations is at the forefront of this campaign. Whereas, as discussed in several places above, there is no evidence that these modern events are different or more common from those in times past. A variant of this is claims that temperatures for particular places are: "the highest on record". But when one looks in detail at this usually there are other higher temperatures previously, often many decades ago. And what is important is long term trends, not isolated events, as discussed already.

There are many other examples of data being presented to convince the observer to agree with the cause rather than being a fairer representation, especially when the uncertainties and great complexity of the Earth, poorly understood, are considered.

Many people, busy in their lives, must look at the powerful dominant lobby of politicians, various vested groups, educational authorities, wealthy business people, scientists, global organizations, all pushing for fossil fuels to cease to save the planet, and think: "This is simply too big to oppose – what is the harm is just going along with it ?"

There are two massive problems with this approach:

The first is that reduction of human output of carbon dioxide will be futile in controlling the Earth's surface temperature as discussed.

The second is that the bedrock of the great advances that humans have enjoyed in their lives for the past two hundred years is the use of fossil fuels for energy. Life before that was universally hard, short and brutal by modern standards. Lacking the energy of electricity powering so much of our lives, and fossil fuels enabling transport, our forebears were only able to use muscle power, either theirs or animals, or natural wind or water power, all very limited.

Without the ready availability of electricity and fossil fuels everything we take for granted in modern society would collapse. Including life in our cities, hospitals, homes, access to food, water, other essentials and goods, travel, transportation, mobility – all gone.

It is not possible for alternatives to fossil fuel use to adequately replace them. So abandoning fossil fuels, should it ever happen, would result in reversion to the previous state of a short brutal life for many people.

It is very unlikely to truly happen because as the deadline for cessation of fossil fuels looms up, and the consequences become apparent, the western democracies populations will respond by voting for politicians promising to continue fossil fuel use and thus maintain the living standard.

The next chapter focuses on why the alternatives are inadequate.

Chapter 13

Alternatives to Fossil Fuels

Introduction:

Wind and Solar Electricity Generation:

It is common knowledge these are intermittent, they only generate electricity when the Sun is shining and the wind blowing, and then only significantly if weather permits and there is little if any cloud cover concerning solar. There is no Sun energy at night. Nights are very long in winter, and this affects big populations living in the northern hemisphere away from the tropics. There are long periods when the wind does not blow. This means that electricity usage, which continues at a high level in modern society 24 hours per day especially at night or in poor, cold weather **must have another source** capable of meeting this demand.

Batteries are pushed as a solution to this by some. But the biggest batteries are miles away from solving this problem. They only provide a few minutes of peak load power and are very expensive, complicated, thus potentially unreliable.

Pumped Hydro is another solution pushed. But it is also very expensive to construct and also requires a large amount of electricity to pump the water to a height. Then the water runs downhill to power turbines and generate electricity, a process which is not totally efficient in that about 70% of the input energy is finally available as electricity. Many places do not have suitable sites for pumped hydro. Currently world-wide pumped hydroelectricity storage is about 1.6 terawatt hours (1.6 X 10^9 kilowatt hours) per year. Human use of energy is about 125,000 terawatt hours (1.4 X 10^{14} kilowatt hours) per year as explained in Chapter 3, 70,000 times greater. It is again not practical as a large scale global alternative to fossil fuels. And those pushing it say the energy to pump the water uphill could come from wind and solar. Given the comments above and the huge requirements of energy for human use that is an impossible ask considering requirements.

Another problem with wind and solar is that the windmills and solar panels are very widely scattered over huge areas – often of remote countryside or coastal seas. This means they need long transmission lines to feed the electricity to where it is needed. This, and the variable nature of the electricity output, make it very difficult to organize inputs into nations' grids compared to constant output sources close to cities like coal, gas and nuclear. Many countries are having substantial difficulties upgrading their transmission network grids to cope with far-flung wind and solar installations, including many scattered small units, contrasting with a few big coal of gas plants comparatively close to major cities and industries. This upgrade is proving very expensive. Some examples:

In 2022 the UK energy regulator approved US$26.6 billion on network grid expansion to cope with renewables for the next 5 years. The network operators asked for US$30.2 billion.

A survey of EU grid upgrades for renewables published in the "Financial Times" stated that increased expenditure on that up to €39 billion per year would be required by 2030.

In Australia the Australian Energy Market Operator (AEMO) has stated that A$25 billion would need to be spent on network grid upgrades for renewables in the next 5 years.

In the US a cluster of 15 midwestern states are planning to spend US$10 billion on this in the next few years. And there are substantial delays for new solar and wind projects connecting to network grids preventing their operation; some 8100 are significantly held up currently; for example:

- New Jersey State has frozen all new renewable connections until 2026 because of grid limitations.

- Silicon Ranch, a Kentucky and Virginia 300 megawatt solar project, bought the land in 2021 but can't get grid connection until 2029; the land is still empty. Other nearby projects have been cancelled.

In addition in the US solar and wind developers are being asked to pay millions for grid connection; for example a small 100 megawatt Minnesota project was cancelled when asked to pay US$80 million. Another smaller 40 megawatt project was asked to pay $50 million for connection. There are about 320,000 kilometers of grid transmission lines in the US and this needs to double to meet this renewable need.

There are plans in some countries for even home solar generators to pay some of the transmission network costs. It is all very well to encourage solar and wind development – but that is not being matched by the huge infrastructure development needed. This is understandable in view of the massiveness of the requirements. And inevitably there are big environmental and aesthetic prices to pay for large areas of countryside being criss-crossed by transmission lines.

A related problem is that wind especially takes up huge areas of land. Ideally the windmills should be well spaced apart to avoid turbulent airflows to adjacent windmills – 10 to 15 times the blade diameter is commonly recommended. For example if the U.S.A. keeps to its expansion plan by 2030 50,000 square kilometers of U.S land will be occupied by windmills, and 11,000 square kilometers of nearby ocean. In 2021 only 6% of world electricity was produced by windmills; if the aim of 50% by 2050 is realised there will massive areas of land and sea occupied.

A further problem is that the windmills, solar panels and batteries have a limited life. They deteriorate and need to be replaced, probably about every 20 years. That is very expensive. And, as they are made of expensive composites including glass, sometimes carbon, fibre, complex electronics, they contain many toxic compounds harmful to the environment and humans making safe disposal difficult. Currently most of this old equipment is dumped in landfills.

There is a further problem with *Stable Grid Electricity.* When electricity is generated from high pressure steam spinning turbines, as occurs in coal, gas and nuclear plants, they can be made to spin nationally at the same speed generating reliable AC (alternating current) power at 50 cycles per second, the common frequency. And the large number of turbines spinning in a country provide **backup inertia** to even out power demand fluctuations and keep the national grid smoothly operating despite variations in demand which are very common.

By contrast electricity generated from wind and solar is initially DC (direct current), then is converted to AC in inverters. There is no inherent stability and little capacity to respond to demand changes; grid and supply failure can occur. This requires large expensive machinery to counteract including very large capacitors and large expensive **synchronous condensers** to try to ensure a reliable power supply.

Yet another problem is that wind, solar, batteries and pumped hydro are very expensive to make and install, especially when compared to fossil fuel alternatives for electricity generation. It is not possible anywhere for them to be economically profitable in their own right based on costs versus electricity production. Those countries which have installed appreciable amounts have only done so because of massive government subsidies funded by taxpayers, especially in Australia, Europe and North America.

Some other issues with wind and solar:

Wind

The *area* of land or sea occupied by wind turbines is potentially huge if it becomes the major source of electricity. This is because turbines have to be well spaced to avoid interfering with the neighbouring turbines airflow and power. Generally the spacing should be 15 times the diameter of the turbine. This ranges from 80 metres to 320 metres, so the spacing should be from 1200 metres to 5 kilometres but is often less, which will impact performance. In the U.S.A. power output is approximately an average of 2.8 megawatts per square kilometre of windfarm according to a big study referenced below; in Austria and Denmark, where detailed studies have been done, 4 and 1.7 megawatts per square kilometre. A big study in the U.S.A. suggested an annual output of 214 terawatt hours – 214 billion kilowatt hours – of wind power electricity annually from farms occupying 12,800 square kilometers of land or sea.

In 2021 1870 terawatt hours of electricity energy was produced by about 500,000 wind turbines world-wide. That implies an area of approximately 100,000 square kilometres currently occupied by wind farms globally. That is about the size of the U.S. state of Maine, or the North Island of New Zealand, or England.

Total global electricity production in 2019 was 24,000 terawatt hours – 24,000 billion kilowatt hours – more than 100 times the amount in the U.S. study mentioned above. That implies that to produce the world's electricity from wind power would require at least 1.2 million square kilometers of land or sea holding over 6 million turbines – providing the wind keeps blowing! That would mean the equivalent of all of France, Germany and Britain covered in wind turbines. Or all the equivalent of all of Victoria, New South Wales and Tasmania in Australia.

It can be argued that not as much as this would be required to make a big impact on CO2 production from fossil fuels, and that solar electricity generation would help. But the world's demand for electricity is growing rapidly, and wind is the dominant renewable source now, solar is smaller globally; so to meet alarmist demands something like these expansions must take place.

This is only for electricity generation which is only 20% of human energy needs. Another 40,000 terawatt hours are used for transport and 46,000 terawatt hours for industry. If this was to come from wind, as some idealists propose, another 4.3 million square kilometers would be required making 5.9 million square kilometers altogether. This is two thirds of Europe, or the same of the USA, or three quarters of Australia. Clearly this is ridiculous.

Returning to just electricity generation this huge area of land and sea occupied would be spread world-wide. But it would be concentrated near big population centres especially in the northern hemisphere including nearby oceans. The electricity from the turbines needs a big network of cables to dispense it and the shorter these are the better.

But it is very well known that many people hate wind turbines. They find them visually ugly and the rhythmic low frequency whooshing noise going day and night very disturbing, even impacting their health. There are intense objections to wind farms. As a result wind farms have generally been approved in agricultural land with human population densities of 14 per square kilometre or less in advanced countries. And the land should not be mountainous to avoid destructive turbulence. These factors limit the available land area for wind farms.

An additional major concern is *the effect on wildlife especially birds*. Birds usually fly within a few hundred metres of the ground, usually are looking down for food or prey; they are vulnerable to be scythed by the huge slim blades which are hard for them to see. There have been a number of studies especially in the U.S.A. on wind turbine bird deaths.

A major U.S. one by Loss and others referenced below using 2012 data where humans searched for dead birds near turbines concluded that an average of 234,000 deaths occurred per year in the U.S.A from turbine strikes. It ranged from 140,000 to 328,000 in different years studied up to 2012, when there were 44,577 turbines there.

The authors found that when dogs were used to look for dead birds between 1.6 times and 2.7 times more were found. And in 2021 there were many more turbines, 65,067, many bigger and more lethal generating 111,000 megawatts (thousands of watts) compared to 60,000 megawatts in 2012. An estimate of the number of birds killed by turbine strikes in 2021 is therefore about 1.2 million in the U.S.A.

A crude estimate from this data about the possible number of birds killed by wind turbine blades each year world–wide is 9 million. If the great expansion occurs to 6 million wind turbines discussed above this could rise to more than 100 million, including many sea birds over coastal wind farms. Especially in the northern hemisphere where many of the wind farms would be. This would arguably have an impact on all life in the area. Studies by Loss and others have shown that birds respond to these problems by leaving the area of wind farms; for example a study of Dakota grassland country found that where previously 9 bird species lived only two remained. Few other studies have been done and generally wind farm companies do not release any data they have, the authors say.

In addition to direct blade strikes killing birds, they are also killed by colliding with overhead power lines, and by electrocution from these. Another U.S.A. study by Loss and others estimated between 12 and 64 million birds killed each year there by overhead power lines, about 10% by electrocution and 90% by collision. Given that wind turbines and farms are spread out in country areas, and are usually connected to national grids by many overhead power lines, it is clear that a big expansion in their numbers would result in a big increase in these deaths; usually transmission lines approximately double in amount when a change to wind or solar occurs. Unless the additional huge expense of running long distance high voltage lines underground was incurred.

Noise is a very well-known problem with wind farms. The continuous low frequency "whoosh – whoosh" of the turbines day and night when the wind is blowing, which is a great deal of the time where the wind farms are located, is deeply resented by many nearby residents, some claiming important health effects. Noise levels of around 53 decibels are common, the same as a noisy office. It is a major reason for regulators only allowing wind farms in sparsely populated areas. And they are usually not allowed in heritage wilderness areas such as national parks in advanced countries for that reason as well their visual ugliness and effects on birds and wildlife.

These problems result in the area of land suitable for wind farms being severely limited in many advanced countries. F. Nitsch and others conducted a thorough study in the Czech Republic in Europe of potentially suitable land for windfarms. Of a total land area of 78,000 square kilometers for the country they concluded only 500 square kilometers was suitable. This is about a tenth of the area found suitable by earlier studies. They concluded that the potential for great land based expansion of wind farms in advanced populated countries, where after all the electricity is needed, has been greatly exaggerated.

There is a move to locate wind farms at sea in view of these restrictions. The wind turbines are either mounted fixed to the seabed in shallow waters on steel and concrete towers, or floating structures moored in deeper waters. In addition to noise in the air these produce noise underwater. Sound travels far better in water for much longer distances as students of whale sounds know. There have been many studies on the effect of this on sea life, especially seals and dolphins which have advanced hearing. One by Stober and others referenced below showed that seal and dolphin behaviour altered for 6.4 kilometers from a single large 10 megawatt wind turbine if it had a gearbox. More modern turbines without a gearbox caused an effect for 1.4 kilometers. Other studies by Tougaard, and Nedwell, have shown sound levels in the water 20 meters from the wind turbine base of 120 to 178 decibels, very noisy indeed. They compared the sound to that of passing large ships concluding it was 10 to 20 decibels less; but it goes constantly. There have been few published studies on noise levels in big offshore wind farms like the largest so far, the Walney Extension in the Irish Sea with 87 wind turbines over 149 square kilometres, generating capacity 659 megawatts. But it is clear the sea here must be very noisy indeed for these animals resulting in them most likely leaving the area. If the massive size increase discussed above takes place to 6 million turbines, many at sea, there must be major effects on wildlife from underwater noise.

The wind farms then would take up huge areas of ocean and be a problem for navigation. Major storms would cause damage no matter how good the engineering including breaking loose from mooring and uncontrollably drifting into nearby turbines. The maintenance requirements for millions of huge offshore wind turbines, their 300 metre fans mounted over 200 metres above the sea, moored in stormy seas, would be massive and continuous, as anyone with experience caring for moorings in sheltered harbours would know. This would cost many billions of dollars inevitably.

The turbines would need to be connected to land users by a huge network of *undersea high voltage transmission cables* which would bring their own set of massive construction and maintenance issues, as well as further effects on wildlife. Many sea creatures have complex nerve systems sensitive to electric and magnetic fields, inevitably present with the cables.

In addition to these difficulties the *working life* of a wind turbine is limited to about 25 years, or 120,000 hours use, then at least the blades have to be replaced. If the area is rainy often, which many are, raindrop impact on the fibreglass blades results in rough pitting which is evident after 2 or 3 years, and results in a 25% loss in power generation by 14 years according to a study by A. Casterrini and others referenced below. Recycling old fibreglass blades is currently not practicable so they are usually cut up and buried in landfills. Some landfills already have many thousands buried. There is concern about toxic chemicals leaching from the old blades including Biphenol A.

The limited working life means that about every 25 years the world's wind turbine blades would need to be replaced by newly manufactured ones and the old ones cut up and buried. If there are 6 million turbines after the great expansion that is a huge ongoing industry using a great deal of energy, much of it from fossil fuel use no doubt. The great majority of blades are now made in China, with its abundant electricity from coal.

In addition concerning offshore wind turbines inevitably some will become damaged, perhaps the blades fragment and fall into sea; there they will slowly break up and leach their debris and toxins.

In addition there are huge construction costs for all the steel and concrete required.

And, finally, as said above, because the wind frequently does not blow there has to be a backup source of electricity equal in size to meet society's needs.

It is suggested that any reasonable person looking at all these problems would conclude that wind power has many severe, threatening problems to understate. The wind itself may be free, but there are massive costs and problems with the equipment.

Solar

There are two types of solar energy devices:

Solar Thermal devices collect the heat from sunlight and use it to directly heat something, usually liquid water. They are very commonly employed to heat household hot water systems in hot sunny countries such as Australia. They are low tech, simple and very effective. As about 40% of household electricity consumption is for water heating, and as the heat can be easily stored in the insulated hot water tanks when the Sun is not shining, they achieve their purpose well. Usually they are accompanied by a gas or electric back up to cover long cloudy periods.

A theoretical variant of solar thermal are large *solar chimneys.* These could be located in remote sunny areas and could have a large area of heat collecting panels heating air. They slope upwards towards the centre of the device, guiding the hot lighter air there by convection. At the centre there is a high conical chimney; the hot air rushes up this and is compressed, going faster. It is then used to drive a turbine electric generator. Theoretically 200 megawatts of electricity could be generated in hot sunny areas like deserts by a one kilometre high chimney at the centre of a feeding area of panel 7 kilometres square. But, while there have been plans, and small prototypes built, no large scale develops have occurred. Problems include very high building costs, the ability of the very high towers to stand up to high winds, the big area of land involved, all adding to the risk of investment.

Solar Photovoltaic devices, solar panels are now very common. They exploit a simple phenomenon which occurs when sunlight strikes a *semiconductor*; this results in the sunlight energy agitating electrons (the minute parts of an atom which whizz around the nucleus and carry a negative electric charge) which then move towards one side of the semiconductor creating a negative electric voltage there and a positive voltage on the other side. If this two sides are connected by wires an electric current flows. So electricity is generated from sunlight.

Semiconductors weakly conduct electricity in contrast to metals, which rapidly conduct it, as we see in electric wires often copper. There are many semiconductors but by far the most common in solar panels is *silicon.* This is one of the most common elements on Earth, very profuse and easily accessible in sands and rocks. For this purpose the silicon has to be made into a *crystalline* structure which greatly increases its electricity generation, then is fashioned into paper-thin large sheets. This requires heating the silicon to 1410°C, usually in factories using energy from fossil fuels.

The sheets are then fashioned into *solar cells* which are usually 156 mm by 156 mm squares. The cells incorporate a metallic grid to conduct electricity produced. One cell produces 0.46 volts. The cells are then organised into modules of between 32 to 96 cells, commonly 60 or 72. They are connected in electrical series so each cell adds to the voltage of its neighbour. A 60 cell module produces about 27 volts, and a 72 cell one about 33 volts. Concerning electrical power output a 60 cell module produces about 300 watts in bright sunlight and a 72 cell one about 400 watts.

A 60 cell module is usually arranged in 10 rows of 6 cells, making a panel one metre by 1.6 metres; and 72 cell one has 12 rows of 6 cells measuring one metre by 1.9 metres. The modules have a protective cover, usually transparent plexiglass, and an aluminium frame.

The electric current produced is Direct Current, DC, whereas all common electrical devices in society are alternating current, AC. So the DC has to be changed to AC by an *inverter* device. About 10% of the electrical energy is lost in this process.

The panel are organised in *arrays* of several panels. For households it is common to see arrays of 10, 20 or 30 panels with 60 cells each on roofs. An array of 10 panels produces about 3000 watts, 3 kilowatts electricity in bright sunlight. One of 30 panels produces about 9 kilowatts.

Solar Panel Manufacture and Disposal:

As said above quite considerable energy is needed to manufacture solar panels; for every kilowatt panel capacity 2000 kilowatt hours are required. The high temperatures required to process silicon have been mentioned and there are several other energy intensive processes involved including aluminium and plexiglass production.

The world electrical capacity of solar panels in 2021 was 849,473 megawatts, or about 850 million kilowatts. To make these panels would have required 1700 billion kilowatt hours of energy, which is 1700 terawatt hours. This would have been spread over several years and is not very significant in global terms, humans producing and using about 24,000 terawatt hours of electricity each year. But the great majority of solar panels are produced in China using cheap energy from fossil fuel, usually coal, power generation and cheap labour in that authoritarian non-democratic state; only a few are made elsewhere, mostly in Canada and the U.S.A.. As a result the cost of solar panels is lower, about $1000 U.S. for each kilowatt capacity.

The ethics of people in western democracies rushing to buy these solar panels encouraged by their governments offering tax and other financial incentives, and preaching about environmental benefits, is obviously suspect to understate. But most people in the western democracies probably give no thought to that, lured as they are by their governments. To be fair this problem of buying cheap goods from an authoritarian state is an old, common one for the democracies.

In 2021 1000 terawatt hours (one thousand billion kilowatt hours) of electricity was produced by solar, 3.6% of total production. This is projected to triple by 2030 and perhaps reach 40% of global electricity production, about 9 million megawatts capacity to make 10,000 terawatt hours, by 2050 as part of "nett zero" ambitions. To make all the solar panels required, about 20 billion, up from 2.4 billion currently, could take 16,000 terawatt hours of energy, a big chunk of total global human annual electricity production and usage of 24,000 terawatt hours as said above, admittedly spread over several years. The *area* occupied by these would increase from 3400 square kilometers to 120,000 square kilometers. And it is likely the great bulk of this manufacture would occur in China where there would be a huge industry making solar panels for the world at a very rapid rate. An industry which would be very secure given the working life of panels and the need to renew them all every 20 to 30 years. Anyone who thinks that all this energy required to make the massive number of panels demanded by advanced countries as they rush towards "nett zero" is going to come from renewables such as wind and solar in China, given the known rate they are expanding coal generation, is truly an optimist.

An interesting side issue is the idea of *floating solar panels.* A multinational group under Sandia National Laboratories Research have proposed that globally there are 114,000 bodies of water – lakes, reservoirs, protected water – suitable for floating huge platforms of solar panels. The area totals 550,000 square kilometers, and the proposal is to occupy one third of them with the panels, 183,000 square kilometers, bigger than England and Wales, or Florida or Missouri states in the US. That would produce 9400 terawatt hours annually, according to the researchers, about a third of human global electricity production. About 40 countries could produce all their electricity plus some by this method because they have large areas of suitable water including Brazil. But when one considers the working life of panels and the disposal issues discussed below the proposal becomes truly frightening. To say nothing of the local effects of huge areas of floating solar panels occupying sheltered water in many countries, especially scenic ones, or those with marine commercial and recreational interests. Aging and damaged panels would clearly be toxic. The researchers calculated that there would be a reduction in water evaporation globally of 106 billion cubic meters liquid water, a massive interference locally regarding climate and rainfall, and also to some extent globally given water evaporation is vital for cooling the planet and for climate as discussed extensively already. And if one third of large bodies of water is covered in panels life in the water beneath would be severely affected, including cessation of photosynthesis in aquatic plants. In addition panels convert only about 30% at best of the Sun's energy striking them, the rest is emitted as heat energy back to the atmosphere. This is in contrast to the ability of water surfaces to absorb up to 80% of the Sun's energy as discussed, and would likely lead to more local effects. The proposal illustrates the extreme lengths some groups will advocate to achieve their ideas.

It follows that if a similar area on land was occupied by panels there would be consequences affecting plant life and the climate also; but many panels on land are on rooftops or city areas without that effect. Some large "solar farms" are on rural or wilderness land, and then these problems needed to be considered.

A huge looming problem with solar panels concerns their working life of 21 to 30 years, and subsequent disposal. The elephant in the room is what happens to the old panels.

Currently world-wide virtually all of them are dumped in landfills. In 2022 there were one million tonnes of old solar panels dumped in the U.S. and 1.6 million tonnes in China. This is projected to increase in the U.S. to 8 million tonnes in 2030 and 80 million tonnes by 2050, given the aim to greatly expand solar power. Less than 10% is recycled in the U.S. This is because there are big problems with recycling old solar panels.

Because they are made to last for up to 30 years in all weathers they are robust and very difficult to dismantle. Doing so is expensive and labour intensive. In the U.S. it costs $15 to $45 to recycle a silicon module of just 60 cells, and only $1 to $5 to dump it. So recycling seldom happens.

In the EU there are regulations forcing some recycling of old solar panels; but in practice just the aluminium and glass is recycled as that is comparatively easy, the rest includes the silicon cell and its components is crushed into tiny fragments and incinerated at high temperatures usually using fossil fuels.

In China, with apparently one third of the world's solar panels mostly in the remote northwest there is no recycling occurring as far as is known; estimates of the amount of solar panel waste, given the country has three times the U.S. quantity of solar panels, would be perhaps 20 million tonnes annually by 2030 and perhaps 250 million tonnes by 2050.

Solar panels contain lead, heavy metals and other toxins which leach into groundwater from landfills. As expected there is a lot of talk about expanding recycling but the real difficulties, coupled with the headlong rush to expand solar power, make achieving it remote. One idea is to make the panels easier to dismantle and recycle, but there is no practical evidence that is happening. It would increase the chances of a shorter working life, and increase costs, arguably. The emphasis is on plentiful cheap panels aiding the rush to "nett zero" and these major disposal issues are simply ignored in most countries.

Solar Panel Performance Issues:

Solar panels only convert a fraction of the Sun's energy striking them into electricity. The theoretical maximum is 32% but in practice the best achieve 20% when new. As they age this drops considerably. Because the solar cells are wired *in series* to add the tiny 0.46 volts made by each cell to its neighbours as explained above any fault in a cell severely impairs the electrical production of the whole unit. Faults may occur due to damage, a manufacturing defect, or aging. The great majority of solar arrays are made in China as said. This makes it more difficult for purchasers in other countries to ensure quality which is crucial for good performance.

The intensity and duration of sunlight is of prime importance. That depends on the season, the length of daylight for the location, whether the Sun is high shining directly on panels, or low shining obliquely. Cloud cover is very important and may reduce output by 80% commonly. These factors are important especially for heavily populated northern hemisphere countries well away from the Equator including northern Europe and North America where winters are long, dark, cold and stormy resulting in high demand for electricity for heating, light and many other purposes. Solar performance will be low then. Cloudy rainy weather can go on for weeks stopping performance.

Shading makes a big difference and may be difficult to avoid in built up populated areas, cities. If one part of an array is shaded it impairs the performance of the whole installation markedly.

Many promoters use 5 hours of sunlight per day as an average when explaining expected performance. Then the calculation concerning expected generation often assumes the installation will put out the rated full capacity for, say, 5 hours. For example a typical household 5 kilowatt installation is calculated to produce 25 kilowatt hours per day for 365 days of the year resulting in over 9000 kilowatt hours of electricity generated. That is pretty close to the average annual electricity consumption for households in wealthy western countries of about 9000 kilowatt hours making it very attractive.

But the world-wide reality from the current solar generating capacity of 850,000 megawatts apparently generating 1000 terawatt hours per year as said above is an average of just over 3 hours equivalent sunlight resulting in full capacity generating per day. So there are problems with these optimistic estimates. The main ones are:

1:

Solar panels are rated on their output in full sunshine from directly above concerning the raw direct current (DC) current produced by the panel; but this has to be changed into alternating current (AC) to be useful as explained above. This results in a 10% loss.

2:

The electricity then has to transferred by cables to the points of final use, often via intermediate stations changing voltages with the common result of a further loss of 5 to 10%.

To help explain the next points figures 19, 20 and 21 from Chapter 3, the section on the Sun's radiation reaching Earth, are repeated below together with a summary of total Sun radiation received for different latitudes:

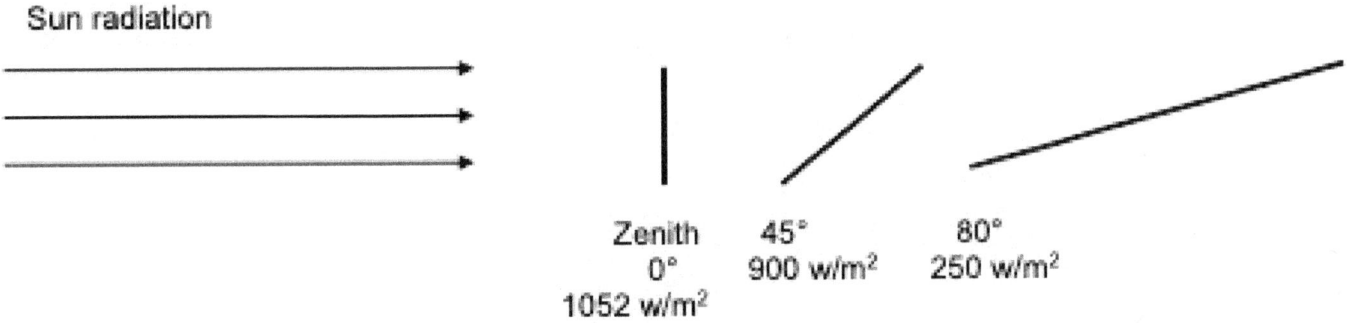

Figure 19 - repeated: showing the effect of shifting away from where the Sun is directly overhead (at its Zenith) towards the Poles. The Sun is lower in the sky, its radiation is spread out more when it strikes the surface and is weaker.

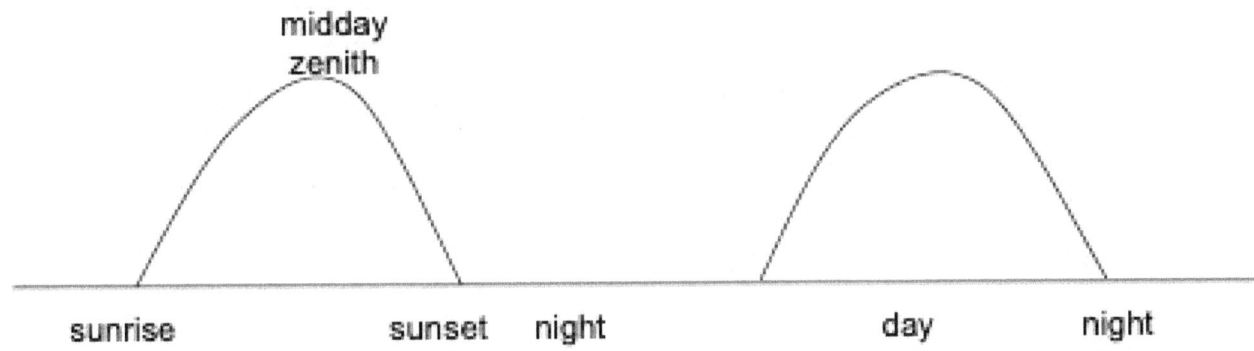

Figure 20 - repeated: showing the effect of the Earth's daily rotation on the rate of Sun radiant energy striking the surface. There is no Sun radiation at night obviously. Then after sunrise the Sun is very low in the sky, the radiation is spread out as in Figure 19. The radiation gets more concentrated into a smaller area as the Sun rises. In places where the Sun is directly above at midday (at its zenith) it gets to about 1000 w/m² on a clear day as explained above.

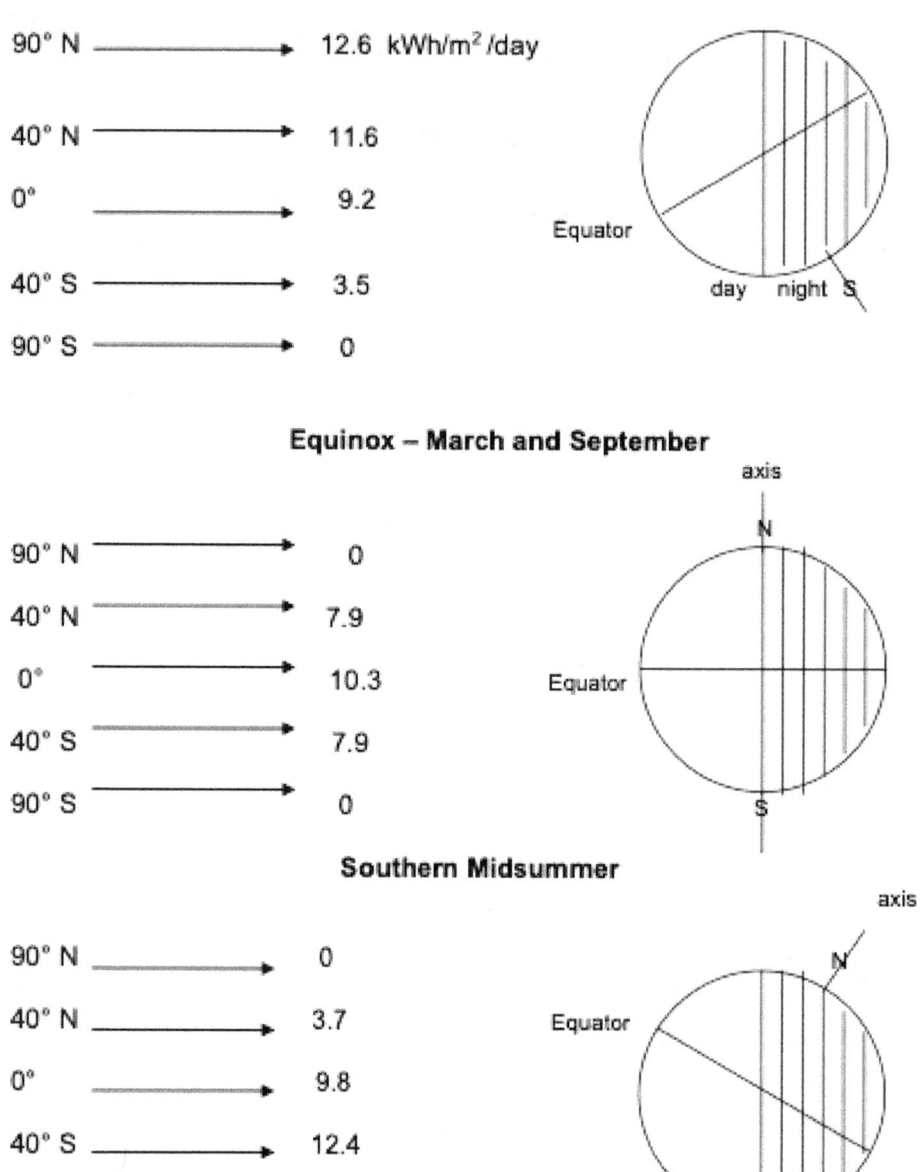

Figure 21 - repeated: showing the total Sun radiation energy received each day per square meter at northern midsummer, the equinoxes in March and September, and southern midsummer for the Poles, 40° North and 40° South, and the Equator. The Equator receives a steady high amount throughout the year. The Poles receive very high amounts at midsummer, but nothing from equinox to equinox, 6 months of the year. The South Pole receives a little more than the North Pole at midsummer because the Earth is 5 million kilometers closer to the Sun then as discussed already. The mid-latitudes receive a varying amount from high amounts at midsummer to low amounts at midwinter.

The *average* received each day over the whole Earth is about 6 kWh/m². For the whole year the *average* amount of Sun radiation energy for the different latitudes in Figure 21 is described below:

- North Pole, 90° N: 3.1 kWh/m²
- 40° N: 7.8
- Equator: 9.9
- 40° S   8.8
- South Pole, 90° S: 3.4

So the Equator region receives the most over the year as we expect. And the Polar regions the least because of the long dark winters.

To continue:

3:

Figure 19 shows that the amount of solar energy striking an area of one square meter drops substantially when the Sun is lower compared to when it is high – about a 75% drop if it is just above the horizon compared to directly overhead. This means that early in a day and later solar panels produce far less that the rated capacity of electricity.

4:

How far a location is from the Equator, its *latitude,* makes a big difference as discussed above concerning Figure 21. This is important as many large advanced societies are a long way from the Equator, especially in Europe and North America, but also Japan, Asia, China and Russia.

5:

The *seasons* also play a big role in reducing solar power output especially concerning the northern winters obviously.

6:

Weather variations, especially cloud, which often reduces solar output by 80% as discussed, are very important, again obviously.

These last three factors, latitude, seasons and weather explain big differences in the practicality of solar installations in different places; for example below are listed the average daily hours of cloud free sunlight in these locations:

United Kingdom   3.8 hours cloud free sunlight per day

Dunedin, New Zealand   4.6

Shanghai	4.8
Beijing	5
Victoria State, Australia	5.4
New South Wales State	6.4
Madrid	7.5
Perth	8.7
Los Angeles	9.1
Alice Springs	10.4

Clearly for many places solar power output would be less than half that for sunnier areas.

All of this means that the electric power output of solar installations will always be significantly less that their rated capacity. And as modern society needs electric power continuously in all weathers and at night big backup power sources are required.

The manufacture and disposal of solar equipment is far from clean and low energy as discussed. Solar has only become popular in western democracies because of low cost equipment from China, cheap because the high energy needed for manufacture comes from coal power stations, and because of cheap labour compared to the western democracies, combined with large government subsidies.

Batteries:

As we all know batteries store electrical energy. They all have a positive and negative terminal. Electric current flows from the positive terminal through external wires and devices, powering them, whether it be lights, motors, various electric devices, to the negative terminal of the battery.

What really happens is that *electrons*, the negatively charged tiny parts of atoms whizzing around the nucleus as explained before, travel from the negative battery terminal to the positive one through the wires and devices being powered. The idea of *current* flowing from positive to negative is simply a human idea used to explain what is happening.

Historically the first observation which ultimately led to batteries was made by a French research physicist, Nicolas Gautherot, in 1801. He was doing experiments on electric current passing through salt solutions and noted that the positive and negative terminals had a charge, a *voltage difference*, from one another even after the current had stopped for a few minutes.

The first practical battery was invented by another French physicist, Gaston Plante, in 1859. It was the lead-acid battery so very common for many uses ever since. The way this battery works in shown in Figure 70 below.

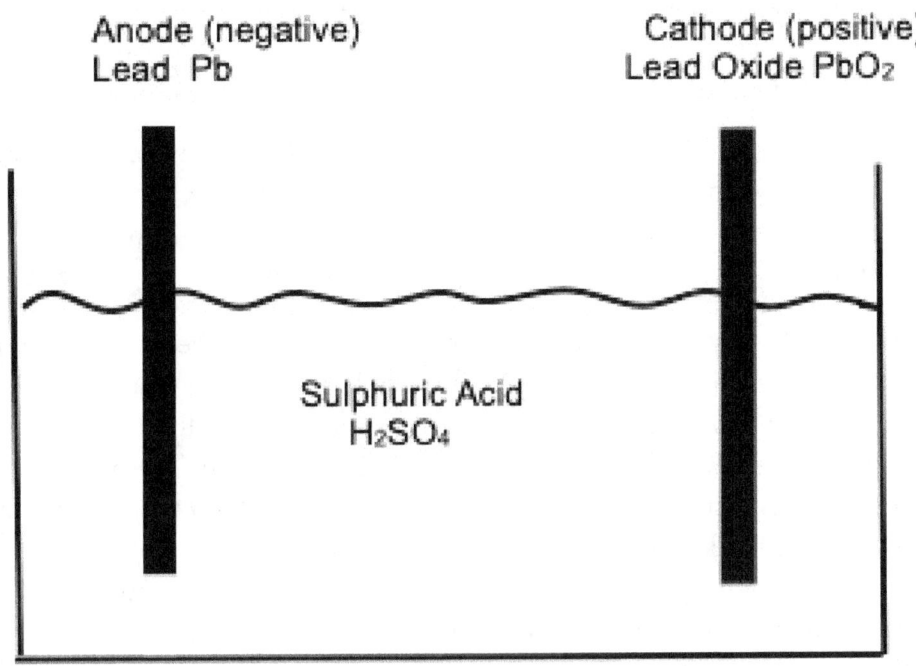

Figure 72: a simple Lead Acid battery.

When the battery is used, and current flows from the positive terminal through wires to make a device – commonly the starter motor to start an engine – work, the Lead in the anode negative terminal changes to Lead Sulphate, reacting with the Sulphuric Acid. The reaction produces electrons which result in the anode becoming negatively charged.

At the cathode, positive terminal, the Lead Oxide also reacts with the Sulphuric Acid solution to form Lead Sulphate but takes up electrons in the process, so becomes positively charged.

The reactions can be written scientifically like this:

At the Anode – negative terminal:

$Pb + HSO4$ goes to $PbSO4 + H^+ + 2e^-$

At the Cathode – positive terminal:

$PbO2 + HSO4^- + 3H^+ + 2e^-$ goes to $PbSO4 + 2 H2O$

The sulphuric acid solution gets weaker as the battery is used. Pb is the scientific code for Lead. H2 SO4 is Sulphuric Acid.

The battery can be charged by applying a positive current with a higher voltage than that of the battery to the cathode – positive terminal, the current flows through the battery to the anode – negative terminal and back to the charger – which is often a vehicle alternator. When this happens the above reaction reverses and the Lead Sulphate at the anode goes back to lead, that at the cathode goes back to Lead Oxide, and the Sulphuric Acid solution becomes more acid. If the battery is overcharged hydrogen gas and oxygen can form from the water in the battery.

Figure 72 shows a single lead acid battery *cell*. This can generate 2.1 volts when fully charged. Commonly 6 cells are linked *in series* together, each adding its 2.1 volts, making 12 volts altogether. This is virtually universal in vehicles to work the starter motor. Once the vehicle engine is running the battery is recharged by the engine alternator.

Over the 160 years since this battery was first developed there have huge improvements. It is extremely common with at least US$15 billion sales annually world-wide. One estimate suggests there are 2.6 million tons of lead in lead acid batteries world-wide. There are perhaps three billion lead acid batteries globally, one for every three people. In advanced societies there are almost one per person. The reasons for this are that these batteries are very cheap to make and pack a big voltage punch for their weight for engine starting, other tasks. They are very versatile and safe, can operate from minus 40°C to plus 60°C, rarely give trouble, or catch fire. There are different types for various purposes ranging from starting batteries, to deep cycle, maintenance free versions, other special purpose batteries for aircraft, ships, submarines. And every electric vehicle has a Lead Acid battery to run its control aspects.

But they have a limited life, about 500 cycles of being discharged, then charged on average. This is due to hard crystalline lead sulphate forming on the anode and cathode amongst other wear problems. And if discharged below 50% charge they are usually ruined. It is not possible to rapidly charge them. They lose about 3% of their charge each month even if idle. Hence, as every motorist knows, they need replacing every few years.

A bright aspect of this is that recycling lead acid batteries is common and comparatively easy. Between 60 and 95% of lead acid batteries are recycled globally and in advanced societies, the USA and western democracies, it is close to 100%.

Lithium Ion Batteries

Lithium batteries have clearly revolutionised modern life. They power phones, computers, multiple other devices. Their basic science is not too dissimilar to lead acid batteries and is shown in Figure 73 below:

At the positive electrode, the cathode, there is an Oxide of Lithium, similar to the situation in Lead Acid batteries where there is an Oxide of Lead. The type of Lithium Oxide at the cathode varies commonly from mixtures of cobalt, manganese, or iron fluorophosphate with subsequent variations in cost and performance. In addition nickel, titanium, copper, and other metals can be present for complex reasons.

At the negative electrode, the anode, there is solid carbon in the form of graphite. When the battery is fully charged the carbon has a lot of lithium ions in it, is fully "lithiated", LiC_6, one lithium ion to every 6 carbon atoms.

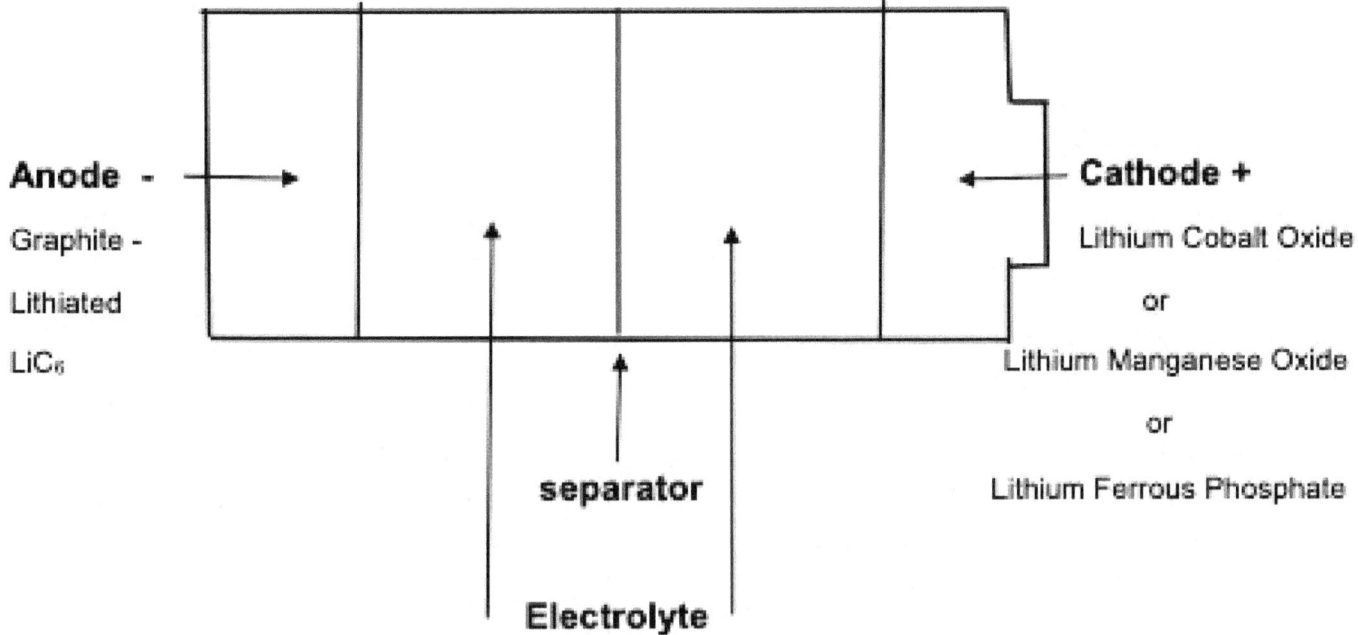

Figure 73: a simple lithium ion battery.

During charging a strong positive voltage is applied to the cathode and lithium + ions flow from that through the electrolyte solution past the separator, which lets lithium + ions pass, to the anode where the lithium ions are taken up by the carbon – graphite to form LiC6.

During discharge when the battery is being used to power something the reverse happens. The lithium ions leave the anode, made of carbon – graphite, and travel back through the electrolyte past the separator to the cathode where they are incorporated as an oxide of lithium.

In this process electrons (tiny with a negative electric charge as explained) are released at the anode and taken up at the cathode, so they *flow* from the anode to the cathode through the wires, device being powered – perhaps a phone or computer – forming an electric current which powers that device. And that current is *described* as being positive voltage at the cathode and negative voltage at the anode.

It can be written scientifically like this:

At the Cathode – positive terminal:

Li Co O2 goes to Co + Li$^+$ + e$^-$

charging

discharging

At the Anode – negative terminal:

C6 goes to $Li^+ + e^-$ Li C6

charging

discharging

Each simple battery like this is called a *cell*. The voltage developed varies from 4.2 volts for lithium manganese cathodes to 3.2 volts for lithium iron phosphate batteries when fully charged. The cells are connected to many others **in series** to add these small voltages up to produce a useful bigger voltage – perhaps 12 or 18 volts for phones, laptops, small tools to 300 volts or more for electric vehicles, whose batteries may have 7000 or more cells.

The credit for masterminding the original research which led to these incredible batteries belongs to John Goodenough who worked at Oxford University in the 1970's and 80's. He was Professor of Mechanical Materials Science and Electrical Engineering at the University of Texas and awarded a Nobel Prize in 2019, when he was still working on battery research at the University at the age of 98.

These batteries can store much more power for their weight than lead acid. They can store from 80 to 200 watt hours per kilogram compared to 20 for lead acid. They can be quickly recharged with higher currents. They last for much longer, typically up to 2000 cycles of charging and discharging compared for about 500 for lead acid. They store the power for longer, self-discharge is usually less than 2% per month. They can be made very small – the smallest cell so far is 3.5 mm across. They are sealed units so can be used for phones, computers without leaking.

As everyone knows they have revolutionised life. The global production of lithium batteries has rocketed this century. Expressed in terms of millions of kilowatt hours of power storage by these batteries in 2022 almost a thousand million kilowatt hours were made (948 gigawatt hours). Electric vehicles account for about a third of this (274 gigawatt hours). This is projected to increase massively by 40% compound annual growth with the moves to nett zero. By 2030 production could be four times greater than now and 70 times greater by 2050 according to some. In 2020 this industry was worth $US 37 Billion globally. In 2028 it is projected to be worth $US193 Billion. Some projections estimate up to $US400 billion by 2030, other $US 200 billion.

Making Lithium Ion Batteries

This is a highly complex and energy intensive process. Initially lithium and the other metals have to be mined. Australia is the biggest supplier of lithium accounting for 54% of the 85,000 tons mined globally in 2018. There the lithium is extracted from hard rocks by crushing and other processes consuming a lot of energy. The next biggest supplier is Chile where lithium is found in deep underground brines, very salty groundwater, especially in the arid Atacama desert. These are flushed to the surface using a great deal of water. 2 million litres of water are required for every tonne. This is controversial as it has resulted in severe water shortages for nearby communities and farmers in this very dry area, as well as contamination of soils and groundwater with toxic waste.

Chile accounted for 24% of global lithium supply in 2018. Other big suppliers were China and Argentina. All of the Australian lithium and most of the rest is shipped to China, which is by far the biggest manufacturer of lithium batteries.

There are concerns about the global reserves of lithium which amount to 43 million tons. But only one third, 14 million tons, can be mined. 87% of these reserves are in underground brine waters in South America. At the rate of mining in 2018 of 85,000 tons that year that would be enough for 165 years. But that rate is projected to very substantially rise. Some projections consider a 70 fold rise by 2050 to be needed. Lithium shortages are predicted, and already there has been a ten-fold increase in the price of Lithium Carbonate to $US 74,000 per tonne from historic prices; in 2021 the price rose 123%.

Cobalt is another metal commonly used in lithium batteries, although there are moves to replace it with iron fluorophosphate as mentioned. But batteries containing cobalt still perform very well, they are the most common type so there is an ongoing need for it. The price of cobalt also is rising, in 2022 it was $US27,000 per tonne. Rapidly increasing prices are occurring for the other materials including nickel, now $US12,000 per tonne.

Very large amounts of electrical power energy are required by lithium battery factories; about 65 kilowatt hours for every kilowatt hour battery storage capacity produced. As said in 2022 948 gigawatt hours (948 million kilowatt hours) were made; that would have required 60 thousand million kilowatt hours of electrical power in the factories (60 thousand gigawatt hours, which is 60 terawatt hours). Compared to the World total electricity production and usage of 24,000 terawatt hours annually that may seem small; but it is projected to increase 70 times by 2050, when perhaps 4200 terawatt hours would be required by the factories.

Currently 77% of global lithium battery production takes place in China where there is cheap electricity from coal power stations, which China continues to build in large numbers. This enables them to supply cheap lithium batteries world-wide.

But the electrical power needs of the *factories* are just a small part of the energy requirements to make these batteries: mining, processing and transporting the raw materials is a far bigger part. When all these are taken into account up to 500 kilowatt hours of energy is required for every kilowatt hour of battery made. So, even if a lower estimate of 300 kwh per kwh of battery made is used to estimate energy requirements the 60 terawatt hours of *factory energy needs* for 2022 lithium battery production becomes 300 terawatt hours. And if predictions of 70 times more lithium battery production by 2050 are realised then 21,000 terawatt hours of energy would be required then, almost 87% of the current total world electricity production.

What is very clear is that making lithium batteries consumes huge amounts of energy, especially electrical power. That is all supplied from fossil fuels currently, be it for mining machinery, heavy trucks, shipping, more trucks, largely using diesel, then the factories using electricity from coal fired power stations virtually all in China. It is clearly unrealistic to believe that battery manufacture can expand sufficiently to meet 2050 ambitions concerning nett zero. Especially when the potential shortages of lithium and other raw materials are added to the huge energy requirements. Lithium batteries are only "cheap" at present because of low cost electricity for the factories in China, and diesel usage for the other costs.

Lithium Battery Disposal and Recycling

Lithium batteries have a limited life, usually about 2000 cycles of charge/discharge. This may be about 4 – 5 years for many devices including electric vehicles. Some claim they will go for 10 years. Their capacity to store electrical power declines with age. They then have to be replaced.

The vast majority of old lithium batteries are dumped in landfills. In 2020 an estimate of the amount in Chinese landfills was 500,000 tonnes. Over 95% of old lithium batteries in North America and Europe are dumped in landfills, 97% in Australia.

The amount of lithium batteries is projected to massively increase with the shift to more electric vehicles. In 2019 2.1 million electric vehicles were sold world-wide, this is projected to rise to 24 million sold in 2030. Old scrapped lithium batteries are predicted to amount to two million tons per year by then, with an accumulated total of four million tons.

Obviously it would be much better to recycle old batteries. But there are big problems with this:

They are difficult to recycle. The batteries are sealed units with many components and many different toxic metals and materials. They are not designed to be easily dismantled, rather they are made as safe as possible, able to take a big charge quickly, and as cheaply as possible.

The toxic substances include the electrolyte in them, lithium fluorophosphate LiF_6. Even with minimal physical contact workers dismantling lithium batteries risk respiratory failure, cardiac arrest and death. The solvents sealed in the battery cells include highly flammable alkyl carbonates which have caused big fires in landfills. Other toxic substances include the metals cobalt and nickel, highly toxic to animals. If these leach from landfills into ground water aquatic animals in rivers, lakes are especially at risk.

So far the only practicable way to dismantle the batteries is by manual labour with many precautions, a time consuming and expensive task. There are many research projects looking at better dismantling by robots, automation, ultrasound, but so far they are just small pilot research programmes.

There are two main methods of "recycling" in limited use:

Pyrometallurgy: the lithium batteries are crushed into small particles by heavy machinery, then heated at 1500°C. this results in everything burning except the contained copper, nickel and cobalt which is recovered. All the lithium, aluminium, chemicals and solvents in the batteries are incinerated at this very high temperature and lost. The resulting gases and ashes are processed further to remove most toxins before being released into the atmosphere or landfills.

Hydrometallurgy: again the batteries are crushed, then treated with hydrochloric acid, nitric acid, sulfuric acid at 100°C, with other powerful solvents. This leaches the metals including lithium, aluminium, copper, nickel, manganese which are recovered. The process produces big volumes of highly toxic liquids which need further processing before being released, usually into landfills.

Both of these methods are very complex, require a lot of electrical energy, and only recover a small proportion of the expensive substances in the batteries for reuse. In 2018 100,000 tons of old lithium batteries were "recycled" world-wide, including 67,000 tons in China and 18,000 in South Korea. It is hoped to be able to "recycle" 1.2 million tons by 2030. This is a small fraction only of the predicted lithium battery waste then.

Plastics in the batteries are degraded by these processes producing toxins including dioxin. Graphite in them is burnt to CO_2, it is not possible to recover the graphite without a highly expensive process keeping out oxygen and water.

Economically even this primitive "recycling" is far from cost effective compared to just buying new batteries. For this reason there are big government subsidies for research and pilot recycling projects in the advanced democracies, many of which have regulations endeavouring to force "recycling". But these are far from effective as shown by results from Australia where 3% of old lithium batteries are "recycled' by shipping them to South Korea and China. In the EU and North America 95% of old batteries end up in landfills, as do the great bulk in China and Asia.

Another problem with the economics of recycling is uncertainty about the future technology of vehicles – will they be hydrogen powered? Is another type of battery going to replace lithium? Is the future price of materials used including lithium, cobalt going to drop undercutting the recycling economics? Are lithium powered electric vehicles going to expand according to current predictions? These are some of the uncertainties.

What is certain is that if the aims to greatly expand the lithium powered electric vehicle world-wide fleet are realised there will be huge mountains of old lithium batteries in landfills or stockpiled in many countries releasing highly toxic substances into the environment, including rivers and the oceans, catching fire, resulting in a great deal of anxiety for the human community. This certain outcome is in marked contrast to the blatantly optimistic propaganda about "recycling" in many proponent websites.

Electric Vehicles

Electric vehicles are being extremely heavily promoted by a variety of powerful groups, including governments, environmentalists, manufacturers and enthusiasts. They have a number of attractions compared to petrol and diesel fuelled vehicles:

- They do not directly emit exhausts containing CO_2 and other gases, soots, resulting from petrol and diesel use.

- They are very quiet, an advantage in big cities. But this advantage can be overstated as a lot of noise from vehicles is tire noise; and most electric vehicles are replacing small petrol ones which are quiet anyway.

- They need less maintenance than petrol and diesel vehicles. Especially there are no oil and filter changes, or other regular maintenance required by internal combustion engines. Rather they have electric motors which last for ages with little maintenance. They do not need complex gearboxes.

- There is no comparatively large heavy engine and gearbox in an engine bay. Rather the battery is under the vehicle floor and the motors spread near the wheels. This results in more space for passengers and luggage, and leads to design opportunities to make very attractive "new age" vehicles.

- They do not need refuelling with petrol or diesel, a big cost saving. Recharging the battery is cheaper usually per kilometer travelled.

- They often have stunning acceleration and performance from a standing start compared to petrol and diesel vehicles because the electric motors quickly develop full power from very low engine speeds. Whereas petrol and diesel engines need to be spinning faster before developing a lot of power. At higher speeds petrol and diesel variants are more powerful.

- In many countries especially Europe, China, North America, there are substantial government grants and tax incentives encouraging purchase of electric vehicles; and in some penalties for purchasing petrol or diesel vehicles. Increasingly large cities have parts were only electric vehicles may go, and offer free parking for them.

In 2012 there were 118,000 electric vehicles globally, in 2021 this reached 16.5 million and 6.6 million were made that year. More than half, 3.3 million were made in China. In Norway, where there are very strong subsidies, tax concessions and regulations, 86% of new car sales in 2021 were electric vehicles. Some countries including the UK, the EU, some US states including California, are outlawing petrol and diesel cars by 2035. Car manufacturers Volvo and Mercedes have announced cessation of petrol and diesel cars by 2030.

But there is a need for perspective. In 2021 there were 1.446 ***billion*** vehicles world-wide (1,446 million) excluding motorcycles, which are very common in Asia (Taiwan alone has 28 million motorcycles). This means that for each electric vehicle there are 87 petrol or diesel ones. Each year 80 million vehicles are made globally, almost 90% are petrol or diesel. Concerning trucks 99.7% of those made globally in 2021 were petrol or diesel, only 0.3% electric. And they were mostly small short haul trucks.

Most of the electric vehicles made are small and used for short distances only. Of the 3.3 million made in China in 2021 more than half were 2 or 3 wheelers – motorcycles or tricycles. Only a tiny proportion of large cars, "sports utility vehicles" (SUVs) and light trucks are electric. And virtually no heavy trucks.

About a quarter of the electric vehicles are plug in hybrids (PHEVs) – they have a small petrol motor to charge the battery and run the vehicle if the battery is flat, increasing range and reliability. They can be plugged in to a power source to charge. They have a much smaller battery than pure electric vehicles, usually 16 to 20 kwh. The combination of petrol engine and battery power, electric motors makes them complicated compared to pure electric or just petrol vehicles.

Non-plug in hybrids also have a small petrol motor which charges the battery and runs the vehicle if the battery is flat. Their battery is also small. They are less popular now than plug-ins, about 5% of electric vehicle sales.

There are several reasons for this continuing dominance of petrol and diesel vehicles:

Cost:

Electric vehicles are more expensive than petrol or diesel. In the US the cheapest new car is a petrol Mitsubishi Mirage at $17,000. This is very economical to run. The average small new electric vehicle there is $66,000. The large lithium batteries in electric vehicles cost at least 40% of the total price. These are usually between 40 and 100 kwh for electric cars.

Electric cars cost about 40% more than the equivalent petrol model. For example in Australia an MG ZS small car costs A$22,000 for the petrol version and A$44,000 for the electric. A Kia EV6 small car costs A$37,000 for the petrol version and A$72,000 for the electric, a big 94% margin.

In the US the cheapest petrol vehicle is at US$17,000 as said. And the cheapest electric is a Nissan Leaf at US$20,000. But the actual cost of this is US$27,000, the government gives a tax credit to bring the price down so electric vehicles can compete economically with petrol as part of its "nett zero" push. The most popular US vehicle, the Ford F150 light truck, costs US$33,000 for the petrol version and US$55,000 for the electric, including the government tax credits.

In Norway petrol cars are subject to a 100,000 Norwegian Kroner (NOK) flat fee on purchase, plus VAT at 25% on the whole price. Electric cars purchasers, however, do not pay the flat fee, and they used to not pay any VAT either. But recently the government has made electric vehicle purchasers pay 25% VAT on amounts over NOK 500,000, under NOK 500,000 there is no VAT. A big tax concession explaining why 86% of new cars sold in Norway are electric. In addition electric vehicles in Norway do not pay parking charges in cities, nor do they pay toll road charges, in contrast to diesel and petrol vehicles.

There is great variation amongst countries as to how they encourage electric vehicle purchase by subsidies and tax credits. Politicians clearly feel this is what their electorates want, based on ambitions to achieve nett zero. In the meantime the government loses revenue and the taxpayer pays.

But there are some ironies in the Norwegian situation, which has been held up by nett zero enthusiasts as an example of the future. Some of these are:

Despite the number of electric vehicles rising from 9900 in 2012 to 647,000 in 2022, now 20% of Norway's car fleet of 3.2 million, Norway's oil consumption has barely changed. The long term average from 1965 to 2020 is 199,000 barrels per day (24 million litres). In 2021 it was 199,000 barrels each day. The highest year was 2010 when it was 226,000 barrels per day, so there has been some reduction, generally considered to be about 5%. This has disappointed many considering the big effort to promote electric vehicles. There is a reduction in petrol consumption, as expected as almost all electric vehicles are replacing small petrol ones. But diesel consumption has barely changed at about 220 million tonnes per year.

The huge government tax concessions subsidizing electric vehicles cost 39 billion Kroner (US$4 billion) in 2022, for a population of 5 million people. This is an ongoing increasing cost if nothing changes; but the government is planning tax changes in response to pressure from taxpayers. A tax on vehicle weight is planned, which will hurt heavy electric vehicles. And decreases in the tax concessions are mooted.

Norway is very wealthy because of it's fossil fuel production; 49% of it's export income is from oil and gas exports from the huge North Sea oil fields. Norway produces 3% of global natural gas and 2.3% of global oil. A lot of this is exported to other European countries. It has done very well out of the big increases in gas prices. So the subsidies for electric vehicles are paid for largely by profits from the oil and gas production, which are kept for the taxpayer there, not distributed to shareholders as in the US. In addition Norway imports 600,000 tonnes of coal, 120 kilograms per person, each year. So the country, despite the hype about its electric vehicles and renewable electricity largely from hydropower, is very hooked into fossil fuels. And that is not going to change.

2. Difficulties recharging:

There are three ways to recharge an electric vehicle:

Level One Charging:

This is "at home or business" charging. The vehicle is simply plugged into a domestic power outlet. This allows up to 16 amps power, or 2 kilowatts, to run into the battery. In one hour two kilowatt hours (kwh) of electric energy would be added to the battery.

Level Two Charging:

This involves going to a charging station, which will usually be publicly available or may be installed at a business, and plugging in.

Single Phase Level Two charging occurs at 3, 6, 4.6 or 7.2 kilowatts, the current flowing at up to 50 amps.

Three Phase Level Two charging occurs at up to 22 kilowatts, a current up to 80 amps.

Level Three Charging – Fast Charging:

There are two levels of this:

The lower level is up to 80 kilowatts at 50 to 1000 volts, current up to 80 amps.

The upper level is up to 400 kw at 50 to 1000 volts, current up to 400 amps.

Fast chargers are usually found on highways and operate by direct current (DC), as opposed to level one and two chargers which work on local AC current.

Concerning level one charging at home overnight or at a business, up to 2 kw. This is very slow. To fully charge a small Nissan Leaf car battery with 40 kwh capacity takes 20 hours. More practically if recharging starts when the battery is 20% charged, and stops at 80% taking 24 kwh, it would take 12 hours, overnight. The range of this vehicle fully charged is 170 km at best, so for short daytime commutes that is fine. Except that it is adding to the household power bill in the same way as leaving on a 2 kw heater all night every day!

But for bigger vehicles with bigger batteries, say a Volvo XC40 with a 78 kwh battery doing 200 km each day using, say 51 kwh power to do so, level one charging becomes problematic. To replace 51 kwh at 2 kw charging would take 25 hours.

Level one charging has other problems. It is easy if the home has a garage or dedicated parking space where power leads can be easily run to the vehicle. But many people live in multistorey apartments and park their vehicles on the street, sometimes a distance away. To charge at home they have to run extension leads across footpaths, impractical if they are high in the building. These are unpopular with foot traffic and not secure. So for these people level one charging is not practicable.

Concerning level two charging the vehicle has to be driven to the charging outlet, then plugged in. The driver has to have the appropriate APP on the mobile phone to access the power in the charger and pay for it, then logon again when charging is completed to allow the plug to be disconnected from the vehicle so it can be driven away. The fastest level two chargers deliver 7.2 kw as said by local AC power at local voltages, say 240 volts in Australia or 110 volts in the US. So for a small Nissan Leaf needing 24 kwh, as in the example above, it would take 3 hours to deliver that. The larger Volvo would take 7 hours to receive 51 kwh. These are very long times to stay plugged in. If that was all that was available locally and there were many electric vehicles obviously there would be long queues for the charger and many upset people.

Hence Level Three Fast Chargers are considered essential if electric vehicles are to be practical for all but very short distance use. The Nissan Leaf could get its 24 kwh in 18 minutes from an 80 kw charger, and in 4 minutes from the 400 kw one – except that only a few electric vehicles can use these fast chargers. Most are not able to take more than 50 kw. If they do the batteries overheat and catch fire or explode. The vehicles have computer systems to detect battery overheating and slow the charge rate. So the fastest the Nissan Leaf can get its charge is about 30 minutes. And that is the case for most electric vehicles. Only very expensive high performance electric vehicle like the Tesla S, Jaguar I Pace, can charge at higher rates.

As a general rule level one charging can deliver enough power to travel about 7 km per hour of charging; level two – about 70 km; and level three – about 250 km.

So a fast charging level three station with six outlets could theoretically service 12 vehicles per hour if each only stayed for 20 minutes plugged in, receiving enough power for 100 km. This would allow time for connecting, disconnecting, changing to a new vehicle. But many people would stay plugged in for longer to receive a bigger charge to go further. They might go shopping, have a coffee or a meal. It is obvious a big problem with long queues and angry waiting motorists would result. Then they have to go through it all again 100 kms further on their journey.

In marked contrast a petrol or diesel station with six pumps could easily process 24 vehicles per hour, allowing 15 minutes each turnaround; and they would receive enough power to travel 500 to 700 km.

The more electric vehicles there are the worse this major problem will get.

Plug in hybrid vehicles have greater range – when running on petrol. Their range on battery is only about 50 km usually.

These problems are worse for larger electric vehicles. For example an American GMC Hummer SUV has a 212 kwh battery. If it is at 20% charge and needs 127 kwh to get to 80% that will take one and a half hours on an 80 kw fast charger. If a 400 kw fast charger could be found and the vehicle accepted that rate without overheating the battery it might take only 20 minutes. Clearly level one – taking 60 hours - and two – taking 18 hours - charging would be impractical. Large SUVs and light trucks account for more than 50% of all vehicles. When they carry large families, loads, or tow trailers, caravans and boats these problems are insurmountable as their power requirements are greatly increased.

Level three fast chargers are expensive to install and require a special three phase heavy duty power supply. Just the charging unit averages $40,000 in the US. They are already in short supply. An additional problem is that different makes have different plugs, not all electric vehicles can use them. Tesla has its own charger network in the US and only Tesla vehicles have the plug which fits. But there are moves to standardize plugs, and Tesla is under pressure to open its network to others by allowing plug adaptors.

In Australia there are 1581 level two and 291 level three DC fast chargers. In addition there are 40 Tesla 400 kw "superchargers" for that make only. In 2022 there were 83,000 electric vehicles in Australia, or about 285 per fast charger.

In the USA there were in 2022 128,000 public chargers, 28,000 of which were fast. There were 1.7 million electric vehicles, or about 60 per fast charger.

In the European Union there were 225,000 public chargers, 25,000 of which were fast. There were 5.5 million electric vehicles, or about 220 per fast charger.

This great shortage of fast chargers is only part of the storey; the **distribution** of fast chargers is far from even and very patchy. There are some places, like California and Germany, with a large number; and many other places with none. This applies to all the countries with fast chargers.

It is likely that most people with electric vehicles have small ones used for short runs, suitable for level one charging at home. In the US, for example, 92% of electric vehicles are charged at home. It is clearly difficult if not impossible to use electric vehicles even in these advanced countries for long journeys especially if they are larger and carrying big loads, given the shortage and poor distribution of fast chargers, without even considering the certainty of long queues.

A further set of problems surrounding public chargers are equipment failures. They are complicated computerized devices with sophisticated electronics which need internet access to function. A survey of 657 public chargers in the California Bay region found that 25% were not operating. Problems included:

- Non-responsive screens.
- Electronic payment failures.
- Charge initiation failure.
- Network disconnection.
- Broken electric cables.

On the Eastern Seaboard of Australia 40 fast chargers were out of action for at least several months. Delays accessing parts, which are in short supply world-wide, were part of the problem.

It is clearly very frustrating to drive up to a charger with a low battery looking forward to getting on with the journey, only to find the charger is out of action. Especially if effort has been expended finding it!

Using public level two and three fast chargers is also much more complicated than buying diesel or petrol; in Norway electric vehicle users need 15 APPs on their phones to effectively use the vehicle. This is reported as an adverse factor hurting sales as they are prone to not work. They need internet access, not available in many country remote areas.

All this is in marked contrast to our universal experience of driving into a petrol station, filling up the tank in about 3 minutes, paying by cash or card to a human being at the counter, chatting about the weather, maybe going to the loo or getting a drink, and driving out within 15 minutes. And no more fuel needed for 500 to 700 kms! Even when towing the caravan and carrying all the family.

Another issue with charging is the variable cost of electricity for level one home charging, level two public chargers and level three fast chargers. For level one home charging it is common knowledge that electricity prices are rising quickly, especially in Europe. But the price is higher to recharge at a level two public station. For example in The US the home electricity price is 10 cents per kwh, at a level two charger around 22 cent per kwh, and at level three fast charger between 55 cents and $2 per kwh. All the predictions regarding the cost of electricity say it will increase dramatically as it has in Europe; this must impact on the cost to charge an electric vehicle.

And the difference in price between home charging and the more desirable fast charging is very likely to increase following the law of supply and demand. More people with electric vehicles are going to want fast charging. There is very limited erratic supply, they are expensive, government subsidies are not likely to last forever, so it is a given that fast charging will get more expensive.

A more general issue with charging electric vehicles is that, as said before, 32% of human energy production and use occurs in vehicles and transport. Currently this energy is supplied by fossil fuels, oil largely. If a big chunk of it is to powered by electricity then electricity production is going to have to increase substantially, some estimates say by 40% to meet "nett zero" ambitions. This is a very difficult task. And the resulting increase in demand for electricity must increase the price if these forecasts occur.

A local issue for many countries concerning this need for locally provided electricity by local electric vehicles is that their energy must come for the local electricity supply; whereas currently for their petrol and diesel vehicles the energy is imported with the fuel for the vehicles. Clearly this imposes a load on the local electricity supply of those countries, which is likely to be difficult to deal with considering society's multiple electricity needs and difficulties supplying it, especially where coal power stations are no longer permitted.

This is a potential big problem in the third world where electricity is scarce and largely provided by diesel generators and coal currently. Only 0.3%, however, of vehicles there are electric.

More broadly the aim of "nett zero" proponents is that all electricity should come from renewables. The impracticality of that has been already addressed. The current reality is that electric vehicles get their electricity from the local supply of the country they are in. Usually that is supplied by coal or natural gas power stations, with a small proportion from renewables when the wind blows and the Sun shines, as discussed already. Then irony is that these vehicles are using energy made by coal and natural gas power stations largely and there is no evidence that will change in the foreseeable future. In areas of China where electric vehicles have been "mandated" by the autocratic government pollution and particulate matter in the atmosphere is much higher than areas using petrol or diesel vehicles.

A further irony is that electric vehicle take up in advanced countries has largely been small cars used for short distances suitable for home charging; these vehicles didn't use much fuel anyway when they were petrol; in contrast to the ongoing big fuel users consisting of long distance heavy load carriers which continue to be petrol and especially diesel. Globally only 0.1% of trucks are electric – and virtually all of them in China using coal power station electricity. China has a big number of electric buses, 670,000, 4% of the world's total buses, also largely using coal provided electricity. In the advanced world and third world virtually all these vehicles are diesel.

So the irony is that while proponents of "nett zero" point to increasing numbers of electric vehicles in Europe and North America, and demand more government funding for these combined with punitive regulations against the real workhorses in society – petrol and diesel vehicles – the reduction in CO_2 emissions resulting is much smaller than suggested by these numbers; especially when consideration is given to the source of the electricity used, mostly from gas and coal. Only "low-hanging fruit" is being picked, some of it arguably just a gesture by proponents.

3. Range:

Most people buying electric vehicles intend to use them for short distances daily, perhaps a short commute to work or for local city travel. As said 92% of US electric vehicle owners charge it at home, implying only short distance use. There is a good reason why the majority of electric vehicles are for short distances only and that is the *range*, distance they can travel before recharging, is far shorter than petrol or diesel.

There is a basic reason for this. Petrol and diesel contain 12 kwh of energy per kilogram of weight. So for an average car with a 50 litre fuel tank which takes 37 kilograms of petrol, the energy available is 444 kwh. But, as petrol cars have a *efficiency* generally of about 37%, meaning that percentage of the fuel energy is finally used to propel the car, the rest is lost as heat, the final energy available to shift the car is 164 kwh.

Lithium ion batteries in electric vehicles can store 80 to 200 watt hours (0.2kwh) per kilogram weight. Petrol and diesel store 60 times as much per kilogram. Electric cars are more efficient, generally 75%, meaning that 75% of the energy in the battery is finally used to propel the vehicle. So to have an equivalent amount of energy to the petrol car above with a 50 litre tank, the electric car would need to store 218 kwh of electrical energy.

Even if it is the very best type of lithium battery storing 200 watts per kilogram the battery required would weigh 1090 kilograms. A more average quality lithium battery in a cheaper vehicle storing 100 watt hours per kilogram (0.1 kwh/kg) would weigh 2180 kilograms, over two tonnes.

So the weight of the battery in electric vehicles is between 30 and 60 times as heavy as the equivalent weight of petrol for the same vehicle range ignoring that the electric vehicle is heavier because of the battery and therefore range will be less. Comparing with diesel, which is more efficient, around 41%, the equivalent battery needs to be heavier again.

It is common knowledge electric vehicles have a much shorter range than petrol or diesel. Some examples:

- MG ZS small car. The petrol version has a 48 litre tank and 800 km range; the electric version has a 50 kwh battery weighing 417 kilograms and range is about 260 km.
- Ford Focus medium car . The petrol version has a 47 litre tank and range of 700 km. The electric version has a 33.5 kwh battery weighing 417 kilograms and range of about 220 km.

- Ford F150 light truck. The petrol version has a range of 880 km. The electric version has a 98 kwh battery weighing 1100 kilograms, the size of two double mattresses, and range of about 360 kilometers.

It is also well known that the load a vehicle is carrying and how vigorously and fast it is driven impact on fuel consumption markedly. The same is true for electric vehicles. It is clear that and electric light truck which may carry very heavy loads or tow caravans may have a very short range, perhaps less than 200 km, especially in hilly country.

4. Cold

Cold temperatures also impact on electric vehicle range; and very cold temperatures, below freezing may stop the vehicle and prevent charging. Lithium batteries should be charged at temperatures between 5°C and 45°C. If the temperature is less than 5°C the charging rate will be very slow indeed. If it is less than 0°C, a very common event in many places, the battery will be irreversible damaged by build-up of metallic lithium on the anodes. This has to be an important consideration for those living in regions with very cold winters.

5. Fire and Explosion

The cells of lithium batteries contain highly inflammable liquids in a sealed aluminium container. During charging and discharging heat is produced. In addition oxygen can be formed at the cathodes under certain conditions. If a cell is faulty, old or is damaged "thermal runaway" or increasingly high temperatures may result in the cell fluid catching fire, fuelled by the oxygen, and perhaps exploding. A large amount of energy is then released suddenly, 5 to 10 megajoules per kg of battery. This is more than explosive TNT has, 4.6 megajoules per kilogram.

If this happens to a cell usually spread occurs to adjacent cells and the whole battery is in danger of exploding with huge force, up to double that of TNT.

Causes of this include:

- Overcharging, high rate charging, recharging a battery which has been over-discharged, charging in cold temperatures.

- External heating or fires.

- Electrical short circuits either internal in the battery or external in circuits supplying the battery.

- Physical damage.

- Corrosion, for example from immersion in sea water.

- Old batteries with deterioration.

Because of the risk of fire and explosion due to overheating during charging electric vehicle batteries have cooling systems, often electric refrigeration, to prevent this; and computer systems to limit the rate of charging if the battery gets too hot. And similar systems control the rate of discharge during vehicle use. But rarely these fail.

A fire near an electric vehicle risks the battery catching fire and exploding. This is important if a house or building with vehicles stored in it catches fire.

Most electric vehicle batteries are part of the floor; they are vulnerable to damage from "bottoming" on speed bumps or obstacles, rocks on the road. Damaged parts are then vulnerable to "thermal runaway" and battery fire.

Vehicles involved in accidents may suffer battery damage risking fire and explosion. Passengers have been killed by these outcomes. Rescue crews are now warned about this, considering the explosive force possible, and usually keep everyone well away and isolate the vehicle. Disconnecting the battery's 12 volt supply from the vehicle lead acid battery helps shut it down. Plenty of cold water cools it hopefully.

If an electric vehicle is immersed in seawater corrosion may damage the battery and cause later battery fires; this may be up to 3 weeks after being flooded. If the vehicle is stored in a house garage then the house may go up. This happened to eleven houses in Florida after Hurricane Ian there. Florida authorities now warn people about this and recommend removing the batteries and immersing them in water to cool them and decrease the risk. Many consider seawater immersion is not safe.

Tricky features of these battery fires and explosions are that they may occur sometime after the damaging event and catch everyone unawares; they tend to reignite after they have apparently been put out, often days later; the smoke is highly toxic.

To put the risk in perspective one study suggested there are 25 fires per 100,000 electric vehicles per year; versus 1500 for petrol vehicles. But battery fires are very tricky, concerning delay and power, and difficult to extinguish.

It is not just electric vehicle lithium batteries which are subject to this risk; ***all*** lithium batteries are vulnerable. Some examples:

Aviation Fires and Explosions:

In 2007 Boeing started using lithium ion batteries to power aircraft systems in its 787 aircraft. But there were two serious fire incidents in the first 52,000 hours of use and the fleet was grounded pending FAA investigation. Lithium batteries were permitted finally with many precautions, including being sealed in special compartments away from vulnerable areas, overboard vents to get rid of toxic fumes in the event of a fire, adequate backups.

Airbus also started using lithium batteries about the same time but reverted to other types. They now use lithium also with similar precautions, as do other manufacturers.

There have been many lithium battery fires in cargo carried by aircraft including:

- Asiana Flight 991 in 2011 developed a fire in the cargo hold in lithium batteries being carried there. The aircraft crashed.
- A Japan Airlines 787 developed a cargo hold fire in lithium batteries shortly after landing in Boston in 2013.

- On 28th January 2011 a B744 aircraft over the China Sea developed a fire in the cargo hold in lithium batteries; it crashed and broke apart.

- On 16th January 2013 a B788 aircraft experienced a short circuit in a lithium battery which caught fire and made an emergency landing.

- UPS Flight 1307 developed a fire in the cargo hold in lithium batteries and crashed at Philadelphia killing two crew.

In a report in 2015 the American FAA noted there had been 158 "incidents" of fire in lithium batteries carried in cargo holds. Precautions were recommended included density of stacking, venting, containment to prevent damage. But continuing risk was acknowledged despite precautions and the explosive power of lithium batteries emphasised.

There have also been many fires in passenger cabins in lithium batteries in phones, laptops, other devices carried by passengers. One report documented 322 such fires between 2000 and 2021 world-wide, including 39 in 2020. Incredibly there may be up to 500 lithium batteries in the cabin per 100 passengers. These fires are difficult to put out, emit toxic smoke, keep flaring up and may reach 900°C. They used to result in aircraft being diverted for emergency landings incurring an extra cost of at least US$400,000; but since 2017 "AvSax" bags have been carried by many aircraft. These were invented by a British company, Environment Defence Systems Ltd, and more than 15,000 have been sold to 80 airlines. They contain a polymer which expands when two litres of water is added to surround and cool the offending device. They have been used 37 times since 2017.

Rubbish Dumps Fires and Explosion:

As discussed virtually all old lithium batteries end up in landfills. There they may catch fire due to internal deterioration as explained. The resulting fires can be difficult or impossible to extinguish, emit and leach toxins, be deep in the landfill.

Grid Storage Batteries Fire and Explosion:

There are now many large grid storage lithium batteries in several countries including especially South Korea, China, USA, and Australia. They include large batteries in homes to store electricity generated by solar in daytime for use at night or cloudy days, as well as larger municipal batteries. There have been serious fires; some examples:

In South Korea between 2017 and 2019 there were 28 difficult fires in large storage batteries. As a result 35% of South Korea's storage batteries were shut done.

In China a 25 megawatt hours large storage battery exploded in fire and two firefighters were killed.

In Arizona on 19th April 2019 a 2 megawatt hour facility on fire exploded severely injuring 8 firefighters.

The fires have mostly been in nickel cobalt manganese type lithium batteries which are popular as they perform well. There is a trend to Lithium Iron Phosphate types which don't perform as well but are safer at higher temperatures.

6. Electrocution

A usual operating voltage for electric vehicles is 300 volts. In addition they are direct current (DC) not alternating (AC) like household appliances, which is safer. The 300 volt DC cables in electric vehicles are orange. If these are exposed in an accident there is a risk of electrocution.

As discussed above voltages during fast charging, which is DC, can be very high, up to 1000 volts. In the event of a fault there is an electrocution risk. Fortunately these are rare.

7. Accidents

Drivers new to electric vehicles are often surprised by the rapid acceleration from a standing start discussed already. This has caused many accidents, crashing into obstacles, other vehicles or people. This type of accident is reported to be 50% more common for electric vehicles.

Electric vehicles are very quiet, clearly a big advantage in cities and generally. But pedestrians may not be aware of them and be struck. Accidents with pedestrians are reported as being 20% more common than other vehicles. Some authorities regulate compulsory noise below 20 km/hour.

8. Battery Replacement Cost

New electric vehicle manufacturers usually cover their batteries under warranty for 7 to 8 years and mileages from 100,000 to 240,000 kilometers. Volvo covers 8 years and unlimited mileage.

Most consider the battery should last about 10 years, but that is going to depend on how it is used. As discussed already lithium batteries usually last about 2000 charge – discharge cycles, so frequent use will shorten battery life. Also use of fast chargers more often as opposed to slow home trickle chargers causes more heat and degradation in the batteries.

Manufacturers are generally not forthcoming about the cost of replacing the battery out of warranty, preferring understandably to assure new vehicle purchasers they will not have to as they will usually replace the vehicle before that.

Predictions used to say the cost of batteries would come down, but in fact they are rising following the law of supply and demand. They are now over US$100 per kwh. Then there is the cost of labour and parts to install the replacement.

In the US experience of replacement costs including the battery and installation shows that:

- Small Nissan Leaf: $3000 - $9000.
- Tesla 3: $15,000.
- Tesla S: $20,000.
- VW eGolf: $23,000.

For older vehicles with higher mileage costs increase; for example a 12 year old Chevy Volt small vehicle with 70,000 miles cost $29,000 for battery replacement.

In Australia costs of A$12,000 to A$20,000 are quoted for battery replacement for small vehicles with less than 50 kwh batteries, and A$50,000 plus for prestige longer range vehicles – Tesla S, Jaguar and others – with about 100 kwh batteries.

This is a major consideration for electric vehicle purchasers, especially second hand vehicles. A 10 year old vehicle usually may have travelled about 150 – 200,000 kilometers. Or much less for a city vehicle. Modern petrol and diesel vehicles will usually have much more life to go before major maintenance is required. But the electric vehicle at that age is facing this big battery replacement cost.

9. Obsolescence and Disposal:

An aspect of this is older electric vehicles becoming obsolete because they are superseded by newer models with bigger better batteries and greater range. It is then not worthwhile spending large sums to replace batteries as the second hand value of the vehicles is very low. In China there are large fields full of abandoned old electric vehicles. The government there ordered vehicle manufacturers to make large quantities of cheap electric vehicles and forced their purchase by subsidies, and penalties on competing fossil fuel vehicles. A similar problem is emerging in the western democracies because of poor resale value of old electric vehicles.

Old or damaged petrol and diesel vehicles are often crushed to small cubes and their metals retrieved and recycled, a relatively cheap process explaining why there are not huge areas of dumped old vehicles. But electric vehicles cannot be crushed, that would result in huge fires and explosions from the lithium batteries. And removing the batteries is an expensive time consuming non-economic process. So the old electric vehicles get parked in fields and left to be someone else's problem – or to catch fire and explode emitting toxic fumes!

In summary anyone looking at all these problems would be inclined to think that at the very least the jury is still out on the future of electric vehicles compared to diesel and petrol ones. Especially as their market share is heavily dependent on government support in the democracies, in turn dependent on taxpayers continuing to tolerate subsidies. And that the heavy lifting in transport, earthworks, shipping, construction and industry is done by diesel vehicles and will not change for the above reasons. While they have a following in big cities, uptake in country areas where greater distances are travelled is much less.

Grid Storage Batteries

These are batteries which complement wind and solar renewable electricity sources which only supply when the wind is blowing or the Sun shining. They aim to continue supply when the renewables cannot.

Home Storage Batteries have an increasing following. They may either be ***lead acid*** batteries or ***Lithium.***

Lithium batteries are most popular because they last longer, up to 2000 cycles, usually 10 to 15 years. Most are warranted for 10 years. They can be discharged to 20% of their capacity, meaning a 10 kwh battery could be discharged without damage to 2 kwh, 8 kwh used. They require less maintenance and are more efficient, releasing about 70% of the electricity put in. But they are more expensive. Often up to $2000 US or Australia per kwh installed is experienced, say $20,000 for a 10 kwh battery. This is about double the cost of lead acid.

But cheaper lead acid will be lucky to last 500 cycles, 5 to 7 years. It needs more maintenance. And it can only be safely discharged without damage 30 to 50%, so a 10 kwh battery bank could deliver 5 kwh electricity for use only. And they are about 50% efficient.

Home storage lithium batteries have all the risks discussed for electric vehicle batteries including fire; recently a large manufacturer did a world-wide recall of 10 kwh batteries for this reason. Lead acid does not have this risk.

And old lithium batteries are hard to recycle as discussed; most end up toxic in landfills. Lead acid is largely recycled as discussed.

Many governments are pushing home storage with subsidies. That and increasing power prices are encouraging home owners to install more. The marketing push is in a "honeymoon phase" and little is said about future disposal problems if big numbers are installed, nor the huge energy cost of making the batteries, mostly in China using coal fired power station electricity as discussed, nor big future costs of replacement.

Generally homes are advised to have battery capacity of about half their daily usage. In the US, Canada and Northern Europe this is about 30 kwh per day, in Australia 20, Southern Europe and Britain 10, the third world much less. So in wealthy countries 10 to 15 kwh battery banks are advised, costing up to $30,000 US and Australian.

The presence of an electric vehicle needing to be charged at home makes a big difference to the battery capacity required; for example if it is a small vehicle with a 50 kwh or less battery, that still means a much bigger home battery to recharge it at night when the solar would not work – if it is not going to be recharged from the normal home power supply from the grid. A larger vehicle doing a higher daily mileage with a 100 kwh battery obviously poses big costly problems recharging at home from a home battery at night, generally impractical.

Large Community Grid Storage Batteries are widely promoted as being part of the shift to renewable electricity. There are a number of different types of large grid storage batteries:

The most common are ***Lithium ion batteries.*** Some examples of these in various countries are:

The largest US grid storage battery is at Vistra Moss Landing, California where the battery has 1600 megawatt hours capacity (1.6 gigawatt hours or 1.6 million kilowatt hours).

Other big installations in the US include the 250 megawatt Gateway, California; the Manalee Energy Storage Facility 409 megawatts in Florida.

Currently the US has 9.2 gigawatts or 34 gigawatt hours installed; it aims to have 30 gigawatts or 111 gigawatt hours by 2030. A large number of projects are planned.

South Korea has many lithium storage batteries installed and there were many fires from 2017 as discussed resulting in shutdowns and new regulations.

China has 8.7 million kilowatts (8.7 gigawatts) of lithium battery storage, giving about 18 million kilowatt hours energy (2.1 hours). There are plans to increase this greatly. But most of China's energy storage is in pumped hydro facilities. There are 23 of these in operation and plans to build 31 more. This will bring China's total energy storage from 45 gigawatts to 65.

Europe has 2.25 gigawatts installed giving 4.45 gigawatt hours at present . There are plans to increase this 20 fold by 2031 to 45 gigawatts, 89 gigawatt hours. Europe's largest battery is at Wollonia, Belgium, a 100 megawatt facility.

Australia has a big battery installed in Victoria, 300 megawatts giving 450 megawatt hours; it is called the "Victorian Big Battery". There is another big one at Wollgrove, South Australia, 50 megawatts, 75 megawatt hours. Another notable one is at Bulgarie Green Energy Hub in Victoria, 20 megawatts, 34 megawatt hours. Australia has over 1000 megawatt hours of grid storage under construction and is aiming to reach 50 gigawatt hours (50,000 megawatt hours) capacity eventually.

But the largest method of energy storage world-wide is pumped hydroelectricity, discussed above. In 2020 850 gigawatt hours (850 million kwh) was stored by pumped hydro and only 10% by other methods including batteries making a total of about 930 gigawatt hours stored.

The stated aim is to expand this very greatly to complement wind and solar allowing the combination of wind, solar and storage including batteries to replace fossil fuel electricity, which currently provides 76% of global electricity.

Total global electricity is 24,000 terawatt hours, which is 24 million gigawatt hours. If one very optimistically argued that wind and solar could expand to supply half of that when the wind was blowing and the Sun shining, that would leave 12 million gigawatt hours to be supplied by storage, providing world electricity consumption did not increase. If that was to be done by batteries then assuming an efficiency of the batteries of 70%, and 80% discharge as discussed above, 21 million gigawatt hours of battery would need to be installed. That is at least 250,000 times the amount of current battery storage and is clearly impossible. Even a much more modest contribution by batteries presents insurmountable difficulties

And the greater stated aim is to move to electric vehicles powered by renewable energy, wind and solar largely, with battery storage for night and no wind times. Global energy for transportation annually is about 40,000 terawatt hours, 40 million gigawatt hours. To have battery back-up for this following the same reasoning would require 35 million gigawatt hours battery storage, 400,000 times the present battery storage. Again even a modest contribution by this technology presents insurmountable problems.

The 40,000 terawatt hours of energy for global transportation is currently supplied by: diesel 36%; petrol 36%; fuel oil 9%; aviation fuel 9%; natural gas 6%. The only renewable source is liquid biofuel, 4%. Electricity is less than 1%.

Enthusiasts for electric vehicles argue that the total transport energy would be less because electric vehicles are more efficient than petrol or diesel: about 75% versus 35 – 40%. But this ignores the reality that batteries themselves are about 70 – 80% efficient, there are transmission losses around 5%, then delivery losses due to inverters and other technical issues, so at the end of the day the energy requirements are very likely to be similar.

A feature to notice about grid storage batteries is **how long they can supply the stated electricity wattage (power) for.** Often this is only an hour or two. For example the South Australian battery mentioned above can supply 50 megawatts for one and a half hours. The US current total of 9.2 gigawatts can only go for just over 3 hours before the 34 gigawatt hours has gone. The US uses about 4 million gigawatt hours electricity per year (3900 terawatt hours). Meaning that the current US grid storage would be gone after 5 minutes! Even if the US achieves its aim to expand to 111 gigawatt hours storage that would be gone after 14 minutes.

And in Australia the Victorian "Big Battery" with 450 megawatt hours (0.45 gigawatt hours), storage would supply Victoria, which has an annual electricity consumption of 43,000 gigawatt hours, for 5 minutes. Similarly the South Australian 75 megawatt hour battery (0.075 gigawatt hours) would be exhausted after 3 minutes given that state's annual electricity usage of 12,000 gigawatt hours.

These considerations are rarely aired publicly. Rather emphasis is placed on the rate of growth of battery capacity in countries, that it is the future, that there is no alternative given the "climate crisis". Many billions of dollars are going into grid storage batteries. The same concerns already discussed regarding manufacture of lithium electric vehicle batteries apply even more to these much larger items – the huge energy requirements, the limited global supply of raw materials, the massive disposal problems when they are old – their life is usually 10 to 15 years. But the basic impracticality of the aim for them to replace existing electricity production is paramount.

Flow Batteries

These batteries are large tanks full of electrolyte solutions with cathodes and anodes usually made of carbon. The brilliant concept was pioneered by Charles Bradley who patented a Zinc Bromide battery in 1885. More brilliant work was done by Maria Skylles-Kazucos of the University of New South Wales, Australia in the 1980s on Vanadium batteries and patented. These now account for more than 50% of installed flow batteries world-wide. There are more 78 different types of flow batteries being investigated using many different electrolytes in solutions of water.

Her battery was a **Vanadium Redox Battery.** This has separate solutions of vanadium salts and vanadium oxide salts in different tanks. The solutions are pumped into a tank where a reaction takes place producing an electric voltage of about 1.2 volts. Many tanks can be wired in series to produce higher voltages. These can be used to produce a working electric current like other batteries. And the system can be recharged like other batteries by applying a stronger voltage to the electrodes.

Vanadium Redox batteries do not store as much electricity as lithium-ion; they store about 20 watt hours per kilogram weight versus up to 200 for lithium. They are also not as efficient, about 65%, meaning they release only 65% of the electricity used to charge them. But they are cheaper to make, may last for 10,000 cycles versus 1000 to 2000 for lithium, can work at high temperatures compared to not more than about 40° C for lithium, and cannot catch fire. They can be made very large to store a lot of electricity, about 230 megawatt hours per acre size.

A number are being built around the world in China, the US, Australia and other countries. A 400 megawatt Australian one is at Maryborough in Queensland. A US company is building five 50 megawatt hour units. There is a big one in Japan storing 60 megawatt hours. And in China an 800 megawatt hour battery is being built.

Vanadium redox batteries have been much slower to catch on than initially anticipated. A main reason for that is uncertainties about the future price of vanadium which makes investors hesitate. Plus the competition from lithium.

Another type of flow battery is the *Iron Redox Flow Battery.* This is similar in principle to the vanadium one but uses iron (ferrous and ferric) salts instead of vanadium ones. The iron salts are in a chloride solution and are pumped into a chamber where the reaction takes place producing about 1.2 volts also. Many cells can be connected in series to produce larger voltages, and the battery can be recharged. It can also work at a wider range of temperatures than lithium, minus 5°C to 60°C. It is a little more efficient than vanadium, 70%. It also can last for 10,000 cycles or more. It has other problems including air causing iron oxide formation like rust, excess hydrogen formation at the cathode leading to fire risk, and others. It is at an earlier stage of development to vanadium types and there are fewer installed. It does not suffer from the uncertain high price issue which affects vanadium. But iron flow batteries are rare still compared to Lithium, the market size was just $US 2.5 million in 2020. There is a lot in interest, however, and this is projected to grow to $US 15 million by 2028, still tiny against lithium.

ESS Inc is a prominent company manufacturing iron flow batteries and is based in Oregon, USA. It is expanding rapidly, including into Europe. It markets a 600 kwh battery in a shipping container amongst other products. The container weighs 38 tonnes including 22 tonnes of electrolyte solution. Multiple containers can be incorporated in an area storing 74 megawatt hours per acre. A Spanish solar company is using these 17 of these to store 8.5 megawatt hours. ESS Inc reports huge customer interest and Bill Gates is an investor, so future growth appears certain.

In general flow batteries do not suffer the huge disposal and recycling problems of lithium, especially iron flow batteries which are easily recycled. Vanadium batteries also can be maintained and recycled much more easily than lithium, potentially a massive advantage.

But there is no escaping the futility of trying to replace current energy supplies with batteries as explained. Flow batteries can usually last longer delivering power than lithium ones, for example the 600 kwh ESS Inc battery in a shipping container discussed above can deliver 90 kilowatts for about 6 hours. But as for combining with renewables wind and solar to provide *all* the world's electricity, 24,000 terawatt hours annually, the arguments are the same as discussed above for lithium: about 34 million gigawatt hours (34,000 terawatt hours) of battery storage would be required, assuming 70% efficiency. If iron flow batteries were used occupying land at a rate of 74 megawatt hours per acre one million square kilometers would be needed – equal to the whole of the UK plus Spain covered in batteries; or over 10% of the USA. Or if vanadium batteries could be used, assuming enough of that comparatively scarce metal could be found and financed at its high price, about 220,000 square kilometers would be needed at the usual 384 megawatt hours per acre, the whole of the UK or New Zealand.

And if the energy involved in transportation is added, another 40 million gigawatt hours (40,000 terawatt hours) annually, the total battery requirement following the arguments in the above discussion for lithium would be another 56 million gigawatt hours again assuming 70% efficiency, making a total requirement of 90 million gigawatt hours. For iron flow batteries three million square kilometers would be needed, virtually all of Europe, or one third of the USA. If vanadium, about 800,000 square kilometers, all of France or Texas.

And these arguments are optimistic in favour of batteries in that no account is taken of distribution losses, or losses from converting DC to AC, for example. These could easily add a further quarter to the requirements.

Clearly this is a ridiculous hypothesis but it underlines the complete impracticality of aiming to supply the world's energy needs by batteries. The land area needed is only a small part of the story, the industry and materials required would be gigantic. Yet this is a key plank for those pushing renewables wind and solar to replace current energy sources; because the wind often isn't there and the Sun only shines part of the time batteries are essential to the aim. Pumped hydro cannot do it as discussed already. But despite reality governments and vested interest groups successfully grasp billions and billions of dollars world-wide dangling the impossible dream before the people.

Batteries are useful for short term backup in special locations; but not to largely replace current energy sources.

Hydrogen

This is promoted vigorously as an ideal replacement for fossil fuels for electricity generation and especially for vehicles, ships and aircraft. When "burnt" it combines with oxygen to make water vapor, H_2O, releasing heat energy in the process like burning hydrocarbons, fossil fuels in air containing oxygen. But no CO_2 is produced.

In Fuel Cells hydrogen and oxygen are mixed in a solution and the reaction to produce water results in electricity production with a voltage of 0.7 volts per cell. The cells can be "stacked" in series to produce higher voltages. Electric vehicles powered by hydrogen/oxygen fuel cells are available in many countries including Europe, China and California especially. This is a more efficient use of hydrogen that in an internal combustion engine, 40 - 60% compared to 30%. **Efficiency** here means the proportion of energy generated by the fuel cell reaction for actual propulsion of the vehicle. But fuel cells are expensive as is hydrogen, refuelling station are scarce so the technology is at a very early stage practically.

That water vapour produced by hydrogen use is a much more powerful "greenhouse gas" than CO_2 is not mentioned in these polite circles!

Hydrogen Production:

There are some big problems with hydrogen for these purposes. These include that it is very hard to make requiring a great deal of energy. Hydrogen does not occur naturally on Earth because it is a very light gas which escapes quickly to space, and because it readily combines with oxygen to make water. To reverse this process and make hydrogen from water is therefore expensive. Especially if the huge quantities needed to replace fossil fuels are contemplated. There are three main methods currently in general use:

Steaming Methane: at over 700° C and in a pressure vessel at about 20 Bar (20,000 hectopascals, 20 X atmospheric sea level pressure) and also with a nickel catalyst. Steam and methane react to form hydrogen gas and carbon dioxide – $CH_4 + H_2O$ goes to $CO_2 + H_2$. A lot of energy is required and also methane. This the most common way of making hydrogen for industrial use, over 50% is made this way. This called *grey hydrogen* because fossil fuels are used both as the raw material – methane – and to make the electricity required, and a lot of CO2 is released. If the CO2 is *captured*, prevented from escaping to the atmosphere, it is called *blue hydrogen*. A lot more energy is required to achieve this.

Methane Pyrolyis: here methane is bubbled over a catalyst at 1000°C to 1900°C and breaks into hydrogen gas and solid carbon (CH_4 goes to H_2 and C). It requires a lot of energy from fossil fuel use and so still releases large amounts of CO2. There are several other methods using coal and fossil fuels. These account for about 46% of global hydrogen production.

Splitting water into hydrogen and oxygen: this requires electricity to pass through water and usually a catalyst. Only 4% of hydrogen is currently made this way. There are two main types of equipment used:

- Polymer Electrolyte Membrane (PEM) is most popular for using renewable energy from wind or solar because it works at less than 100° C and at varying input electricity voltages. But the equipment is expensive as is the platinum catalyst required. If renewable electricity is used it is called *green hydrogen.* Less than 1% of global hydrogen is made currently using renewable energy.

- Alkaline Electrolyte Cell (AEC) equipment is cheaper including a nickel catalyst but it needs 200° C plus and a very strong alkali, so is not suitable for use with renewables.

A lot of electricity is required. To make one kilogram of hydrogen 50 kilowatt hours are needed. Then if it going to used in vehicles it needs to be compressed using at least another 15 kilowatt hours. As discussed below when one kilogram is burnt or used for energy 32 kilowatt hours are produced, about half the energy needed to make it. So using hydrogen involves a lot of energy input. And the energy output is half the input making it very wasteful and expensive compared to fossil fuels. And also compared to using the electricity for other purposes than hydrogen production, given only half the energy put in is finally available when the hydrogen is used.

If the hydrogen is to be used in vehicles the final efficiency, that is the proportion of the energy finally used to move the vehicle compared to that needed to make and deliver hydrogen, is less than 15% for conventional internal combustion engine vehicles and less than 30% for fuel cell vehicles. If the intention is to use just renewable electricity to make hydrogen by electrolysis of water for vehicles then between 3 and 6 times the amount of energy used by the vehicles to move is required from renewable sources such as wind and solar.

Currently in 2021 94 million tonnes of hydrogen was made world-wide. More than 99% was made using fossil fuels for the energy input, less than 1% was made using renewable energy. Hydrogen is used widely in industry:

- Oil refining: 39.8 million tonnes.

- Ammonia production for industry use: 33.8 million tonnes.

- Methanol production for industry: 14.6 million tonnes.
- Steel production: 5.2 million tonnes.
- All other uses including vehicles: less than a million tonnes.

Hydrogen production for use in industry is projected to increase to 180 million tonnes by 2030. Hydrogen production from renewable electricity, wind and solar, is increasing by about 9% per year as this expands but is still only a very small proportion of total hydrogen production as said. Some predictions aim to produce 30 million tonnes from renewables by 2030 - green hydrogen – and 30 million tonnes from fossil fuel use with carbon capture – blue hydrogen. But that would involve a huge expansion in renewables and carbon capture.

Hydrogen Storage:

Liquid hydrogen has a very low boiling point of minus 252° C (20° Kelvin). Further its **critical temperature** – which is the maximum temperature a liquid can exist even with very high pressures – is 33° Kelvin, minus 239° C. So to get liquid hydrogen it has to be cooled to usually minus 253° C. That is a demanding energy intensive process involving the use of liquid nitrogen, and between 6 and 10 kilowatt hours of electricity per kilogram hydrogen liquified. Then the liquid hydrogen has to be stored in very strong very insulated containers at this very low temperature. Finally when the hydrogen is used it has to be carefully turned into gas again. This is all very different to the ease of storing and using fossil fuels. Currently globally 355 tonnes per day are liquified.

Hydrogen is often stored as a gas. If it is going to be used in vehicles it needs to be in very strong high pressure containers to enable a useful volume to be stored because at low pressure only a very small amount of hydrogen is storable in a big volume – one kilogram at sea level atmospheric pressure occupies 11 cubic meters. These high pressure tanks are heavy and dangerous if damaged, say in an accident. The pressures used are between 100 Bar (one Bar is the atmospheric pressure at sea level) and 700 Bar, very high. In hydrogen vehicles 700 Bar is often used.

Hydrogen can be stored at lower pressures in underground caverns left over from oil extraction especially if they are lined with salt which helps prevent leakage. Storage in long steel pipes at low pressure is another method. But the hydrogen causes the steel to deteriorate as a complication.

Another method of storage attracting attention is as **Ammonia, NH3**. Ammonia is commonly made for use in fertilizers often by electrolysis of water at 300° C and high pressure, 200 Bar. It can be stored as a liquid at high pressure or dissolved in water. About 200 million tonnes is made globally each year. When it is heated to about 1000° C it breaks up into nitrogen and hydrogen gases. The nitrogen can be separated making the hydrogen available. The whole process requires much energy and care.

When "burnt" in an engine or used in a fuel cell hydrogen releases 32 kilowatt hours of heat energy per kilogram. This compares well with other fuels:

Diesel   12   kwh/kg

Petrol   12

Propane   13

Methane   14

Hydrogen   32

But:

This has to be looked at comparing *a litre* of each fuel to be practical – and then the picture changes:

Diesel	12	kwh/litre
Petrol	9	
Jet Aviation Fuel	8	
Propane	6	
Methane	6	
Liquid hydrogen (minus 253° C)	2.5	
Hydrogen gas at 700 Bar	1.6	
Hydrogen gas at 350 Bar	0.9	

This means that a vehicle using hydrogen gas at 700 Bar, a common pressure for hydrogen vehicles, instead of diesel needs to store 7.5 times as much volume in its fuel tank for the same available energy. Instead of, say 50 litres diesel for a car, it would need 375 litres hydrogen. A petrol vehicle would need a hydrogen tank of 280 litres to replace 50 litres. A Toyota Corolla, which has a 50 litre fuel tank and a 361 litre luggage capacity would need virtually all this occupied by the hydrogen tank for the same vehicle range capability. A Land Rover Discovery SUV with an 82 litre diesel fuel tank would need a 615 litre 700 Bar hydrogen tank, more than half its luggage capacity of 1124 litres with the third seat row folded flat. And a truck, which commonly carry 1000 litres of diesel, would need 7500 litres in 700 Bar hydrogen tanks.

Currently hydrogen small cars often have 6 kilograms of compressed hydrogen gas at 700 Bar in a 150 litre tank; this gives them 240 kilowatt hours of energy to use. Compared to 450 kilowatt hours from a common 50 litre petrol tank. This means the range and/or load carrying ability is about half.

Current hydrogen larger trucks and buses carry about 100 kilograms of hydrogen gas in a 2500 litre (2.5 cubic metres) high pressure tank. This gives them 3200 kilowatt hours of energy. Commonly these vehicles carry 1000 litres (one cubic metre) of diesel giving them 12,000 kilowatt hours. Again range and/or load carrying ability are affected.

Hydrogen powered aircraft have been proposed. A Boeing 737 small passenger jet carries 20,000 litres of aviation fuel and would need 700 Bar hydrogen tanks totalling 160,000 litres for the same range. That would occupy 160 cubic meters of space (a cubic meter is one meter by one meter deep by one meter high). The volume of the fuselage of a 737 is about 400 cubic meters so a third of that would be high pressure heavy hydrogen fuel tanks.

A large long-haul passenger jet like the Airbus A380 carries 300,000 litres of aviation fuel, so would need 2.4 million litres of compressed hydrogen in 700 Bar tanks. That is 2400 cubic meters. The total volume of the fuselage of an A380 is about 1500 cubic meters. So hydrogen powered long-haul aircraft would need to be much larger than those powered by fossil fuels. That in turn means they would need more hydrogen fuel, a "catch 22" situation.

It could be argued that the *weight* of the hydrogen required would be just less than half the weight of equivalent aviation fuel for the same energy, but when the weight of the high pressure tanks and other equipment required for hydrogen is considered the advantage lessens. Then there are risks with carrying big volumes of hydrogen in very high pressure tanks. Aircraft are built to last for many years which means the tanks would be cycled from very high pressure to low pressures many times, increasing risks.

And the above estimates of the volume of tanks do not take into account the extra volume due to the very high pressures in the equipment. Not only are tank walls much thicker than petrol, diesel of aviation fuel tanks but also it is likely several smaller tube or ball shaped tanks would be used as opposed to large rectangular tanks; so the space occupied by these would be greater.

Even if liquified hydrogen at minus 253°C is able to be stored in, say, a ship, five times the volume of fuel would need to be there for the same energy output. It is very unlikely liquified hydrogen at minus 253°C would be practical or safe for a commercial aircraft given the complexities and weight of the required equipment, including that to turn the liquid hydrogen back to gas again for use.

Hydrogen Delivery and Fuel Stations:

In 2021 there were 400 hydrogen vehicle refuelling stations world-wide servicing 30,000 hydrogen fuel cell vehicles including trucks, buses, cars. In Germany there are 81 stations for 530 fuel cell vehicles. Korea has 30 stations for 4000 vehicles. Japan has 110 stations. China has 8000 fuel cell vehicles including 2000 buses and 1500 trucks. Switzerland has 50 fuel cell trucks.

A common factor for all these places is **very high government tax and other subsidies** to make hydrogen compete economically with fossil fuels which otherwise it could not do. For example in Switzerland vehicles pay $1 per kilometre travelled but hydrogen vehicles pay nothing. In Japan, Korea, China, the U.S. there are big subsidies. These counteract the reality that hydrogen costs much more than fossil fuels. For example US$12 per kilogram is commonly stated as the true cost of compressed hydrogen delivered to a vehicle versus US$2 for diesel. Even if it is assumed the more energy is available from the hydrogen it is still at least double the cost of diesel. Hence the government subsidies.

In addition some governments are proposing to reduce the number of diesel and petrol stations to increase the incentives to use hydrogen.

Hydrogen is delivered to stations by pipelines, or, more commonly, trucks. If by pipeline it is at 20 Bar pressure. If by truck, at 200 to 500 Bar. The trucks contain large tubes with the hydrogen gas at those pressures.

At the stations the hydrogen is initially compressed to 950 Bar (950 times sea level atmospheric pressure). This causes the hydrogen gas to be heated. It is then kept in a high pressure tank. Before it can be safely delivered to a vehicle it has to be cooled to minus 40°C. It can then be delivered to the vehicle with a special hose and connector at 700 Bar.

The vehicle tanks have special safety systems to vent hydrogen gas and reduce excess pressure if they heat up; in open air this dispenses quickly but in closed spaces like garages it can accumulate and become an explosion risk; only a small amount of hydrogen in air can explode with a small spark.

Less commonly liquified hydrogen at minus 253°C is delivered in special trucks. It then has to be changed to high pressure hydrogen gas at 900 Bar by special pumps and equipment. Again it has to be cooled to minus 40°C before going into a vehicle at 700 Bar.

Hydrogen vehicles have to have special equipment to reduce the pressure from 700 Bar to use in the engine; commonly it is dropped to 10 Bar, occasionally as high as 300 Bar for special engines.

These factors mean that hydrogen fuel stations are much bigger, more complex and expensive than diesel and petrol stations. And because they store explosive hydrogen gas at very high pressures there are more government regulations concerning them. Nearby residents are more concerned. So they are much more expensive to build, take up more space, more construction time, and more time for government approval.

Another problem is the ***delivery capacity per day*** of hydrogen fuel stations. A usual capacity is 600 kilograms per day. The biggest in China is 2000 kilograms per day. Small fuel cell cars often have a 6 kilogram tank, so 100 cars per day could be refuelled at a normal station. But large trucks and buses may carry 100 kilograms so after just 6 trucks the station is empty. Small trucks commonly have 20 kilograms, so after 30 of these the station is empty. These are very small numbers compared to diesel and petrol stations. This impacts not only supply issues but also the earning capacity of these expensive complex stations.

And the decreased range of these vehicles means they have to refuel far more often than petrol and diesel ones; small cars two or three times as often and large trucks and buses 15 times as often – unless they have much bigger tanks.

In addition the ***engines*** using hydrogen have problems: fuel cell vehicles are very expensive, and to use that technology means a huge re-tooling of the vehicle manufacturing and support industries. Conventional hydrogen internal combustion engines with cylinders and pistons like ordinary diesel and petrol cars and trucks are much cheaper than fuel cell, their construction and use fit into existing systems more easily. But they are very special with increased engineering to deal with problems including pre-ignition, flame back and explosion, crankcase explosions, increased wear due to water from the burnt hydrogen getting into lubricant oil, and others. So they are at least 50% more expensive than a diesel of petrol engine.

Hydrogen fuel cell vehicles are about twice the cost of electric vehicles, which in turn are about twice the cost of conventional petrol or diesel vehicles purchased new.

Considering all this it is understandable that the sale of hydrogen fuel cell vehicles has slowed and some major truck manufacturers, such as Scania of Sweden, have stopped further development of hydrogen trucks.

Looking more broadly at the question: could hydrogen replace fossil fuels for all global transportation including cars, trucks, ships, planes – as is urged by some ?

In 2020 40,000 terawatt hours, or 40 trillion kilowatt hours of fossil fuel energy was used for vehicles world-wide.

To replace that with hydrogen would require 1.25 trillion kilograms (1.25 X 10^{12}) of hydrogen annually (1250 million metric tonnes). To make that with renewable electricity would require at least 80 X 10^{12} kilowatt hours of electricity (80 trillion).

But total world electricity production was 24 X 10^{12} kilowatt hours in 2020 (24 trillion)

So to make the required hydrogen to replace fossil fuels *just for transportation, vehicles* would require *more than three times* the total electricity currently made for all purposes and using all methods of electricity production including fossil fuels world-wide.

This is clearly a completely impossible task. It therefore will not happen. What will happen is the continuing "boutique" use of hydrogen on a small scale for smaller short range light vehicles and other "showcase" uses.

In 2021 94 million metric tonnes of hydrogen were made world-wide, most of it by methods needing high inputs of fossil fuel energy especially the steam – methane method described above, and other methods with a big input of fossil fuel energy. Only 3.4 million tonnes was made by water electrolysis, and virtually all the electricity involved came from fossil fuel use. Less than 1% of hydrogen production came from renewable energy use. To multiply the amount of hydrogen made by thirteen times to raise output to the 1250 million tonnes required *just for transportation, and* only use renewable energy electricity, which would need to be increased at least a thousand times greater than current use for this purpose, is clearly impossible.

A further consideration is that the great bulk of hydrogen produced now is used in industry. These industrial uses are projected to continue to increase. All other uses including transportation in 2021 totalled 0.9 million tonnes. So hydrogen for transportation has to compete with these much greater industrial demands. Given the ready availability of fossil fuels for transportation with many competitive, practical and cost advantages compared to hydrogen makes it even more unlikely hydrogen will dominate transportation in future.

And some argue that hydrogen could replace fossil fuels for global industry – another 46,000 terawatt hours, the: "elephant in the room" compared to electricity generation and transportation. All the above considerations make this aim impossible.

When these realities are studied the contrast with the utopian view promoted by some that hydrogen is: "the easy fuel of the future" is brought into perspective to understate.

Biofuels

Biofuels are inflammable fuels which can be burnt in air to produce heat energy, and which are made from plant or animal material. They can be used in a similar way to oil products, diesel, petrol and aviation fuel, ideally.

Because they are manufactured from recently grown plants or from animals that have recently died they are regarded by those concerned about carbon dioxide production as having come from sources which recently took up CO2 to make them; in the case of plants by photosynthesis and plant growth using up CO2; and for animals, they have eaten plants or other animals which ate plants when growing.

This is in contrast to fossil fuel use, oil products and coal, when the carbon dioxide uptake from the atmosphere was millions of years ago, and then it is released by recent use.

Those concerned about carbon dioxide production say that the CO2 released into the atmosphere by burning biofuels does not count, as it was recently absorbed from the atmosphere in forming the plants and animals used to make the biofuel. And if the biofuel was not used by humans it would break down naturally with decay eventually, 80% of the CO2 would be released into the atmosphere anyway, only 20% would be incorporated in soils long term.

So biofuels get a free lunch as it were concerning their CO2 output by those counting emissions when burnt as fuel, even though they are producing carbon dioxide in a similar way to fossil fuels.

The principle biofuels are:

Bioethanol:

This is made from plants by fermentation, like making beer and alcohol; indeed the product is ethanol, C2 H5 OH. The ethanol is added to petrol to supplement performance in **ethanol blend petrol**. The process of making it usually requires considerable heat, usually supplied by natural gas. Ethanol blend petrols can cause many problems with engines, principally petrol leaks from seals being destroyed, unless they are made to take them.

Biodiesel:

This is made from oils and fats of animal or vegetable origin by chemical processes, again involving energy input, often from natural gas. It is most common in Europe. Common plants used include rapeseed, canola, palm oil, soy. Again it is usually an additive to fossil fuel diesel as by itself it results in performance and maintenance issues, and a big increase in particulates (ash, soot) production, as well as, ironically, CO2 production. In France it makes up 8% of some diesel fuels, in Europe generally about 5%.

Biofuels contributed 4.3% of world transportation fuel energy in 2021. As part of "nett zero" ambitions the International Energy Agency states that biofuels need to expand greatly to meet 64% of all transport fuel by 2050. Most consider this completely unrealistic.

There is also an aviation biofuel to replace aviation gasoline and kerosene but it is at a very early stage of development and accounts for well less than 1% of aviation fuels.

Biofuels are very controversial as often high quality land is diverted from food production to growing biofuel crops. There is then much less biodiversity – plant variation – in that region. In addition there is a trend to troublesome ground water acidification as a consequence. Many forests are being cleared to grow biofuel crops due to economic motives resulting in, ironically again, increased CO2 production, especially in poorer third world countries. .

Many consider biofuels the only potentially practical "renewable" energy sources for aviation, marine and shipping, heavy freight and industrial vehicles, which currently run on aviation fuel, fuel oil or diesel, are biofuels; given that batteries and hydrogen are impractical finally for these purposes.

In 2020 1.3 trillion litres of petrol and 1.5 trillion litres of diesel was used world-wide. In addition 255 billion litres of aviation fuel was used. This totals about 2.8 trillion litres (2.8 X 10^{12} litres. A trillion is a million millions, and a billion a thousand millions).

64% of that is 1.8 trillion litres hopefully to be biofuels to meet "nett zero" aims .

Concerning **Cost** currently fossil fuel petrol and diesel costs about 0.64 cents (US) per litre for the crude oil and refining, that is the manufacture; it is taxed heavily in many countries accounting for the higher retail cost. So for 1.8 trillion litres to be hopefully replaced by biofuels the current cost for fossil petrol and diesel is 1.2 trillion US dollars.

There are many estimates of the cost of manufacture of biofuels but as a generalization it ranges from 0.98 cents (US) to $1.27 (US) per litre, according to a study referenced below. Other studies range within that band usually. They are certainly more expensive than fossil fuel diesel and petrol. If one assumes $1.1 dollars per litre manufacturing cost for biofuel petrol and diesel the 1.8 trillion litres would cost two trillion dollars, a difference of 800 billion dollars (US), 66% more than the current cost. Because petrol and diesel are basic costs in society there would be many flow-on increases in living costs for populations in affected countries. Already in Europe the effect of the limited biofuel initiatives there has increased fuel costs by about $17 billion dollars (US) per year; and there are many protesting, including some calling for a return to fossil fuels without the addition of biofuels.

In addition to this manufacturing cost there is a big potential cost to governments and taxpayers. Fossil fuel diesel and petrol is heavily taxed especially in the advanced democracies where taxes, duties, other government charges are usually well over 100% of manufacturing cost, at least doubling the price. This tax revenue is used for all kinds of government expenditure including health, education and social welfare. Whereas biofuels are usually subsidized heavily to help make them competitive and promote their use. So a major switch to biofuels would result in a big loss of government tax revenue of trillions of dollars globally. If government services were to be maintained at the same level there would need to be more taxation.

In addition to these financial costs there are substantial **Energy Costs.** Biofuel manufacture uses a lot of energy at every step from farming feedstock, transporting it to plants for manufacture, the manufacturing process which needs heat usually to 500°C plus energy to work machinery. The final energy available from the biofuel when used is **Output Energy**; the energy used to make it is **Input Energy.**

For biofuel manufacture the ratio of output energy to input energy is usually between 1.16 to 1 and 2.16 to 1. If the International Energy Association's proposal is followed to have 64% of global transport fuel from biofuels that means 950 billion litres of biodiesel, 750 billion litres of bio-petrol, and 150 billion litres of aviation biofuel. The calculated *output energy* from that at 12 kilowatt hours per litre for diesel, 9 for petrol and 8 for aviation fuel is about 22,000 terawatt hours, 22 trillion kilowatt hours. So the ***input energy*** to make that using the ratios above is between 7000 and 10,400 terawatt hours. This is a big chunk of the world's energy considering total global electricity production is 24,000 terawatt hours.

Most of this energy to make biofuels currently comes from natural gas and electricity largely from coal fired power stations. The stated aim is to supply this from renewables but considering the huge amounts required and the problems discussed already that is clearly impractical. This large energy requirement is in marked contrast to fossil fuel diesel, petrol and aviation fuel which already have the great bulk of their energy in them.

But an even bigger issue with large scale use of biofuels is the **Land Area** occupied to grow the feedstock crops. This varies greatly depending on the type of crop and the quality of farming and manufacturing processes, but generally for diesel from soybeans about 200 litres per acre is expected; for rapeseed and canola between 170 to 800; other feedstock crops are usually less. An exception is water based algae crops in large ponds which can produce between 6000 to 60,000 litres per acre. But the production energy requirements, which include the removal of the water usually by heat, are huge; so this is not a practical very large scale option.

So to return to the proposal to make 1.8 trillion litres of biofuels annually: allowing a reasonable production rate of 250 litres per acre means this would require 7 billion acres of arable land suitable for cropping. But of the total global land area of 32 billion acres only 1.7 billion acres are arable, able to be farmed in crops; and, considering the many demands including food on arable land, most suggest that the maximum possible for biofuel crops would be one billion acres and even that is clearly massively pushing it, 70% of all arable land. Thus there is a huge discrepancy between the land needed to meet this proposal, 7 billion acres, and the reality, at extreme best one billion acres. It is therefore not surprising that a major report on the future of biofuels stated: "Biofuels will never be a major transport fuel as there is not enough land to grow plants for all vehicles - but they can be part of the renewable mix."

Yet another problem with biofuel expansion is investor nervousness. Global investment in it has declined 80% since 2008. This is largely due to the high cost of production versus the potential market price given the cheapness of fossil fuel competitors – even with their very high tax levies. Building and running biofuel plants is very dependent on government taxpayer funded grants, subsidies and loans. Investors worry that these may not, even will not, last forever. And there are instances of companies going bust despite all the government help to increase uncertainty. For example in 2015 a Kansas biofuel plant failed despite an initial government grant of US$97 million and a government guaranteed loan of US$132 million. In addition there are allegations of dumping of cheap government subsidized product from South America, where costs are lower, into higher cost countries making profits from biofuels even more uncertain. The push for some third world countries to replace food crops with biofuel feedstocks driven by more profit and encouragement from advanced countries increases this problem.

Considering all these problems it seems unlikely biofuels will expand much beyond their present 4.3% of world transport fuels, but remain essentially a gesture for the "renewable lobby" to point to. Countries where they are used, principally Europe, subsidize them substantially to encourage consumer use, as does Australia. Left to find their own place in the market without government subsidy it seems likely they would fail, especially considering potential engine and performance issues. Then there is the "conjuring trick" of not counting their CO_2 production as discussed above; if users were truly sincere about this they might instead bury the plants deep in the ground to avoid the carbon dioxide production from processing and using them. Or else leave forests intact and continue food production. But these proposals are certain to be not appreciated in those quarters!

Wood

Some reference has already been made in Chapter 8 concerning carbon dioxide production from wood burning. Human use of wood for fuel for cooking and heating is almost universal especially in third world countries. Estimates of the amount of CO_2 produced by that range from 4 to 10 billion tonnes annually as said, compared to total CO_2 production of from 535 to 901 billion tonnes, and that from wildfires 8 to 50 billion tonnes. It is a difficult amount to measure as discussed. This type of low tech burning has positive effects on soil fertility and plant growth due to nutrient rich ash falling on the surface; large algal blooms occur where ash fall on the sea as discussed. Many consider it environmentally advantageous and part of the natural world.

200 years ago when steam engines first appeared wood was used as a fuel for transportation, powering trains, traction engine tractors and ships. It was soon replaced by coal because of the much greater heat output of that fuel. Wood fires commonly burn at about 600°C whereas coal ones can achieve 2000°C. A given weight of coal will usually produce almost twice as much heat energy as wood. And of course gas and liquid fuels are much more energy productive and practical than these fuels for transportation.

But there is a move to use wood to generate electricity in power stations especially in Europe, usually in stations which previously used coal. Again those counting CO_2 emissions give this a free lunch for the same reasons discussed above for biofuels. But if the wood came from trees grown in Europe the "counters" would include those felled trees as being part of the tally of CO_2 production because they were not left in the ground to keep growing absorbing carbon dioxide. That tally is neatly sidestepped by the European wood burning power stations by importing wood from trees felled in American forests, especially from the western US states of Washington, Oregon and California where there big forests of suitable trees for mechanized felling and chipping on an industrial scale; also there is an expanding industry using hardwood forests in North Carolina, Louisiana and other US states as well as more broadly internationally driven by profit. The European counters do not include this in their tally's, even though they purport to be concerned about global emissions.

The process of felling forests, making the logs into small pellets about 8mm by 20mm, and transporting them to power stations including by sea requires considerable energy usually supplied by diesel and fossil fuels. The cleared forests no longer grow absorbing CO_2. The wood pellets when burnt in the power stations quickly emit CO_2 into the atmosphere at a rate of about 4 kilograms of CO_2 for every kilogram of wood.

This in contrast to the situation if the trees were left to grow to maturity, then fall and rot into the soil, the "natural process". Then the living tree continues to grow absorbing CO_2 by photosynthesis and producing oxygen for often many years or decades; and when it finally dies at the end of its lifespan it slowly rots away, in the process providing food and a home for many organisms. It slowly emits CO_2 as it does and a lot of its substance is incorporated into soils facilitating regrowth of the forest.

How much CO2 is emitted by wood rotting naturally compared to being burnt quickly is controversial; some say 80%; others say only about 15%, the rest incorporated in organic soil components. What is certain is that burning quickly produces a lot more CO2 than the natural process.

In addition the cleared forest resulting from pellet production for power stations is bare earth; soil formation has been stopped by the tree removal. So there is a cost in terms of future CO2 absorption by slower forest regrowth.

Also when the wood pellets are burnt in the power station they produce less energy per kilogram of CO2 emitted than coal; Coal produces about 36,000 kilowatt hours of heat energy for every kilogram of CO2 emitted. Opinions vary about how much is produced by wood from about 25,000 to about 32,000 kilowatt hours per kilogram of CO2 emitted. Most believe it is around 25,000, meaning 50% more CO2 than coal for the same heat output. Some say it three or four times as much CO2 than coal for the same heat energy produced. It is clearly more.

There are some vested interest groups who say pellet burning produces only 20% of the CO2 produced by coal burning in power stations but that claim does not square with scientific studies of power station emissions.

These groups also claim that up to 80% of the feedstock for pellets is waste wood or residue left over from crops, like rice husks, which would otherwise be unused. But the reality is that demand for pellets far exceeds the supply from those sources, hence forest trees are largely used, including large areas of hardwood trees from US forests. It is more expensive to make pellets from waste.

Globally 52 million tonnes of pellets were made in 2018, and that is projected to increase by 5% each year. The trade was worth US$8 billion in 2022 and is expected to reach US$15 billion by 2026. Pellets sell for about US$500 per tonne. The price varies greatly with market fluctuations. It is highly profitable.

In Europe and North America there is an increasing market for using wood pellets to fuel heating stoves for homes and buildings replacing oil, gas, which is in short supply, and coal, which is plentiful, cheap but often banned. This has the potential to further greatly increase wood pellet burning.

On a weight basis wood produces less heat energy than coal and other fuels. 4.5 kilowatt hours is produced per kilogram burnt. This is because only half of dry wood is carbon concerning weight. Black coal produces about 7, brown coal about 5. By comparison natural gas is 12 to 15, diesel 12 as discussed already.

A further consideration is the amount of **Energy Input** needed to make pellets compared to the **Energy Output** burning them, similar to the biofuels energies discussed above. The input energy includes felling, processing of wood into pellets, then transport to the power stations. It varies depending especially on whether the feedstock wood is dry or wet. If dry it is about 150 kilowatt hours per tonne of pellets, if wet up to 1000 kilowatt hours. As the output energy is about 4500 kilowatt hours per tonne burnt it follows that between 3 and 20% of output energy is needed to make the pellets. This energy is largely supplied by diesel for machinery and transport, and electricity, often from coal fired power stations, for the manufacturing plants. As demand increases there is a trend to use wetter feedstock woods. So it is reasonable to assume about 10% of output energy will be required to make and deliver pellets.

The International Energy Association has said that by 2035 at least 6.5% of world electricity should come from biomass, which is essentially wood pellets. That would be at least 2,000 terawatt hours of electricity generated by wood pellet burning, and the input energy to make the pellets would be about 200 terawatt hours (200 billion kilowatt hours). That is a further substantial demand on global electricity and transport energy. But there are plans to greatly expand this beyond 2035 to meet "nett zero" demands by 2050; so that input energy for wood pellets would also increase greatly.

Many government authorities in the advanced democracies have pronounced that wood burning for power generation is "Carbon Neutral" meaning emissions of CO2 from using it to make electricity get a free lunch. But many scientists and "renewable lobbyists" strongly disagree with this, pointing out all the aspects discussed above from forest destruction through to high CO2 emissions. Those in the UK point to the wood fired power stations there emitting big amounts of CO2 and yet not only getting this not counted but also gaining huge government subsidies as "renewables". For example the biggest wood power station in Britain, the Drax station using wood pellets from Louisiana forests largely, has received over 5 billion pounds subsidy and emits 16 million tons of CO2 from burning over 5 million tons of pellets annually, the biggest single carbon dioxide emitter there; but that does not count. Nor does the 480,000 tons of wood pellets burnt annually at the Stevens Croft station at Lockerbie. There are similar situations in Germany and other European countries.

Concerning home heating there is clearly an irony that people are forced to stop using coal and instead use pellets which produce at least 50% more CO2 !

There is also concern and debate in US areas where industrial forest destruction for pellets is taking place. Many are poor regions and there are allegations of regulations being flouted with large scale pollution and environment destruction.

It is hard to escape the conclusion that this is an activity facilitated by profit, government subsidies politically driven, and, to be very kind, unwise regulation.

Geothermal

Geothermal energy uses hot water from deep underground to generate steam and drive turbines generating electricity. Also the hot water can be used directly to heat homes and buildings in cold climates. Iceland is a country where 90% of heating is from geothermal.

To tap into this energy for electricity generation deep wells usually have to be drilled from one to ten kilometers down. Then the power plant is built to harness the steam. This costs between 2 and 5 million US dollars per megawatt. So a 1000 megawatt station – and there are none operating – would cost two to five billion US dollars to build. This is expensive especially when compared with gas, but cheaper by far than nuclear or wind or solar with battery backup.

Despite the geothermal steam not having a direct cost there are significant costs operating geothermal plants and the electricity produced finally costs between US$80 and US$220 per megawatt hour. Coal is 40 to 80, gas 70 to 120, solar 90 to 170, wind 65 to 115, nuclear 110 to 190 as discussed in detail in the section on coal below.

In 2019 15 gigawatts (15 billion watts) of electricity came from geothermal globally. This is a minute percentage of world-wide electricity, 0.00006%! Only 7% of global potential for this has been tapped so far. Some consider it could meet 3% of global demand by 2050, and maybe 10% by 2100, but that seems to be pushing it very hard.

Six countries currently use geothermal for more than 15% of their electricity: Iceland, New Zealand, Kenya, El Salvador, Philippines, and Costa Rica.

The United States has the most geothermal electricity production; 3889 megawatts (3.8 gigawatts). This includes the Geysers Field group in California, 15 plants generating 725 megawatts. Electricity from this slightly cheaper, US$50 to 80 per megawatt hour. The US has great potential to expand this source as only 0.7% of potential sites have been used. Some say 100 gigawatts, 10% of the country's use, could come from geothermal.

Indonesia is second globally with 2200 megawatts (2.2 gigawatts).

Historically the first geothermal electricity was generated in Italy in 1904 and powered four light bulbs. Italy subsequently built a commercial plant which was the only one operating globally until 1958.

Problems with geothermal include the high capital cost of building; wells can cool and run out of hot water; because of the low temperature of the hot water efficiency is low, 7 to 10%; subsidence and earthquakes at the site can occur. Indeed in Switzerland 10,000 quakes more than 3.4 Richter strength occurred in the first six days of a plant operating and it was shut down – but that is an exception. The hot water can contain toxic chemicals such as mercury, arsenic, boron, antimony and salt which need to handled properly.

An advantage is that geothermal is constant day and night, when the wind and Sun are not present.

Geothermal has some future potential but is not the answer for the world's massive electricity needs.

Tidal

There is a large amount of natural energy used to make the ocean surface rise and fall every six and a quarter hours, in some places five to eleven meters. And there have been many efforts by humans to use this.

The Romans used sea-tight walls to trap the incoming tide, then it flowed out through sluices driving water wheels. This was widespread in Europe during the Middle Ages.

The first electricity generation from tides occurred in the 19th century. But currently there are very few commercial plants. The largest is at Sihwa Lake, South Korea. This has a peak capacity when the tide is running fast at 254 megawatts, an average overall of about 55. It is a sea wall with 10 turbines underwater.

The next largest is at La Rance, France, a plant with 27 turbines which opened in 1966. It has a peak output of 240 megawatts, and average of 57. The third largest is in Nova Scotia, Canada, a 20 megawatt unit.

There are several other pilot small plants globally in China, Russia, South Korea and the US. A project in the East River near New York has 30 35 kilowatt turbines 10 meters deep and is monitoring effects on fish life, underwater noise, other environmental effects.

Many projects have been planned but later abandoned. Examples include a 1977 Canadian plan to build a 3800 megawatt plant; very high costs resulted in cancellation. In Vancouver a tidal plant was opened in 2006 but shut because of high costs in 2011. In South Korea an 812 megawatt plant was cancelled in 2013, and a 1320 megawatt plant was halted for environmental reasons in 2013. In India a 50 megawatt plant was cancelled due to high costs. In the UK a 320 megawatt plant at Swansea was cancelled in 2018. In the US the Snohamish PUD project was halted due to ballooning costs in 2014.

Most plants use either a water tight sea wall to trap the tide and force it through narrow passages where the turbines are – so called **Tidal Barrage.** Or a **Tidal Fence** is used where the turbines are sited in a fast flowing tidal current.

Clearly the resulting power is *intermittent*: when the tide is not running, slack water, there is no electricity. Problems include:

- Very high cost to build.

- A lack of suitable sites: the plants can impact on shipping and recreational boating, fishing, coastal environments, water quality.

- There is concern about fish being struck by blades, the effects of underwater noise on marine animals, electromagnetic effects from underwater cables taking power to the shore, effects on bird life if fish leave the area because of the above problems, and resulting effects on marine plants.

- Underwater turbines are subject to corrosion and need to be made of expensive stainless steel and other resistant materials. They need lubricants which may leak and cause effects on sea life. Maintenance is difficult requiring divers compared to surface equipment.

- Fouling of turbines by marine growth, barnacles etc is another problem needing ongoing maintenance.

- The installations need a complicated network of cables to deliver electricity to users. And the resulting power can lead to instability in national grids requiring Synchronous Condensers, large capacitors or other equipment for stability like power from wind and solar as discussed above and in the section on coal.

So despite over 100 years of effort tidal power remains very tiny globally.

Waves

Again there is a huge amount of natural energy used to make waves world-wide. The natural energy is from the Sun causing winds resulting in waves. Humans have been seriously attempting to use this since 1890.

Devices generally try to use the rise and fall of the waves to work machinery which drives electrical generators. These include anchored buoys, rafts, jointed long cylinders, and water columns where the rising and falling water drives turbines.

There are no commercial plants operating currently. In 2000 the world's first commercial plant, the LIMPET Islay station off Scotland, was opened and fed power to the national grid; it has now closed.

In 2008 a plant was opened off Portugal; it has also closed.

There is a great deal of ongoing research but problems include:

- Very expensive to build

- The electricity is difficult to distribute as discussed for other methods, is expensive, and has the same problems with national grid instability.

- Suitable sites away from navigation, commercial and recreational activities are not easy to get agreement on.

- The power is not constant; often the sea is calm.

- And often it is very rough with confused wave coming from several directions; this can result in breakages and the engineering required has to be very robust, expensive.

- The equipment has to be anchored to the sea bed, difficult often. If it breaks loose that is a major issue.

- Equipment maintenance is much more difficult and expensive than land based machinery.

- Corrosion and fouling are other problems.

It is clear, given there are no commercial entities, that everyone who has looked hard at this over the 130 years since the first projects has moved away to other electricity sources for the above reasons.

Nuclear

Nuclear energy is heavily promoted by some as the ideal solution to achieve nett zero. They say it offers a huge electricity supply for virtually no carbon dioxide emissions.

Nuclear energy from uranium was first developed practically by scientists in Nazi Germany in 1939. During World War Two atomic bombs were pioneered by Manhattan project scientists in the United States, culminating in their use on Japanese cities in 1945 ending that war.

In 1954 the first nuclear power station was opened at Obninsk in the Soviet Union. This was followed in 1957 by a US plant in Pennsylvania. These were followed by many government funded plants in several countries including Britain and the US. A utopian world of unlimited energy from nuclear power was promoted. This even included a plan to build a new Panama Canal with 651 hydrogen bombs each of 42 megatons explosive power doing the earthworks! But the Cold War was in full swing and many feared Earth was about to be turned into a lifeless rock.

The utopian scenario was given a reality check when in the 1960's many of the government built US nuclear plants were sold to utility companies for about $60 million, far less than the cost to build. Then in the 1970's there were plans to build about 200 nuclear plants in the US by private utility companies but there were huge cost overruns and many delays. This reflected the fact that the true cost of previous government funded stations was hidden. It was a time of high inflation and interest rates. In 1979 the Three Mile Island nuclear power station disaster occurred and widespread public opposition developed. So in the 1980's more than

120 projects were cancelled, 35 continued but cost overruns were six to eight times original estimates, and delays were six to twelve years beyond expected commissioning times. It was described as the: "Greatest managed business disaster in business history".

Currently there are 99 nuclear power stations operating in the US, and 422 world-wide in 32 countries. Globally 57 are under construction, construction times are about 10 years. The US fleet is aging and approaching the end of the original 40 year licensing. About three quarters have extensions for a further 20 years. Some are being decommissioned, an expensive time consuming process. Many in the US are being replaced by natural gas stations which are much cheaper and quicker to build.

The French also built a large number of nuclear power stations in the 1970's and 80's. These are aging and problems including corrosion cracks in reactor cooling pipes have caused 26 of the fleet of 56 to be closed. France has plans to decrease its reliance on nuclear power from 70% to less than 50%.

Decommissioning old nuclear power stations is a decades long process costing many millions of dollars. This is because after a reactor has been shut down some lower level radioactive decay takes place creating heat; so cooling still has to occur for many years to prevent fire and explosion. Old stations are large reinforced concrete and steel buildings with many pipes and contaminated areas, therefore difficult and very expensive to dismantle.

Uranium is a rare element comprising about one millionth of the Earth's crust. Large open cast mines produce huge quantities of crude ore which is processed into concentrated uranium. This requires large amount of energy for machinery and vehicles, and big quantities of toxic waste are produced. The open cast mines destroy the natural environment.

The uranium is mostly Uranium 238, a very stable form with a half-life of over four billion years, as much as the Earth's age. About 0.7% is a less stable form, Uranium 235 which has a half-life of 700 million years. Atoms of this form tend to break into smaller particles which then strike other U 235 atoms causing them to break into smaller nuclei, for example Krypton 36 and Barium 56. A tiny amount of mass is lost in this process and that is converted into huge amounts of energy usually as heat. The amount of energy follows Einstein's law $E = MC^2$, meaning the mass times the square of light speed equals the energy; truly huge. One gram of uranium 235 reacting this way has the same heat energy output as burning three million grams (3 tonnes) of coal.

In nuclear power stations the uranium is in the form of *fuel rods*. These contain **enriched uranium**, where the concentration of U 235 has been increased from 0.7% to 3% by special methods; this results in much more heat energy being produced.

The nuclear reaction is controlled by **control rods**. These are made of a material which absorbs the small fragments from breaking up uranium 235 nuclei and prevents them striking other nuclei continuing the reaction. Graphite and Boron are amongst the different materials used. The rods are passed between the fuel rods to slow or stop the reaction.

The nuclear reactors of fuel and control rods are in containers immersed in **coolant fluids** which are pumped in pipes to take away the huge amounts of heat generated. The fluid most commonly is water, but liquid sodium, carbon dioxide or Helium gas, or oils are sometimes used.

These fluids in turn pass the heat energy on to pipes full of water, which boils to make high pressure steam. This is aimed at the propeller like blades of **steam turbines** which then rotate at high speed driving **electricity generators** producing electricity.

Only about one third of the heat energy produced by the nuclear reactors finally ends up as electricity; the rest has to be removed from the power station, which would otherwise continue to heat up and eventually explode. This is usually done by putting the station alongside a lake or the sea, the circulating that water with pumps and pipes to cool the station. In that process large amounts of the nearby lake or sea are heated with adverse effects on marine life.

All these components are housed in a very strong steel reinforced concrete building, a **containment building.** This is designed to hopefully stop radioactive superheated fluids and gases escaping in the event of a breakage or explosion.

There are many excellent diagrams of nuclear power stations in the websites referenced below. There are many variations and complexities in the design of nuclear power stations.

As a nuclear power station ages the fuel rods gradually lose Uranium 235 as it changes to lighter elements and the heat energy generation declines. Old rods then have to be removed and replaced with new. The old rods are then **nuclear waste** and have to be stored safely as they are still very radioactive. The 99 US reactors result in about 2000 tons of waste each year. The waste is radioactive because it contains many unstable isotopes which decay giving off small harmful particles including alpha (two protons and two neutrons like Helium nuclei), neutrons, other radiation which can damage living cells. Examples are Caesium 137, which has a half-life of 30 years; and Strontium 90 with a half-life of 28 years. All the waste isotopes have half-lives shorter than 90 years except seven. These are:

- Technetium 99 – half-life 211,000 years.
- Tin 126 – 230,000 years.
- Selenium 79 – 320,000 years.
- Zirconium 93 – 150,000 years.
- Caesium 135 – 2.3 million years.
- Palladium 107 – 6.5 million years.
- Iodine 129 – 15.7 million years.

Of these Technetium 99 and Iodine 129 are the greatest hazards to living creatures as they are readily taken up by them. And their radioactivity lasts for very long periods clearly. Many of the shorter lasting isotopes are also hazardous. Therefore nuclear waste has to be handled carefully and cannot be allowed to escape into the environment, especially not into the sea or environment water.

In practice nuclear waste is buried deep in stable ground hopefully free from earthquakes, volcanic activity, floods where it can be left for millions of years, hopefully. This is difficult and expensive to do. Getting government approval is one of the difficulties as there is usually considerable local community resistance to have waste nearby, and having it transported through neighbourhoods.

In the UK there are plans to bury waste 1000 metres deep, but there have been delays. In the meantime waste is stored in reinforced steel and concrete containers submerged in large water pools, for example the Sizewell B waste pool. This facility was full in 2015. Another at Sellafield contains 2150 cubic metres of waste. World-wide this is a problem which will increase if nuclear power expands.

In addition there have been many accidents at nuclear power stations causing release of radioactivity. From 1952 to 2014 there have been 100 serious accidents globally, defined as resulting in deaths and/or more than US$50,000 damage. Especially serious ones are when there is **core meltdown**. This where the reactor cooling system fails, massive heating occurs until it gets so hot that the fuel rods melt and destroy the containing structures, the superhot core escapes into surrounding earth where contact with water can result in severe explosion and release of radioactive gases, liquids and solids into the environment.

There have been 15 of these, with three especially serious ones:

In 1979 the Three Mile Island accident occurred. This was due to a core coolant relief valve which was stuck open releasing radioactive coolant and resulting in partial meltdown of the core with release on radioactive material into the local area. Staff behaviour was criticised in later reports. The clean-up took 14 years and cost US$2 billion. Part of the station restarted in 1985 but shut again in 2019 for economic reasons – it could not compete with gas. The site has been decommissioned but that will not be complete until 2080 and will cost a further US$1.2 billion. The episode severely hit public confidence as discussed.

In 1986 the Chernobyl explosion occurred. This resulted in probably 10,000 to 20,000 deaths, mostly from thyroid cancer from Iodine 129 exposure. There were many deaths from direct radiation sickness in the weeks after the accident, mainly in site rescue and clean-up workers. A radioactive cloud spread over Europe containing especially Caesium 137 and Strontium 90. A 30 kilometer wide long term exclusion zone is present for at least 300 years.

In 2011 the Japanese Fukushima disaster occurred. This was due to an undersea earthquake off the coast where the station was located to access cooling water. A large tsunami wave flooded the station causing an explosion when it involved the reactor core. A radioactive cloud with Caesium 137 and Iodine 131 spread widely and heavily contaminated a 30 kilometer wide zone. There has been an ongoing problem keeping the melted core cool and still huge volumes of seawater are poured into the site. There is a 20 kilometer wide permanent exclusion zone. It will at least 100 years before the control rods can be approached and removed. By 2020, 9 years after the disaster clean-up costs were US$300 billion. Plant workers have increased risks of life-shortening cancers. Within 5 years radioactive isotopes from this were found right across the Pacific Ocean. This is considered similar in damage to Chernobyl and greater than Three Mile Island. Subsequently Japan shut down its nuclear power stations, which are all on the coast, and built more gas and coal ones. It has since reopened some nuclear stations. Japan's choice of a substantial fleet of nuclear stations given it is very earthquake prone and maritime has been criticised.

Other notable accidents amongst the 100 include the SL-1 meltdown in 1961 resulting in an explosion killing three, fortunately in a remote area. This was due to human error with the control system. And another involved an explosion in a coastal French station in 1999 due to flood from a severe storm. And another when Israeli and US bombers attacked an Iranian nuclear facility.

Concerning the accidents it is always easy to be wise after the event. But many consider more accidents are inevitable given the extremely complex technology, that it is run by humans, and the stations are sited in our everchanging hazardous environment with earthquakes, volcanoes, storms and other unpredictable events. An authoritative Massachusetts Institute of Technology (MIT) study suggested at least four more serious accidents would occur by 2050 especially if nuclear power is expanded. Obviously the accidents have led to design changes to increase safety with many more backups.

Small Modular Reactors are pointed to by some as offering a practical answer. The concept is to build in factories small nuclear stations less than 300 megawatts power housed in containers small enough to be shipped to their destination. There was initial enthusiasm but that has waned due to the following problems:

- The cost of building them is much higher than large scale nuclear stations for the power output; and large scale stations are much more expensive to build than gas and coal ones.

- They generate 5 times the amount of nuclear waste and 35 times other waste compared to large nuclear stations. This includes radioactive steel from the container walls.

- Uncontrolled nuclear proliferation to irresponsible and/or hostile countries is a concern. The US has slowed down SMR development for this reason.

- Problems with shipping, including loss in the sea in the event of a sinking.

So while there have been more than 80 designs for Small Modular Reactors only two prototypes have been commissioned. The first is floating on a barge near a small Russian town and started operating in 2022. The second is a gas cooled one in China.

The nuclear power saga has many stories of enthusiastic brilliant new developments then disillusionment as troubles occur. For example from 1976 to 1988 the British government built 14 Advanced Gas Cooled nuclear power stations each with two 660 megawatt reactors with great optimism for the future. But they turned out to be very complex to build, were 13 years longer than planned to complete, and far more expensive than budgeted. In the 1980's they were privatized and the true cost revealed. Then unacceptable pipe corrosion was found and the first decommissioned at great expense in 2010. The others were found to be similar affected with corrosion, cracks and multiple defects and started to be closed from 2018. By 2030 they will all have been closed at huge expense. The sites will be isolated for many decades.

The first new UK nuclear power station in 20 years is being built at Hinckley Point. The project is 10 years behind and the latest cost in 2022 was 33 billion pounds (US$40 billion) more than double original estimates. It is being built by a consortium of the French firm EDF and the Chinese one CGN. EDF is in the process of being nationalised (made part of the government) by the French as it is US$17 billion in debt and is primarily responsible for that nations power supply; the government wishes to avoid that failing.

France has a new nuclear station under construction but the cost has already run out to US$17 billion and it is late. Another nuclear station under construction in Finland has costs of US$20 billion so far. In the US only two stations are under construction and costs have risen to US$20 billion each. In South Carolina a nuclear power station project was abandoned after costs reached US$13.4 billion.

Nuclear power stations are most common in wealthy countries, for example France where previously 70% of electricity was from nuclear, and the US where 18% is. Globally 9.2% comes from nuclear. Most countries do not go there.

China currently has 53 nuclear reactors operating providing 5% of its electricity and is building 20 more. They have announced plans to build a very large number and eventually phase out coal in favour of nuclear by 2060. They claim they are able to build plants for US$6 billion each. In addition there are claims there are 30 plants planned for "Belt and Road" countries. If they ever did make good on these stated ambitions it would be arguably a huge worry for people everywhere concerning accidents, waste storage, quality control.

Fast Neutron Reactors are promoted by some. These operate at very high temperatures and are cooled by liquid metals, often Sodium, which explodes in air if there is a leak. They are more difficult to control than water cooled reactors if things go wrong. And more difficult to refuel easily. They are very costly to build and operate.

A new gas power station of similar 1000 megawatt output costs about US$1.2 billion, and a coal one US$3 billion. The only other technology approaching nuclear in cost to build is wind with battery backup; as discussed below a 660 megawatt plants with battery backup for just 3 days is quoted at US$27 billion.

It therefore seems certain that nuclear power will either remain a small contributor to global energy, or else gradually disappear as stations age and are not replaced. If China goes ahead with big expansion plans that is a very serious worry. If such an expansion occurs world-wide it would be a terrible legacy for future generations as they grapple with the problems of multiple aging nuclear power plants with corrosion and many defects, leaking radioactive waste into the environment, and being so difficult to demolish.

Many proponents of "nett zero" advocate massive expansion of nuclear power globally, as do many politicians, pointing to avoidance of CO_2 but plentiful energy. They gloss over the problems discussed above. If this was to happen there would need to be about 4000 nuclear plants world-wide, instead of 400 now, given that nuclear provides 10% of global electricity now. Many more catastrophic accidents would be certain.

And in 50 to 100 years' time they would be an impossible problem for future generations to deal with safely with a high probability of disastrous radioactive contamination. The design life of most plants is about 30 years. That can be extended by a further 15 to 30 years with special treatments especially of the reactor pressure vessel. This tends to get damaged by the constant neutron bombardment inherent in the reaction. For example, the oldest operating nuclear plant at Beznau in Switzerland, built in 1969 and 365 megawatts output, quite small, has 1000 5mm size cavities in the reactor pressure vessel wall. There have been a lot of concerns about its safety, and protests seeking closure. And there are many other problems, some discussed above, due to aging, corrosion in these very high temperature and pressure machines.

Modern plants have a design life of 60 to 80 years. There is a research plan to build one with 100 years, appropriately called the "Centurion". But these slightly longer estimates of life make no real difference to the aging nuclear plant problem for future generations.

References and Further Reading for Chapter 13 – Alternatives to Fossil Fuels

www.entusa.com.au/batteries-vs-pumped-storage-hydropower-are-they-sustainable/

www.en.wikipedia.org/pumped-storage_hydroelectricity

www.en.wikipedia.org/wiki/electric_power_transmission

www.iea.gov/energyexplained/hydrogen/use-of-hydrogen

Nedwell, J et al. Univ Texas Report no. 544 R 0424 May 2003.

www.users.ece.utexas.edu/ling/2A_EU1.pdf

www.thundersaidenergy.com/downloads/wind-and-solar-costs-of-grid-inter-connection/

www.afr.com/energy/huge-grid-costs-blowouts-unnerve-energy-users-20200729-p55gpl

www.sciencedirect.com/science/article/abs/pii/S0360544214003363

Jorge, Raquel et al. "Grid infrastructure for renewable power in Europe: the environmental cost". Energy Vol 69 1 May 2014 pages 760 – 768.

www.ft.com/content/

Alice Hancock article 11/12/2022 Financial Times. Huge EU cost grid upgrades for renewables.

www.oilprice.com/latest-energy-news/world-news/UK-approves-266-billion-power-grid-investment-to-expand-renewable-capacity.html

(Actually $26.6 billion).

www.nytimes.com/2023/02/23/climate/renewable-energy-us-electrical-grid.html

Stober, U et al. "How could operational underwater sound from future offshore nwind turbines impact marine life?" Jn Acoustic Soc Am, 149, 1791 (2021).

https://doi.org/10.1121/10.0003760

Tougaard, J et al/ "How loud is the operating underwater noise from operating offshore wind farms?" Jn Acoustical Soc Am. 148, 2885 (2020).

https://doi.org/10.1121/10.0002453

Tougaard, J et al. Aarhus Univ. Denmark, Jn Acoustical Soc Am.

www.asa.scitation.org/doi/abs/10.1121/1.3117444

https://www.ewea.org>wind-energy-basics>faq

www.science.com/much-power-wind-turbine-generate6917667.html

www.en.wikipedia.org/wiki/wind_turbine

www.en.wikipedia.org/wiki/wind_power

www.en.wikipedia.org/wiki/wind_power_by_country

www.nationalgeographic.org/resource/wind-energy

www.energy.gov/eere/wind/advantages-and-challenges-wind-energy

www.energysustainsoc.biomedcentral.com/articles/10.1186/s13705-019-0234-z

Miller,LM et al. "Corrigadum: observation based solar and wind power capacity factors and power densities." Environment Research Letters 13:104008. 14.079501

https://worldpopulationreview.com>country-rankings

www.iea.org>reports>wind-electricity

www.forbes.com/siles/rrapier/2022/07/04/wind-and-solar-provided-a-record-10-of-the-worlds-power-in-2021/?sh=438803d914aa

www.researchgate.net/publication/348265467

Casterrini, A et al. "Machine learnt prediction method for rain erosion damage on wind turbine blades."

www.abcbirds.org/blog21/wind-turbine-mortality/

Loss, SR et al. "Refining estimates of Bird Collision and Electrocution Mortality of Power Lines in the United States." (3/7/2014)

https://doi.org/10.1371/journal.pone.0101565

www.reseachgate.net/figure/general-description-of-a-wind-turbine-system

www.energysustainsoc.biomedcentral.com/articles/10.1186/S13705-019-0234-z

Nitsch, F. et al. Observation based estimates of land availability for wind power: a case study in Czechia. Energy, Sustainability and Society, 9, No. 45 (2019).

Miller, L.M. et al. Observation based Solar and Wind power capacity factors and power densities. Environment Research Letters. 13:104008. 14:079501

https://worldpopulationreview.com>country-rankings

www.iea.org>reports>wind-electricity

www.forbes.com/sites/rrapier/2022/07/04/wind-and-solar-provided-a-record-10-of-the-worlds-power-in-2021/?sh=438803d914aa

www.ourworldindata.org/wind-generation?time-latest

www.abcbirds.org/blog21/wind-turbine-mortality/ (Loss et al studies and others).

Loss, S.R. et al. Refining Estimates of Bird Collision and Electrocution Mortality of Power Lines in the United States (2014). https://doi.org/10.1371/journal.pone.0101565

www.researchgate.net/figure/general-description-of-a-wind-turbine-system

Stober, U. et al. How Could operational underwater sound from future offshore wind turbines impact marine life ? Journal Acoustical Society America 149,1791(2021). https://doi.org/10.1121/10.0003760

Nedwell, J. et al. Univ Texas Report no. 544 R 0424 May 2003. www.users.ece.utexas.edu/ling/2A_EU1.pdf

Tougaard, J. et al. (2009) www.asa.scitation.org/doi/abs/10.1121/1.3117444

https://www.ewea.org>wind-energy-basics>faq

https://statistica.com>climate

www.en.wikipedia.org/wiki/solar_updraft_tower

www.freedomforever.com

www.ga.gov.au/scvientific-topics/energy/resources/other-renewable-energy-resources/solar-energy

www.energy.gov/eere/solar/how-does-solar-work

www.solvoltaics.com/energy-make-solar-panel/

www.solarchoice.net.au/solar-power/how-are-solar-panels-made/

www.solarcraft.com/solar-energy-myths-facts/

www.axionpower.com/knowledge/power-world-with-solar

www.news.energysage,com/what-is-the-power-output-of-a-solar-panel

www.iea.org/reports/solar-pv

www.sciencedirect.com/science/article

www.en.wikipedia.org/wiki/solar_power_by-country

www.weforum.org/agenda/2022/04/wind-solar-electricity-globa-energy

www.abc.net.au/news/2021-06-06/what-happens-to-solar-panels-after-their-useful-life-is-over/100193244

www.pvindustries.com.au-toxic-leach-holes-in-landfills

www.theguardian.com>environment>oct>aus October 2022.

www.sustainability.uk.gov/recycling-and-reducing-waste/product-stewardship/national-approach-to-manage-solar-panel-inverter-and-battery-lifecycles

www.cen.acs.org/environment/recycling/solar-panels-face-recycling-challenge-photovoltaic-waste/100/i18/

www.currentresults.com/weather/united-kingdome/annual-sunshine.php

https://www.livingin-australia.com>sunshine-hours-aus

www.en.climate-data.org/oceania/australia/nortern-terrory/alice-springs-58b/

www.pv-magazine.com/2023/3/21/global-study-highlights-potential-of-floating-solar/

www.crownbattery.com

www.en.wikipedia.org/wiki/lead_acid_battery

www.electronicnotes.com/articles/electronic-components/battery-technology/how-to-do-lead-acid-batteries-work-tecnology.php

www.visualcapitalist.com/chinas-dominance-in-battery-manufacturing

www.prnewswire.com

www.mckinsey.com/industries/automotive-and-assembly/our-insights/battery-2030-resilient-sustainable-and-circular

www.globalxetfs.com/lithium-market-update-elevated-prices-are-creating-favourable-dynamics-for-miners/

www.nationalgeographic.com/science/article/partner-content-audi-nl-the-cost-of-batteries

www.spglobal.com/mobility/en/research-and-analysis/growth-of-liion-battery-capacity.html

www.woodmac.com/press-releases-global-lithium-ion-battery-capacity-to-rise-five-fold-by-2030

www.globalnewswire.com/en/news-release/2022

www.cen.acs.org/articles/97/i28/time-serious-recycling-lithium.html

www.energy.gov/eere/articles/how-does-lithium-ion-battery-work

www.en.wikipedia.org/wiki/lithium_ion_battery

www.en.wikipedia.org/wiki//fuel_efficiency

www.energy.gov/eere/articles/how-does-lithium-ion-battery-work#

www.en.wikipedia.org/wiki/lithium_ion_battery

www.insideevs.com/news/528346/ev-weight-per-battery-capacity

www.iopscience.iop.org/article/10.1088/2515-7620/ab5e1e "Energy use for large scale Lithium ion battery production." Kurland, S.D. Enviro. Res. Comm. Vol 2 No. 1 20 Dec 2019.

Dai Q., Kelly JC., Gaines L. and Wong M. "Lifecycles of lithium ion batteries for automotive applications." Argonne National Lab, USA. Batteries 2019, 5(2), 48,

https://doi.org/10.3390/batteries5020048

www.mdpi.com/2313-0105/5/2/48

www.csiro/au/en/research/technology-space-energy/energy-in-the-circular-economy-battery-recycling

Zhao, Y et al. "Australian landscape for lithium-ion battery recycling and reuse in 2020." CSIRO Report EP 208519.

www.cen.acs/materials/energy-storage/time-serious-recycling-lithium/97/;28

American Chemical Society 2023.

www.bbc.com/future/article/20220105-lithium-batteries-big-unanswered-question

www.sciencedirect.com/science/article/abs/pii/S03043894210697

Miao, Y et al. "An overview of global lithium-ion batteries and associated critical metal recycling." Jn Hazardous Materials Vol 425 5 March 2022.

www.sciencedirect.com/science/article/pii/S0921344921003712 Kelly, JC et al. "Energy, greenhouse gases and water life cycle analysis of lithium carbonate and lithium hydroxide monohydrate from brine and other resources and their use in lithium ion battery cathodes and lithium ion batteries." Resources, Conservation and Recycling, Vol 174, Nov 2021.

www.link.springer.com/article/10.1007/S11027-019-09869-2 Kelly, JC et al. "Globally regional life cycle analysis of automotive lithium-ion nickel manganese cobalt batteries: mitigation and adaptive strategies for Global Use." 2020.

Yang, S et al. "A review of lithium-ion battery thermal management system strategies and the evaluation criteria." Int. J. Electrochemical Science 14(2019) 6077-6107.

www.electrochemsci.org/papers/Vol14/140706077.pdf

www.carsguide.com.au/ev/advice/evcharging

www.en.wikipedia.org/wiki/Charging_Station

www.marginlewis.com/pubs/2022/06/06/the-importance-of-renewables-for-electric-vehicle-charging

Rempel, D et al. "Reliability of Open Public Electric Vehicle Current Fast Chargers."

https://ssrn.com/abstract=4077554

www.fortune.com/2022/10/17/electric-vehicles-have_charging-access-problem-these-companies-are-working-to-save-it/

www.claytonutz.com/knowledge/2022/September/emerging-challenges-for-australia-electric-vehicle-charging-infrastructure

www.bloomberg.com/news/articles/2022-08-1/the-ev-charging-buildout-has-a-problem-many-stations-dont-work

Mostoi, M et al. "An indepth analysis of electric vehicle charging and station infrastructure, policy implications, and future trends." Energy Reports Vol 8 pp 11504 – 11529 Nov 2022.

www.sciencedirect.com/science/article/pii/S2352484722017346

www.ceicdata.com/en/indicator/norway/oil/_consumption

www.statista.com/statistics/793330/sales-volumeof-auto-diesel-in-norway

www.thedriven.com.10/2022/10/17/norway-ev-sales-high-but-long-road-to-leave-fossil-fuels-behind/

www.carsales.com.au/editorial/details/how-much-does-it-cost-to-replace-an-ev-battery-136621/

www.carexpert.com.au/car-news/ev-battery-prices-increased-this-year-for-the-first-time

www.recurrantauto.com/research/costs-ev-battery-replacement

www.whichcar.com.au/advice/phev-plug-in-hybrid-vehicles-on-sale-australia-january-2023

www.ev-volumes.com

www.iea.blob.core.windows.net/assets/eOd2081d-487d-4818-8c5969b-globalelectricvehiclesoutlook2022.pdf

www.hedgescompany.com/blog/2021/how-many-cars-are-there-in-the-world/#

www.iea.org/data-and-statistics/data-product/global-ev-outlook-2022

www.iea.org/global-ev-outlook-2022/trends-in-electric-heavy-vehicles

www.nrel.gov/docs/fy19osti/74426.pdf

www.iea.org/reports/grid-scale-storage.

www.e360.yale.edu/features/in-boost-for-renewables-grid-scale-battery-storage-is-on-the-rise

www.en.wikipedia.org/wiki/vanadium-redox-battery

www.forbes.com/sites/eriKKobayashisolomon/2022/08/18/we-have-an-energy-problem/?sh=38d35591681

www.en.wikipedia.org/wiki/iron_redox_flow_battery

www.pv-magazine.com/2022/12/09/us-to-deploy-us-2013-30-gw-11-gwh-of-grid-scale-energy-storage-by-2025/

www.blackridgeresearch.com/reports/southkorea-ess-market

www.iea.org/reports/grid-scale-storage

www.solarviews.com/blog/are-lithium-ion-the-best-solar-batteries-for-energy-storage.

www.Igessbattery.com/au/home-battery/news-view.Ig?blcSn=1413

www.clean.energy/reviews.info/battery-storage

www.chinadaily.com.ch/9/202302/24/ws63f81ab3a3105c47ebb0a6b.

www.woodmac.com/news/opinion/european-grid-scale-energy-storage-capacity-will-expand-20-fold-by-2031/

www.euronews.com/my-europe/2022/12/02/europes-largest-energy-storage-facility-begins-operation-in-belgium

www.takonbattery.com/top10-grid-scale-energy-storage-countries-in-europe-2022/

www.energy-storage.news/australia-surpassed-1gwh-of-annual-battery-storage-deploymments-during-2021/

www.fortunebusinessinsights.com/industry-reports/iron-flow-battery-market-101324

www.blackridgeresearch.com/blog/top-flow-battery-companies-manufacturing

www.international.org/tin-iron-redox-flow-battery-looks-promising

www.energy.gov.au/eere/fuelcells/articles/progress-grid-scale-flow-batteries

www.pv-magazine.com/2022/05/24/iron-flow-battery-maker-ess-expands-into-Europe/

www.yearbook.enerdata.net/total-energy/world-consumption-statistics.html

www.iea.org/reports.electricyrinformation-overview/electricity-consumption

www.eia.gov/outlook/ieo/pdf/transportation.pdf

www.en.wikipedia.org/wiki/list_of_countries_by_electricity_consumption

www.ourworldindata.org/energy-production-consumption

www.irena.org/Energy-Transportation/Technology/Transportation/Transportation-costs-renewables-biofuelsall-2016

www.businessinsider.com/this-is-the-potential-of-solar-power

www.en.wikipedia.org/wiki/Solar-energy

www.eia.gov/energyexplained/use-of-energy/transportation.php

www.en.wikipedia.org/Fuel_cell

www.energy.gov/fuelcells/hydrogen-storage

www.en.wikipedia.org/wiki/Hydrogen_storage

www.pubmed.ncbi.nlm.nih.gov/150852731 (hydrogen critical temperature and storage).

www.sciencedirect.com/science/article/pii/S0360319919310195 Andersson. J et al. In Jn. of Hydrogen Energy. Large Scale Storage of Hydrogen. Vol 44, 23, pp 11901-11919, 3/5/2019.

www.en.wikipedia.org/wiki/hydrogen_production

www.iea.org/reports/hydrgen

www.sciencedirect.com/science/article/pii/S2590174520300155

www.pubmed.ncbi.nlm.nih.gov/150852731

www.haskel.com/en-au/blog/how-does-a-hydrogen-refuelling-station-work

https://www.sciencedirect.com/article/pii/S2590116820300436 Greene, D. et al. Challenges in the designing, planning and deployment of hydrogen refuelling infrastructure for fuel cell vehicles.

www.drive.com.au/the-problem-with-hydrogen-fuel-cell-cars/

www.forbes.com/sites/jamesmorris/2021/02/06/why-are-we-still-talking-about-hydrogen?sh=C0978e7f044e

www.en.wkipedia.org/wiki/Hydrogen_internal_combustion_engine_vehicle

www.greencarcongress.com/2009/03/high-pressure-d.html

www.ieafuelcell.com/index.php?id=33

www.eere.energy.gov/hydrogenandfuelcells/tech_validation/pdfs/fcm03r0.pdf

www.sciencedirect.com/science/articles/abs/pii/S0360319922005936

www.energyeducation.ca/encyclopedia/biofuel

www.en.wikipedia.org/wiki/biofuel

www.currentresults.com/weather/united-kingdom/annual-sunshine.php

https://www.livingin-australia.com>sunshine-hours-aus

www.en.climate-data.org/oceania/australia/northern-territory/alice-springs-586/

www.en.wikipedia.org/wiki/biofuel

www.voltaoil.com/what-makes-up-retail-price-for-gasoline/#

www.sciencedirect.com/science/article/pii/S037838201

Patel, M et al. "What is the cost of renewable diesel from woody biomass and agricultural residue based on experimentation". Fuel Processing Technology Vol 191 August 2019 pages 79 – 92.

www.stasta.com/statistics/282774/global-product-demand-outlook-worldwide/

www.farm-energy.extension.org/oilseed-crops-for-biodiesel-production/

https://afdc.energy.gov>files>pdf

www.theglobalist.com/food-fuel-compete-land/

www.energyeducation.ca/encyclopedia/biofuel

www.mdpi.com/2077-0472/7/4/32

Aranjo, K et al. "Global biofuels at the crossroads: an overview of technical, policy and investment complexities in the sustainability of Biofuel development." Stony Brook Univ, NY, USA.

www.bbc.com/news/business-15462923

www.stastica.com/statistics/653066/breakdown-of-the-untied-states-diesel-price-by-expenses/

www.transportenvironment.org/discover/biofuels-aretwice-as-expensive-as-fossil-fuels/

www.tnstate.edu>extension/documents/biodieseleconomics.pdf

Doombosch, R. Steenblik, R. "Biofuels: is thew cure worse than the curse ?" SG/SD/RT (2007) 3/ REVI. OECD Round Table on Sustainable Development, Paris 2007.

Fischer, G. Van Velthaizen, H et al. "Global Agro-Ecological Zones Assessment Report RP06003; IIASA Vienna 2006.

www.world-nuclear.org/information-library/facts-and-figures/heat-values-of-various-fuels.aspx

www.bbc.com/news/business-59546281 (Drax power station).

www.futuremetrics.com-info/wp-conrent/uploads/2013/07/CO2-from-wood-and-coal-combustion.pdf

www.nrdc.org/stories/no-burning-wood-fuels-not-climate-friendly

www.sciencedirect.com/science/article/pii/S1309104215304839

Robinson, D. "Australian wood heating currently increases global warming and health costs." Atmospheric Pollution Research Vol 2, 3, July 2011, pages 267 – 274.

www.theguardian.com/environment/2022/aug/11/burning-imported-wood-in-drax-power-plant-doesn't-make-sense-says-kwarteng

www.mottmac.com/article/2282/stevens-croft-biomass-power-station-uk

www.power-technology.com/projects/wood/burning/

www.worldbioenergy.org/uploads/factsheet%20-%20Pellets.pdf

www.e360.yale.edu/features/wood_pellets_green_energy_or_a_new_source_of_CO2_emissions

www.thehikingauthority.com/is-it-better-to-burn-wood-or-let-it-rot/

www.en.wikipedia.org/wiki/Geothermal_power

www.irena.org/Energy-transition/geothermal-energy

www.css.umich.edu/publications/factsheets/energy/geothermal-energy-fact-sheet

www.eia.gov/energyexplained/hydropower/tidal-power.php#

www.en.wikipedia.org/wiki/Tidal_power

www.en.wikipedia.org/wiki/Wave_power

www.e360.yale.edu/features/why_wave_power_has_lagged_far_behind_as_energy_source

www.en.wikipedia.org/wiki/nuclear_power_power_plant

www.world-nuclear-org/information-library/nuclear-fuel-cycle/nuclear-power-reactors.aspx

www.nei.org/fundamentals/how-a-nuclear-reactor-works

www.ucsusa.org/resources/how-nuclear-power-works

www.britannica.com/science/uranium-235

www.wikipedia.org/wiki/small_modular_reactor

www.wikipedia.org/wiki/nuclear_fission_product

www.wikipedia.org/wiki/long_lived_fission_products

www.en.wikipedia.org/wiki/Environmental_impact_of_nuclear_power

www.nytimes.com/2022/11/15/business/nuclear-power-france.html/

www.en.wikipedia.org/wiki/advanced_gas_cooled_reactor

www.en.wikipedia.org/wiki/Nuclear_power_in_the_united_kingdom

https://www.reuters.com>business>energy>costs-edfs

https://www.smh.com.au

www.afr.com/energy/nuclear-energy-too-expensive-to-replace-fossil-fuels-20220711-p5b0bd

www.en.wikipedia.org/wiki/Three_mile_island_accident

www.en.wikipedia,org/wiki/Beznau_nuclear_power_plant#

Chapter 14

Gas and Coal – a Modern View

Natural Gas

Natural gas (methane CH_4) was used to generate 22% of global electricity in 2022, second only to coal, 35%; and far ahead of intermittent renewables wind (7.5%) and solar (4.5%).

Gas is a very convenient practical fuel for this purpose. Advantages include:

Gas power stations are cheap and quick to build. An 800 megawatt plant in the US costs between US$800 million to 1.2 billion, less than half a coal station and a small fraction of nuclear. As gas is a comparatively clean fuel getting consents is easier, quicker and cheaper.

Gas plants can be turned on and off quickly and easily. They also run efficiently at lower loads compared to coal, which is most efficient at full load, and very expensive to turn on and off. So gas is ideal to meet peak loads or to supplement intermittent renewables.

The exhaust from gas plants is comparatively clean with very low ash or sulphur contents compared to coal. It is mostly water vapour and CO_2. It does not need scrubbing and other cleaning processes used for coal.

For those concerned about CO_2 emissions gas produces about 450 grams of CO_2 per kilowatt hour of electricity made, compared to about 900 grams for coal, half.

Gas is very easy to transport by pipelines and special ships as already discussed.

Gas plants are comparatively quick and cheap to decommission and dismantle should that be necessary.

There are two principle types of gas power plant:

Simple: this type just burns the gas and the hot gas jet resulting is directed at turbines which spin to drive generators making electricity; like an aircraft jet engine.

Combined Cycle: again the hot gas jet resulting from burning the gas is used to drive turbines, but also extra heat resulting from the burning is used to boil water to high pressure steam which is also used to drive turbines, in turn driving electricity generators.

The Combined Cycle plants are much more efficient than simple ones, about 60% of the energy in the gas ends up as electricity. But they are more expensive to build. An 800 megawatt simple plant costs about US$650 million, versus 800 million to 1.2 billion for Combined Cycle. Most plants are the latter.

Some smaller gas plants use piston internal combustion engines. These cost about the same as Combined Cycle for the same power output, but are far less common. It is a convenient method for isolated places just needing a small electricity plant.

World-wide there are almost 10,000 gas power plants larger than 50 megawatts each. The US has the most, 987, generating 39% of the country's electricity. There has been a big increase in gas power generation there over the last 20 years; gas production there was 550 billion cubic meters in 1998 and 950 in 2021. There are plans for 177 more US gas power plants.

Other countries with large fleets of gas power plants include Russia, 177; China, 163; Germany, 131; and the UK with 61. Gas power stations are very widely spread globally including many third world countries reflecting its cheap practicality.

The world's biggest gas plant is in Russia at Surgul, a 5597 megawatt facility. This makes 39 billion kilowatt hours of electricity annually using 10 billion cubic meters of gas. It has six 800 megawatt generators. Another big one is in Japan at Chila, 3996 megawatts. Many countries are planning expansion of gas stations to replace old coal ones; for example Australia plans another 1000 megawatts of gas to cover the closure of the large Liddell coal plant in New South Wales, including a 660 megawatt gas station at Kurri Kurri in the Hunter Valley costing $660 million Australian.

A big potential problem is limited known gas reserves globally. In 2022 these were 196 trillion cubic meters (a trillion is a million millions). As global consumption is about 4 trillion cubic meters per year this is only 52 years supply. Further only a few countries hold about two thirds of known reserves. These are:

- Russia 20%
- Iran 17%
- Qatar 12%
- Turkmenistan 6%
- US 6%
- Venezuela 3%
- Saudi Arabia 3%%

Many other countries have smaller reserves including Australia , 0.4% of global.

This has impacted the price of gas especially for European plants affected by embargoes on Russian gas due to the war in Ukraine. The 131 gas plants in Germany were paying US$2.91 per cubic meter in 2000, but in 2021 this cost was US$8.94, three times higher. This has greatly impacted electricity prices there.

That half of the reserves are in countries hostile to the western democracies is of concern.

The US, by contrast, is self-sufficient concerning gas *at present*, cost stable at about US$4 per cubic meter. US consumption is close to one trillion cubic meters per year and known reserves there are 17 trillion cubic meters. So in about 20 years the US will be importing big quantities of gas. As decades pass and global reserves are used up looming gas shortages mean price increases, and increased electricity prices where gas is a major fuel; which is most places world-wide. Or more reserves may be found; but it is not likely the huge reserves necessary if consumption continues to rapidly expand will be revealed given the highly developed nature of world gas exploration now.

So natural gas is a short term fix only for the world's voracious electricity appetite.

Coal

Coal fired power stations generate 41% of the world's electricity, and in many countries a much larger percentage; for example China 58%, India 70% as do many other Asian countries; Poland 80%; Germany 43%; Indonesia 60%; Australia 51%.

The reasons for this are that coal is very plentiful and widely available in many countries. And electricity made in coal power stations is the cheapest as shown below:

- Coal US$ 40 – 80 per megawatt hour of electricity generated.
- Gas combined cycle 70 – 120
- Gas simple 180 – 430
- Solar 90 – 170
- Wind 65 - 115
- Wind plus batteries as backup when no wind 320 – 640
- Wind plus coal as backup 75 – 120
- Solar with backup is more expensive than wind with backup.
- Nuclear is US$ 110 – 190 per megawatt hour, but this does not take into account the huge cost and time involved in building nuclear power plants as discussed.

Further the cost of building coal power plants is reasonable compared to alternatives. For example costs for a 1000 megawatt plant are shown below:

- Coal US$ 2.2 – 3.5 Billion
- Gas combined cycle 1 – 1.4
- Gas simple 0.8 – 1

- Solar (no back-up) 2.5 – 3

- Wind (no back-up) 1.3 – 1.8

- Wind plus battery backup when no wind 30 – 40

- Nuclear 25 plus

Concerning the cost of building wind with battery backup a 650 megawatt wind plant with battery backup for three days only was quoted at $27 billion in references below. The cost for 1000 megawatt has been estimated from that.

A further massive global use of coal is for industry especially steel and metal manufacture; about a third of coal use is for these purposes. The energy for industry totals 46,000 terawatt hours annually, 37% of human usage, as already discussed.

Coal is a very natural substance found in abundance in the Earth's crust both near the surface and deeper. It has been used by humans for thousands of years for heating and cooking. The earliest record is from China in 3490 BC when surface coal was used in households. In Greece in 370 BC it was used in metal working. The Romans were commonly using coal to heat buildings and baths by 200 AD. The Aztecs widely used coal for metal work and heating.

In Britain in the 13th century coal was so commonly used that it was banned in London as the smoke was considered hazardous to health, and the use of wood and charcoal was made compulsory. But by the 16th century wood was in short supply so coal was again reused. The first known coal mine was built by Sir George Bruce in Scotland in 1575 and was hailed as a wonder of the age. In the 1600's the French in North America were mining coal from seams at the surface and used it widely.

In the 1700's the industrial revolution started in Europe and was powered by coal for more than 200 years. In the 1780's coal was discovered in the Hunter Valley in Australia.

Today about 8 billion tonnes of coal are mined globally world-wide including 3 billion in China and 1.1 billion in the US.

Almost all coal was formed during the **Carboniferous Age** from 360 to 300 million years ago when conditions were ideal for it. The formation of coal needs:

1. Forests of trees. Trees first formed 400 million years ago in China. By 360 million years ago the land was mostly in a huge super-continent called **Pangaea** which had formed by merging of previous land masses by plate tectonics and this was covered with swampy dense forests of trees.

2. Low basin areas which fill with sediment from dead trees. The dense forest trees lived then died and fell into the swampy low basins. As fungi and bacteria which eat and destroy woody trees had not evolved then the dead trees lay there, sank into the swampy mud, and got increasingly covered as more trees lived, died and fell on them. After millions of years the dead trees were deeply buried. The pressure and temperature became very high resulting in coal formation.

The sea level was low at this time increasing the land area, and swampy regions.

At the beginning of the Carboniferous Age the average Earth surface temperature was high, 20°C as opposed to 14°C now. Atmospheric carbon dioxide was also high at 0.15%, 1500 parts per million, compared to 0.04% now. These conditions were ideal for rapid plant growth and photosynthesis. This resulted in massive oxygen formation, atmospheric oxygen was 35% compared to 21% now. So the forests grew very rapidly, died rapidly also, and laid down huge volumes of dead trees to become coal.

At the end of the carboniferous age 300 million year ago this all stopped. The surface temperature dropped to about 12°C, the carbon dioxide had dropped to 0.035% (350 parts per million) and an ice age followed. Why this occurred remains uncertain.

The huge super-continent Pangaea eventually broke up to form precursors of our land masses today by plate tectonics. The coal therefore was widely distributed.

Coal varies in quality depending on the intensity of temperatures and pressures deep in the ground when it was formed. The key factor is the **percentage of carbon** in the coal. Essentially as discussed already photosynthesis results in atmospheric carbon dioxide being made into **hydrocarbon** compounds especially wood, and oxygen using the Sun's energy. Old high quality coal has more carbon than younger lower quality. And when it is burnt as fuel it makes more heat energy because its carbon is the main fuel burnt, forming carbon dioxide after combining with atmospheric oxygen.

High quality coal is called **Anthracite** or **Black Coal.** It is 86 – 97% carbon. It is especially sort after for metal manufacture, also for electricity generation. Parts of the US and Australia have large amounts mined.

The next grade down is **Bituminous** or **Brown** coal. This has 60 – 86% carbon. It is used for steel making and electricity generation.

Next down is **Lignite** coal which has about 50% carbon. It is also used for steel manufacture and electricity generation. It has more water than higher grades and may be dried and washed before use to improve performance.

Below that is **Sub-bituminous** coal with about 35 – 45% carbon, not highly sort after. It is not far away from **Peat**, famously used in Ireland and other cold wet countries as fuel 100 years ago. This is about 30% carbon and contains a lot of water.

When the carbon in coal burns it combines with oxygen to make CO_2 giving off a large amount of heat energy. There are other substances in coal which also burn, combine with air oxygen, forming various **oxides**. While the great bulk of heat energy from coal comes from carbon burning the other substances are important as some cause environmental or health hazards. Their presence reflects the complicated nature of the original trees which made the coal 300 plus million years ago. They are:

- Hydrogen. Most hydrogen from the original hydrocarbons in wood has gone from coal but a small amount remains and can form water when burnt.

- Sulphur. This forms Sulphur Dioxide, SO2, when burnt. In the atmosphere this can combine with water vapour to make Sulphuric Acid, resulting in **Acid Rain** which can damage plants when it falls. Also Sulphur Dioxide forms Sulphate particles in the air which block sunlight and cause smog.

- Nitrogen. When burnt this results in nitrates, which can form nitric acid, H2NO3, in the air contributing to acid rain.

- Fly Ash. This is small particles of unburnt material found in big quantities in the gas resulting from burning coal. Fly ash is responsible for the dense black smoke emerging from old power stations causing most things around them to be stained with black soot, and cities like London to be afflicted with dense fog.

It also contains Mercury, Lead, Arsenic, Cadmium, Chromium, and Selenium in varying amounts. These are all toxic to living organisms to greater or lesser degrees.

Radioactive Uranium is present also as that is part of the Earth's crust as discussed already. About 5 tons of Uranium, containing about 50 kilograms of radioactive Uranium 235, is released annually globally from coal use. This radiation is about a twentieth of the amount released by Chernobyl.

These undesirable substances are largely removed from modern coal plant emissions by a variety of techniques:

Fly Ash:

More complete burning of coal reduces the amount of ash and also increases the heat output, efficiency. This is achieved either by pulverizing the coal into a powder which is then sprayed into the furnace; or by initially pulverizing, then "fluidizing" by mixing it with sand and lime powder, suspending it in an airflow and burning that. This is called a *Circulating Fluidized Bed* system (CFB). It is the newest method and is discussed further below. The pulverizing method is most common in older stations.

Virtually complete removal of fly ash takes place by passing the emissions through *Bag Houses.* These contain thousands of bags of fine filters which last for about 3 years before disposal and replacement. Other technologies including *Electrostatic Precipitation* (ESP) machines which remove ash by charging particles in an electric field; and *Cyclone Precipitators*, which remove the particles by spinning the emissions like a cyclone resulting in a centrifuge action. By these means 99.9% of the fly ash is removed including virtually all particles down to 0.1 microns size. Some less than that do escape this process. These *Dry* processes have largely replaced *Wet* processes using a lot of water in *scrubbers* where the water is sprayed into the emissions to remove the ash, then stored in large ash ponds. The pond solutions should be disposed of carefully to prevent pollutant release into the environment including groundwater, and for modern stations in advanced countries that is happening under government regulations, but is less common in third world countries regrettably.

Removed fly ash can be used for fertilizer and cement. In Europe more than two million tonnes is used for cement manufacture annually.

Many countries have strict regulations to limit fly ash particles in emissions. For example in China only 30 milligrams of particles are allowed per cubic meter of emission gas. In the EU 50 is allowed for black coal and 100 for brown coal. In the US 125 is allowed. In Australia with lax regulations the limit is 260.

Sulphur Dioxide, Nitric Oxide, other pollutants including metals:

Sulphur dioxide causes an increase in cardiovascular and respiratory mortality and increases hospital admissions for these significantly. Birth weights are decreased. Coal plants without emission control measures emit huge amounts; in China this was more than 25 million tons, and over 13 million tons in the US annually before reduction measures were applied.

These are partially removed by **Scrubbers** sometimes combined with **Selective Catalytic Reduction** (SCR) in a process called **Flue Gas Desulfurization** (FGD). The scrubbers remove 90% of the Sulphur Dioxide and most of the other pollutants, which are also partially removed by the fly ash removal methods. Scrubbers use an alkaline water solution to neutralise any acids in the emissions. The solution is then stored in large ponds for later disposal. There are a variety of types of scrubbers including wet limestone scrubbing, wet lime FGD, dry lime FGD, fluidized bed lime injection, dry sorbent injection.

Flue Gas Desulphurization is now compulsory in the US, the EU, Japan, China and many other countries – but is not used in Australia.

Activated Coke is another method being used including in China. The flue gas passes through the coke and sulphur dioxide, most particles and many metals including mercury are largely absorbed by getting stuck on the coke surface. This surface is huge because coke has many crevices. Later the pollutants can be *de-sorbed* and the coke reused.

The quantities of sulphur dioxide which can be removed by these methods are huge; in China alone more than 25 million tons is removed annually, in the US more than 12 million.

Government regulations are essential to force coal power plants to drop emissions. In the EU and China the limit is 400 and 200 mg per cubic meter of flue gas; in the US it is 1500; in Australia up to 2370 for some mines.

Mercury:

Mercury is of special concern because there is a comparatively large amount in low quality brown coals, up to 1.5 grams per tonne. High quality black coal has around 0.01 mg per tonne, much lower. If it escapes in emissions rain will cause it to end up in the sea or land. Fish and molluscs concentrate mercury, and when humans eat them they can get a big dose, which can be toxic to nerves and causes foetal abnormalities. The risk is real where people eat a lot of fish frequently, for example, Japan has had problems. Pregnant women are advised to have only two fish meals per week, and then avoid large predatory fish like shark and swordfish who eat other fish and concentrate mercury. But mercury is part of the natural world and harmless in that context at low levels.

Many countries have strict regulations to limit mercury emissions from coal power stations. Examples are:

- US Black Coal 1.5 micrograms per cubic meter M³ emissions

Brown Coal 14 µg/m³

- EU 30
- China 30

Unfortunately many other countries have no regulations or else very lax ones. A surprising example is Australia with 19 coal stations, none of which have scrubbers, electrostatic precipitation, selective catalytic reduction or other pollution control measures except that 15 have baghouses which reduce ash and particles by more than 99.9%. So four Australian coal power stations do not have bag houses with filters, three in Victoria and one in Queensland. An excuse is that the coal used is very high quality black coal from Australian mines. But some mines there, especially in Victoria, have brown coal. Concerning Mercury Australia allows 200 to 2000 µg per cubic meter individualized for each mine, massively higher than the above. This underlines the importance of government regulations controlling pollution.

That some uncontrolled Australian mines emit 2000 µg/M³ of mercury shows how effective the control measures can be, reducing this to very low levels.

Mercury is partially removed by the methods to remove fly ash and sulphur dioxide mentioned above. An additional method, injection of **activated carbon** particles, often supplements these to get the level low enough to comply with strict regulations in China, the EU and US especially.

Nitric Oxides

Nitric oxides cause severe asthma and decrease lung function, also allergies.

As said these are partially removed by the above methods. In addition three special methods are used to get the level low enough to meet strict regulations again in China, US and EU.

Low Nitrogen Burners (LNB) plus **Selective Catalytic Reduction** (SCR) plus **Overfire Air** (OFA) are the three methods. In Australia with lax regulations stations can emit up to 1500 mg per cubic meter, but China's limit of 200 has recently been dropped to 50. Hence their application of these measures. In the EU it is 200 and 875 in the US.

Selective catalytic reduction is the most common method to remove nitric oxides. This involves injecting ammonia and a catalyst into the emission gas, and a reaction takes place which turns the oxides into water and nitrogen gas, harmless. 85% of nitric oxides are removed.

There are many interesting new methods of pollution control being researched. These include:

Electrostatic Oxidation (ECO): This is being researched by Powerspan Company in a 50 megawatt pilot plant. The method involves a special process combined with electrostatic precipitation and removes fly ash particles, sulphur and nitric acids. It results in Nitric, Sulphuric acids and remnants of ash which is made into fertilizer.

Electron Beam Dry Scrubbing: A low energy electron beam from an electron accelerator is used to "dry scrub" the flue gas removing 95% of sulphur dioxide and 70% of nitric oxides as well as other pollutants. Initially the flue gas is cooled to 70°C with a water spray. Ammonia gas is then added to neutralise the acids and the result is solids for – yes! – fertilizer. The 2015 paper describing it is referenced below. The equipment is large and expensive and little has been heard of the method since.

Circulating Fluidized Bed Burn (CFB) is another coal burning method mentioned briefly above which is moving into more general use. Activated carbon and limestone are mixed with the pulverized coal particles which are suspended in an airflow then burnt at a much lower temperature than that used in the Pulverized Coal (PC) method. Then ammonia gas is added. It results in much lower particulate matter (PM) about 9 mg per cubic meter, and very low nitric oxides because of the low temperature burn, and low sulphur dioxide concentration. Excitingly it can be used for low quality brown coal, biomass, even old tires and waste. And still with good pollutant control. Again China is leading with this as are Japan and many other countries. It is replacing more simple high temperature Pulverized Coal.

Combined Heat and Power (CHB): This method uses the huge amounts of waste heat lost to the atmosphere by coal plants, up to two thirds of the energy generated. The heat is used is special combustion turbines to make more electricity.

High Utilization and High Quality Coal Usage are other more practical strategies to improve efficiencies and decrease pollution. Coal plants work best if operating close to full output. They are slow and expensive to shut down and restart making them a bad mix with intermittent solar and wind, unlike gas. And black high quality coal produces more energy and less pollution than low quality coals. This is clearly demonstrated in Europe where most of the sulphur dioxide emissions come from the southeast where low quality lignite coal is used. These plants have 20 times the output of other European plants concerning sulphur dioxide and 16 times particulate matter.

Another strategy is the move towards High Energy Low Emission plants – **HELE** plants. These use the Pulverized Coal burn method but have much higher steam pressures to improve performance and remove the need to condense the steam back to water. Conventional stations are called "subcritical" and use steam at about 160 times sea level air pressure. "Supercritical" work at 220 times, and "Ultra Supercritical" (USC) at 300 times sea level air pressure. These plants are about 20% more efficient with about 15% less pollutants than subcritical.

Four of Australia's 19 coal power plants are HELE Super Critical, all in Queensland. In China 19% of coal power plants are ultra – supercritical, and 58% of those under construction are. The US has just one ultra – super critical, a 600 megawatt plant in Arkansas which cost US$1.8 billion. This is 30% more than sub critical, which most US plants are; no more ultra – supercritical are being built in the US for cost reasons, but in China where regulations are tougher, the cost structure lower, and the pollution problem far greater, many are.

As well as these very technical methods a simple method is *Coal Washing* before burning; this removal of dirt and dust reduces pollutants including mercury by about 30%.

There is absolutely no doubt that old uncontrolled coal power plants caused very large rates of death and disease, especially lung problems, amongst those living nearby. In the US poor air quality seriously affects two million people each year, and estimated costs are US$100 billion. In the EU it is estimated 3000 premature deaths occur in children alone and 8000 more are affected seriously. The very bad track record of China concerning this is well known and the same goes for many other Asian countries where big populations live in crowded cities near coal power stations. Even in Australia, with its wide open spaces, there is evidence that coal burning causes 800 – 1500 premature deaths each year largely due to lung problems. Historically European cities including London were filthy with soot and the air foul with dense smog due coal use in domestic fires, and later coal power stations nearby.

But a modern balanced view of coal fired power stations must take into account the modern methods described above of practically eliminating serious pollution. The end result from modern plants so equipped is emissions which are visually clean, with no visible soot or brown discolouration, the serious pollutants largely removed, and the emission consisting of about 5% water vapour, about 11% CO_2, about 5% oxygen, and 79% nitrogen gas emerges from the station finally. These emissions are compatible with the Earth's environment given the arguments above about carbon dioxide being irrelevant concerning temperature, and the essential substance for photosynthesis and life.

A technical advantage of coal power stations is that they result in **Stable Grid Electricity** for countries where they are. This is because the turbines generating electricity from high pressure steam spinning them can be made to spin nationally at the same speed generating reliable AC (alternating current) power at 50 cycles per second, the common frequency. And the large number of turbines spinning in a country provide *backup inertia* to even out power demand fluctuations and keep the national grid smoothly operating despite variations in demand which are very common.

This same advantage applies to gas plants and nuclear where high pressure steam drives the turbines generating electricity.

By contrast electricity generated from wind and solar is initially DC (direct current), then is converted to AC in inverters as previously discussed. There is no inherent stability and little capacity to respond to demand changes; grid and supply failure can occur. This requires large expensive machinery to counteract including very large capacitors and large expensive *synchronous condensers* to try to ensure a reliable power supply.

Some have argued that **Carbon Capture and Storage** (CCS), a variant being **Carbon Capture with Underground Storage** (CCUS), so – called **Clean Coal**, is the answer. This offers continued use of coal but the CO_2 is captured, a very expensive energy intensive process, and permanently stored, not released into the atmosphere. It may be pumped deep into the sea, or into old oil and gas field caverns deep underground, or into deep underground salty aquifers. But there are problems with this including:

- It is far more costly to build these plants, at least double or three times, even up to eight times. A 1000 megawatt CCS coal plant could cost US$6 to US$24 billion, compared to US$3.5 billion for a coal plant with other pollution controls as discussed. An example is the Kemper 240 megawatt CCS coal plant in the US which was abandoned unfinished when costs had reached US$7.5 billion in 2017.

- The cost of the resulting electricity is up to 8 times higher than coal alone at US$350 per megawatt hour. This is far more expensive than all other methods apart from wind with battery backup as discussed above; meaning CCS is not economic.

- About 30% of the electricity generated is used for the carbon dioxide storage process. This means that ultimately plants have to be 43% bigger for the same useful output to the consumer, a massive increase in coal consumption and cost.

- The CCS process uses 55% more water than just coal stations, which are already a big user especially for pollution control. This is a further huge financial and environmental cost. One study suggested human water use would double if CCS became widely used.

It is clear that CCS coal power plants are a huge risk financially for these reasons, and cannot compete with any other electricity source apart from wind with battery backup. Especially they cannot compete with gas. Many CCS projects have been started and then abandoned in the last 15 years. Examples are:

- US Petra Nova 240 megawatt plant. This small plant cost US$1 billion, ran for a few years and shut in 2020.

- UK Ferrybridge 5 megawatt pilot plant. A bigger one was planned but the project was abandoned in 2011.

- A Netherlands 250 megawatt project was abandoned in 2017.

- A German pilot plant was abandoned in 2018.

- The EU has spent one billion Euros on research with no plants built.

- In Australia a Queensland pilot plant costing A$240 million to that point was decommissioned and abandoned in 2013. And a large 1320 megawatt CCS plant was started at Vales Point, New South Wales, had massive losses during construction and was abandoned. The remnant was sold for a token sum in 2015.

There is only one CCS coal plant operating globally now, the Canadian 110 megawatt small plant at Boundary Dam. This cost US$1.5 billion to build and provides electricity at double the price of other sources including hydro in Canada. It is abundantly clear that this technology is impractical. In addition by burying CO2 deep underground permanently the normal photosynthesis process is denied, and the oxygen resulting lost from the surface; this imitates the very long term process where all CO2 will eventually finish up in carbonate rocks, and life will be history, as discussed earlier. So it is arguably just as well it is impractical!

Cheap plentiful electricity is the bedrock of modern civilization with its vastly improved living standards, including health, compared to pre-industrial times.

Many countries including China, India, other Asian countries, find it impossible to cease using coal. The living standards of their big populations are very dependent on cheap plentiful electricity and as their industries expand demand increases. Because of frequent cloud cover and little wind much of Asia is not suitable for these renewables, and even if it were the demand is too great day and night for these to be the answer. China uses coal for 58% of its electricity, has built 300 new plants since 2011, and has another 200 gigawatts (say 200 1000 megawatt plants) planned. India has 94 gigawatts being built (say 94 1000 megawatt plants). Many other Asian countries are doing the same. Coal is responsible for 51% of Australia's electricity.

Below is a list of countries, their operating coal power stations in 2022, and the equivalent number of 1000 megawatt power stations:

Country	Plants	Equivalent 1000 Mw stations
China	1118 plants	1074 equivalent 1000 Mw stations
India	285	233
US	225	217
Japan	92	50
Indonesia	87	40
Russia	71	40
Germany	63	37
Poland	44	30

What is very noticeable about China's use of coal is the tough government regulations around pollution to good effect; and that is the case for other Asian, European, North American countries and Japan.

But the movement in Europe and North America has been to move away from coal to gas or renewables as discussed. That is proving very difficult and resulting in big price increases for electricity impacting the living standards of populations. This will get worse as gas reserves run down. And moving to wood, biomass results in more CO2 to worry those concerned about that. Some European countries have moved back to coal including Germany where 43% of electricity came from coal in 2018; others have stayed with it including Poland, 80% from coal.

There are massive coal reserves globally, and more continue to be discovered. In 2021 there were 1041 billion tonnes of minable reserves globally; annual usage is 8 billion tonnes, meaning 130 years supply globally now.

The US has reserves of 250 billion tonnes; and digs 1.1 billion tonnes each year; meaning it has 227 years supply for its own needs. Coal provides 27% of US electricity.

Russia has 160 billion tonnes reserves and digs 420 million each year; 380 years supply. It uses about 90 million tonnes annually and exports the remaining 330 million tonnes for a huge income.

Australia has reserves of 147 billion tonnes, 80% of it close to the surface suitable for open cast mining, and much of it high quality black coal. 483 million tonnes are mined each year of which 382 million tonnes were exported in 2018, much of it to China, for a huge profit funding the "lucky country's" high standard of living. There are 304 years supply.

Ukraine has reserves of 34 billion tonnes, digs 33 million annually, 1000 years supply. Poland has reserves of 26 billion tonnes, 213 years supply. Germany has 36 billion tonnes, digs 170 million tonnes, 211 years supply.

India has reserves of 101 billion tonnes, digs 771 million annually used mainly domestically to supply 70% of its electricity. It has 130 years supply and imports 240 million tonnes each year.

Indonesia's reserves total 101 billion tonnes. It digs 549 million each year of which 80% is exported, the rest used domestically for electricity; it is the world's largest coal exporter. Its reserves will last 67 years at this rate.

China has 138 billion tonnes in reserves, digs a massive 3.5 billion annually and imports 295 million. At this rate its reserves will last 39 years.

Given all these considerations it is completely irresponsible, as many democratic governments are doing, to cheerfully state that coal is finished; and especially that renewables and nuclear can feasibly replace it. There is clear hypocrisy in this view for coal exporting countries like Australia, and the many who continue to need large amounts of coal for their energy. The responsible path is as China is doing: very strict regulation of pollutant emissions combined with expansion to increase the living standards of its population. Many other countries are quietly following this path.

References and Further Reading for Chapter 14 – Gas and Coal – a Modern View

www.wikipedia.org/wiki/Gas_fired_power_plant

www.statista.com/statistics/1281761/number-of-gas-power-plants-by-country/

www.blog.resourcewatch.org/2019/11/13/this-map-shows-29000-of-the-worlds-power-plants

www.usatoday.com/story/news/2019/09/09

www.energy.gov.au/data/electricity-generation

www.theguardian.com/australia-news/2021/may/19/coalitions-600m-gas-fired-recovery-boost-what-you-need-to-know#

www.quora.com/what-is-the-cost-of-building-an-800-MW-natural-gas-power-plant

www.worldometers.info/gas/gas-reserves-by-country

www.statista.com/statistics/557322/installed-natural-gas-generator-construction-cost-in-the-US-by-type/

www.ourworldindata.org/grapher/natural-gas-prices

www.en.wikipedia.org/wiki/Coal_fired_power_station#

www.eia.gov/todayinenergy/detail.php

www.hosemaster.com/pollution-control-in-coal-fired-power-plants

www.nist.gov/news/2022/11/simple-material-could-scrub-carbon-dioxide-power-plant-smokestack

www.pubs.acs.org/doi/10.1021/es102s678

www.theguardian.com/environment/2017/aug/15/australian-coal-power-pollution-would-be-illegal-in-us-europe-and-china-report#

www.iaea.org/sites/default/files/publications/magazines/bulletin/bull56-3/5631213.pdf (electron beam SO2 removal).

www.ncbi.nlm.nih.gov/pmc/articles/PMC9075710/

Asif, Z et al. "Update on air pollution control strategies for coal fired power plants." Clean Technol Environ Policy 2022; 24(8):2329 – 2347.

www.epa.nsw.gov.au/-/media/epa/corporate-site/resources/air/18p0700-review-of-coal-fired-power-stations.pdf

www.envirojustice.org.au/wp-content/uploads/2018/10/EJA-fact-sheet-Emission-controls-for-NOx-and-SO2.pdf

www.sciencedirect.com/science/article/abs/pii/SO269749/2011787 (Bag filters)

www.acarp.com.au/abstracts.aspx

Nielson, P et al. "Measurements of Mercury Speciation from Combustion in Australian Coals." ACARP

www.unnsw.edu.au/news/2021/07/a-tale-of-two-valleys-latrobe-and-hunter-regions-both-have-coal (Bag filters)

www.reneweconomy.com.au/australian-coal-plants-are-producing-less-power-but-more-toxic-pollution/

www.researchgate.net/publication/329984727_Mercury_pollution_from_coal_fired_power_plants_a_critical_analysis_of_the_Australian_regulatory_response_to_public_Health_risks

Bramwell, G et al, ANU.

www.envirojustice.org,au/wp-content/uploads/2018/02/EJA-submission-to-EPA-power-station-licence-review-February-2018.pdf

www.researchers.mq.edu.au/en/publications/speciation-of-mercury-in-flue-gas-from-australian-coal-fired-power

www.unece.org/fileadmin/DAM/env/documents/2011/eb/wg5/WGSR48/informal%/20docs.3_Reduction_of_mercury_emissions_from_coal_fired_power_plants.pdf

www.eria.org/RPR_FY2016_O2.pdf

Economic Research Unit for ASEAN report 2017.

www.en.wikipedia.org/wiki/History_of_coal_mining

www.eia.gov/energyexplained/coal/

www.treescharlotte.org (Earliest trees)

www.science.org/content/article/world-s-tree-grew-splitting-their-guts

www.zmescience.com/feature-post/natural-sciences/geology-and-paleontology/rocks-and-minerals/how-coal-is-formed

www.ga.gov.au/classroom-resources/minerals-energy/australian-energy-facts/coal#

www.pubmed.ncbi.nlm.gov/30949738/

Zhang, N et al. "The toxicological effects of mercury exposure in marine fish." Bull Environ Contam Toxicol 2019 May; 102(5): 714 – 720

www.foodstandards.gov.au/consumer/chemicals/mercury/Pages/default.aspx

www.darwinsdoor.co.uk/timetour/the-carboniferous-period.html#

www.epa.nsw.gov.au/-/media/epa/corporate-site/resources/air/18p0700-review-of-coal-fired-power-stations.pdf Pollution Australian plants.

www.envirojustice.org.au/wp-content/uploads/2018/10/EJA-fact-sheet-Emissions-controls-for-NOx-and-SO2.pdf

www.sciencedirect.com/science/article/abs/pii/SO269749/2011787 Bag filters in Australia.

www.acarp.com.au/abstracts.aspx

Nielsen, P et al. "Measurements of Mercury Speciation from combustion in Australian coals." ACARP

www.unsw.edu.au/news/2021/07/a-tale-of-two-valleys-latrobe-and-hunter-regions-both-have-coal

www.afr.com/energy/nuclear-energy-too-expensive-to-replace-fossil-fuels-20220711-p5b0pd

www.apo.org.au/node/96821

Solstice Development Corp and Minerals Council Australia report cost generation.

www.theconversation.com/factcheck-qanda-is-coal-still-cheaper-than-renewables-as-an-energy-source-81263

www.witpress.com/secure/elibrary/papers/S0907/SDP07088FU2.pdf

Stanciu, D et al. "Exhaust gas treatment technologies for pollutant emission abatement from fossil fuel power plants" WIT Transactions on Ecology and Environment Vol 102 2007.

www.aip.scitation.org/doi/pdf/10.1063/1.5086569 (Coal pollution control methods).

www.linkspringer.com/article/10.1007/S40789-014-0001-x

Wang, A et al. "influence of flue gas cleaning system characteristics of PH2.5 emission from coal fired power plants" Int Jn of Coal Science and Technology. 1 4-12 (2014).

www.sciencedirect.com/science/article/abs/pii/SO269749117305754

Li, Z et al. "Influence of flue gas desulfurization on PM2.5 from coal fired power plants equipped with selective catalytic reduction (SCR)".

www.reneweconomy.com.au/clean-australias-clean-coal-power-stations14224/ (HELE plants Queensland and elsewhere).

www.mdpi.com/2071_1050/13/3/159/

Ali, H et al. "Cost-benefit analysis of HELE on subcritical coal – fired electricity generation technologies in southeast Asia." Sustainability 2021 13(3) 1591 https//:doi.org/10.3390/SU/303151

www.afr.com/politics/hele-coal-cheaper-than-previously-thought-says-mca-20170702-gx2uzc (Minerals council of Australia report).

www.whatswatt.com.au/minimum-demand-managing-energy-supply&demand Re unstable renewables power, need synchronous condensers.

www.whatswatt.com.au/what-is-hele-coal-power

www.en.wikipedia.org/wiki/Fossil_fuel_power_station

www.reneweconomy.com.au/australian-coal-plants-are-producing-less-power-but-more-toxic-pollution/

www.envirojustice.org.au

Coal pollution control.

www.reneweconomy.com.au/clean-australias-clean-coal-power-stations-14224/

www.epa.nsw.gov.au NSW coal plant air emission monitoring.

www.epa.gov Pollution methods activated carbon.

www.sciencedirect.com/science/article/pii/S167498712200 https://doi.org/10.1016/jgsf.2022.101498 Coal pollution. Closure US plants.

www.researchgate.net/publication/329984727_Mercury_poluution_from_coal_fired_power_plants_a_critical_analysis_of_the-australian_regulatory_response_to_public_heath_risks

www.envirojustice.org.au/wp-content/uploads/2018/02/EJA-submissions-to-EPA-power-station-licence-review-February-2018.pdf

www.researchers.mq.edu.au/en/publications/speciation-ogf-mercury-in-flue-gas-from-australian-coal-fired-power

www.unece.org/fileadmin/DAM/env/documents/2011/eb/wg5/WGSR48/informed%/20docs.3_Reductions_of_mercury_emissions_from_coal_fired_power_plants.pdf

www.eria.org/RPR_FY2016_O2.pdf Economic Research Unit for Asean report coal use Asia.

www.sciencedirect.com/science/article/pii/S2405844020310410 Mercury control measures.

www.proest.com/construction/cost/estimates/power-plants/

www.eia.gov/energyexplained/coal/how-much-coal-is-left.php

www.miningtechnology.com/features/feature-the-worlds-biggest-coal-reserves-by-country/

www.world-nuclear.org/information-liobrary/energy-and-the-environment/clean-coal-technologies.aspx

www.climatecouncil.org.au/new-coal-power-stations-polluting-expensive/ Cost and problems with CCS.

www.theguardian.com/environment/2022/oct/15/emissions-capture-carbon-cost-water-electricity CCS water use.

www.en.wikipedia.com/topics/engineering/synchronous_condenser_motor_maintains_voltage_varying_current_in_field_coils_hybrid_systems_wind/diesel_etc

www.sciencedirect.com/topics/engineering/synchronous-condenser-motor-maintians-voltage-varying-current-in-field-coils-hybrid-systems-wind/dieseletc

www.iea.org>reports>executive-summary Coal doubled India since 2007.

www.iea.org/reports/coal-2022/executive-summary Coal 8 billion tonnes global 2022. High gas prices Europe switch to coal.

Chapter 15

Some Consequences of the Forced Shift to Renewables

Government Subsidies for Renewables and Carbon Taxes Globally

In general renewable energy for electricity and vehicles is significantly more expensive than fossil fuel energy as commented on in many chapters already. And the common strategy of governments endeavouring to force massive change to renewables is to subsidise them, and tax the fossil fuel competition endeavouring to make the change attractive to consumers and voters. Some examples:

Europe:

Recently the EU passed a €250 billion renewable subsidy package.

The Netherlands government subsidises renewable by about €12 billion per year. And is also paying €20 billion for a new transmission grid to service electric vehicles by 2035.

The Danish government subsidises renewables by about US$220 million per year (1.5 billion Danish Kroner).

Spain subsidises renewables by €7 billion per year. In addition the government has invested €13 billion of taxpayer funds in renewable development. The government's debt load for solar projects is €30 billion.

The Czech republic previously had big government subsidies for renewable power. But the cost of the power was ten times that from coal and gas, and electricity prices for consumers soared. The accumulated subsidies amounted to €1.7 billion, annually in 2014 €600 million. So later that year the government stopped all support for renewables mainly because of soaring subsidised electricity costs.

France spends about €6 billion each year on renewable subsidies.

Germany spent US$38 billion in the 2020 year subsidising electricity prices for consumers. Some of this was generated by coal and gas. The government has plans to spend US$177 billion on renewable development over 4 years.

Finland has committed US$1.7 billion government spending on renewables over the next few years. It has also just spent over US$35 billion on a nuclear plant as part of "nett zero" ambitions.

Poland has committed US$400 million government funding on renewables.

Norway and Sweden have stopped previous subsidies for renewables; but they continue to fund electric vehicles with tax breaks and subsidies as discussed, and both get the majority of their electricity from hydro.

The UK government spends US$7.4 billion annually on renewable subsidies. In addition it retrieves US$2.3 billion in carbon taxes from UK fossil fuel users. It has committed US$ 30.2 billion for transmission grid upgrades to serve renewables as said.

North America:

Recently the US government committed to US$529 billion investment subsidies for renewables. In 2020 energy subsidies were US$634 billion of which 26%, $165 billion were renewable energy subsidies and the rest, about $469 billion largely to oil and gas. Many of the latter were for general business credits as opposed to cash handouts as occurred with renewables. For the period 2012 to 2017 US$410 billion government subsidies were paid to renewable power plants. There is no carbon tax. To help put this in perspective the US has spent US$70 billion on war help for Ukraine

Canada has government energy subsidies of US$14 billion per year, but only $1 billion of that is for renewables, the rest for fossil fuels. Canada also has a carbon tax levied on businesses which use fossil fuels. From 2019 to 2023 this raised US$23 billion dollars. The stated aim of the tax is to encourage movement away from using fossil fuels.

Mexico is the first third world country to have a carbon tax. This is modest and raised US$275 million in 2019.

Asia:

Japan has a carbon tax which raised US$32 billion in 2020. At that point it was very modest at US$2 per tonne of CO_2, there are plans to raise it to perhaps $80. Government subsidies for renewables were US$3.1 billion in 2021. This compares to US$5.8 billion for fossil fuels. These subsidies are complicated in several forms but that seems to be the essence.

China subsidized solar by US$905 million, a comparatively tiny amount, in 2021. China does not have a carbon tax. As a "command economy" and autocracy the Chinese government does not need government incentives to achieve objectives – and is pushing coal development intensely as discussed. China does not have a carbon tax.

India subsidizes renewable by a modest US$91 million. Total energy subsidies are about US$822 million, most going to fossil fuels. India does not have a carbon tax.

Indonesia, Malaysia, and Vietnam have no carbon tax or renewable subsidies.

Russia has no carbon tax or renewable subsidies. It does get 20% of its electricity from hydro and another 20% from nuclear, about 200 terawatt hours each, but solar and wind are minute. Russia has huge reserves of gas and coal, is extremely rich in energy.

Singapore has a modest carbon tax at US$3.7 per tonne of CO_2. It generates about 50 million tonnes annually, so the tax currently costs business there about US$180 million. But there are plans to raise the rate to US$60 per tonne by 2030, when the tax would reach US$2.8 billion if that occurs.

Pacific:

Australia has committed A$25 billion to renewable energy development in the 2022 and 2023 years. $20 billion of that is going to transmission grid expansion to cope with renewables including hydro. $500 million is for electric vehicle recharging at 117 sites, and $300 million for community batteries to back up solar and wind. The Minerals Council of Australia in a 2017 report pointed to A$2.8 billion each year to 2030 in planned government subsidies for renewables. Government subsidies per megawatt hour of electricity include $214 for solar, $74 for wind, $33 for other renewables including hydro and biofuels, and $0.40 (40 cents) for coal, 30 cents for gas.

New Zealand and other Pacific Islands do not have a carbon tax, nor do they subsidise renewables. There are some government projects funding renewable development, for example a biomass gasification plant in Samoa. Many of the small tropical islands use diesel generators to generate electricity. While fuel tax is charged on the diesel, there is also a rebate refunding it. New Zealand has 70% of its electricity generated by hydro. The small pacific islands are campaigning for large grants from countries with carbon taxes on the basis of sea level rise.

It is apparent there is a huge amount of global government money sloshing around the energy sector generally. While moneys going to renewables have been focussed on above much larger amounts involve the fossil fuel sector, coal and gas. But while renewable subsidies often are cash grants, fossil fuel ones are more commonly general business grants and tax refunds.

Regarding carbon taxes 27 countries now have these. The rate ranges from Uruguay at US$137 per tonne CO_2 to Poland at $1. Several European countries are about US$130. Notable for no carbon tax are the US, China, Russia, India and Australia.

The motives for all this government interference appear to be:

- To make electricity affordable for the countries' people.
- To raise government revenue for general government expenditure.
- To aid development of the energy sector and ensure energy supply.
- To encourage or force renewable development instead of fossil fuels.

The US International Energy Agency considers that in the period March 2020 to November 2022, two and a half years, US$1250 billion was spent by governments globally on renewable energy investment; and another US$600 billion on making energy affordable.

Most of this government funding and energy taxation is new in the last 30 years, much of it a result of the stated aim to move to "nett zero" away from cheap practical coal to expensive difficult renewables.

There are "Catch 22" situations here: renewables are only affordable to people if they are heavily subsidised by governments; but that means that fossil fuel energy becomes uneconomic, not able to compete with subsidised renewables. But because renewable are intermittent and unreliable, governments are forced to ensure the presence of adequate fossil fuel energy. And so are forced to also subsidise fossil fuel sources.

Therefore the massive subsidies are a product of the "nett zero" ambitions focused on carbon dioxide. If those ambitions did not exist energy sources could compete on their own virtues at what their principle task is – to produce electricity and energy for society's requirements in an acceptable manner. And not based on the demonization of a trace gas which is essential to life.

Strategic Military Considerations

Modern warfare requires huge amounts of energy. The World Wars of the last century demonstrated this. Victory went to the side with the overwhelming energy supply – for building many more weapons, ammunition, vehicles, aircraft and ships, and being able to use them, than those defeated. In both the World Wars the dominance of the United States industrial ability determined the outcome. And our lives in the free world have flourished as a result. During World War II the US built 6000 warships, 300,000 military aircraft, 50,000 Sherman tanks and much more. Britain built 144,000 aircraft and 27,000 tanks. The Soviets made 87,000 tanks.

By contrast Japan managed 2000 ships, 85,000 aircraft and 6400 tanks. Germany built 2000 ships, 94,000 aircraft and 28,000 tanks. The losers output of military hardware was dwarfed by the Allies production.

Mao Tse-tung famously said: "All power comes from the muzzle of a gun." And that is still true today in our world with many tyrants and dictators always probing the Western Democracies for weaknesses – witness Ukraine and Taiwan currently.

The world has many huge armies potentially ranged against each other. On one hand there is China, with 2 million soldiers, Russia with 1.1 million, North Korea with 950,000, Iran with 360,000, all ruled by dictators with hostile ambitions. On the other there is the US with 480,000, India with 1.5 million, Indonesia with 390,000, then several European democracies with 60 to 80,000 each including the UK, France and Germany. Japan has 261,000 personnel in its whole defence force.

Concerning army tanks China has 5000, Russia 3000, North Korea 6600, Iran 1400. And the US has 2500, Israel 3700, Japan 1000, Pakistan 2600, Poland 900, India 4500, Taiwan 900, Germany 328, France 400, the UK 160.

Regarding military aircraft China has 4000, Russia 2465, North Korea 900, Iran 343. And the US 5200, Japan 740, India 1645, France 650, Germany 400, the UK 460, Australia 296. All the NATO countries have 20,600 combined.

And China has 485 naval vessels, Russia 290, Iran 60, North Korea 186. The US has 485, Japan 150, the UK 70, all other democracy navies are less than 70.

Virtually all these military hardware items use fossil fuels – diesel, aviation fuel, petrol. And the energy used to make them mostly comes from fossil fuels also.

The adversaries of the Western Democracies, Russia, China, Iran and North Korea, have continued to expand their fossil fuel energy base. Witness Russia with its huge gas and coal reserves and power plants; China building coal power plants rapidly; Iran uses natural gas for 71% of its electricity and oil for 29%;

North Korea uses coal for 80%. None of these countries are closing fossil fuel facilities at a great rate and shifting to renewables. They know war requires industrial might for success.

By contrast the democracies are closing fossil fuel energy sources including for electricity production and vehicles. Because renewables are intermittent and inadequate that has resulted in them becoming dependant on imported energy from the dictatorships, especially Europe which is further along this path which the others are still following. In addition substantial amounts of heavy manufacturing industry have shifted to China where energy and costs are cheaper and practical. Indeed the "deindustrialisation" of Europe has been pointed to by industrialists criticising European rush to "nett zero". In the event of war this lack of industrial capacity in the democracies would be crucial.

An illustration of this problem is world steel production. Steel is obviously the key component of military hardware. In 2021 global production was 1950 million tonnes. Of that 53%, 1032 million tonnes, was made in China. The next biggest was India with 6%, Japan did 5%, the US 5%. The biggest European producer was Germany at 2%. Australia made 0.3% as did the UK. Russia made 4%.

By contrast in the middle of World War 2, 1943, the Allies, the US, Soviet Union and UK, made 77% of world steel, more than three times as much as Nazi Germany and Japan. This is a further illustration of the essential nature of industrial production to victory.

The UK's flight from heavy industry is clear when its path from being the world's largest steel producer in the 1880's, to still significant war-winning production at 21% in 1943, more than either Nazi Germany or Japan, to just 0.3% in 2021, is considered.

It is an unfortunate and lamentable fact from human history that the only way to prevent war is to prepare for it. Many have said that from Sun Tzu of 500 BC China in his "The Art of War" stating: "In war, prepare for peace; in peace, prepare for war.", to George Washington saying: "To be prepared for war is one of the most effective ways means of preserving peace" to Omar Bradley, the commanding general of US troops on D-Day in World War 2: "Peace is our goal but preparedness is the price we must pay."

Aggressors, and there are plenty in the current world, believe they can win because their victims are weaker militarily. History is littered with examples of this from the ancient world where weak city states were sacked and slaughtered by stronger opponents including the Persians, the Greeks under Alexander the Great, the Romans, right through to the 20th century wars; always the aggressors believed they could win.

The pursuit by the Democracies of the patently false hypothesis that the Earth's surface temperature can be controlled by humans altering their production of carbon dioxide is weakening them and risks igniting war amongst all the other consequences.

Life in the Renewable Energy Age:

What would the world be like if the renewable energy drive continues?

First, only the currently wealthy western democracies would change markedly. Third world countries would not be able to afford the huge costs and would stay with fossil fuels for electricity, transport and industry.

And the autocracies would continue to greatly expand their fossil fuel base as they are doing now – especially China, Russia, Iran, North Korea as explained already.

The western democracies – Europe, North America, Australasia, Japan, South Korea especially would experience greatly increased electricity prices and shortages including blackouts. This would result in a flight of industry and business to China especially as is already happening, and to some third world countries. The transport sector would be crippled by the limitations of electric vehicles especially slow and complicated refuelling, impossible queues and limited range. This would severely impact the earnings and quality of life of the people, especially lower socio-economic groups.

Northern hemisphere countries have very cold winters. There would be real problems for many people, especially the poor, heating homes and workplaces. This would cost many lives.

As discussed already a major source of human made carbon dioxide is concrete construction. This would have to decrease substantially at least.

"Lifestyle entertainment" is worth a mention. For example the western democracies all enjoy a big recreational marine industry with many people enjoying boating activities such as sailing, launches, fishing, power boating of all types. This is all heavily dependent on fossil fuels, diesel and petrol engines including many smaller vessels with outboard motors which are massively popular. Replacing these with electric motors and lithium batteries is impossible – for all the reasons given in the electric vehicle discussion including range, fire and explosion risk, weight. Propelling boats take at least four times the energy per kilometer than road vehicles. For large launches and yachts it is many times greater. So these activities would die.

Similarly many western people enjoy going on large cruise ships and seeing the world. These all use large quantities of energy supplied entirely by fuel oil and other fossil fuels. It is not feasible for them to operate on renewables and batteries. So this would go too.

Western people use vehicles for a great deal of recreational travel, ranging from visiting friends, family, going to theatre, films, sporting events, many others. A special group are those towing trailers, horse floats, caravans, other recreational vehicles. Much of this would all stop.

And a very large number of western democracy people frequently fly long distances, even around the world, to see family, friends and other countries. The aircraft clearly use large amounts of aviation fossil fuels. There is again no way this can be done with renewables or biofuels, nor indeed hydrogen as has been discussed. So this great boon would also disappear.

Governments would find it much harder to fund services including health and education due to a decline in tax revenue as business suffered, and the current big taxes from fossil fuels disappeared. The current large subsidies on renewables and electric vehicles would be increasingly difficult.

Wildlife especially birds and marine creatures would be severely affected near huge windfarms on land and sea, their numbers would plummet there. And the massive toxic waste from old windmills, lithium batteries and solar panels in landfills and dumps would be a major concern. The huge increase in grid networks including high voltage overhead land lines and a plethora of undersea cables would take a further big toll on wildlife, to say nothing of the effects of ruining the environment for many humans. Biofuel use in heavy vehicles and for heating and electricity production would mean a shorter more expensive life for engines and vehicles, and, ironically, more carbon dioxide output than if coal and fossil fuels had been used, as well as much greater costs for transporting goods. To say nothing of the environmental effects from massive areas of land used for growing biofuel crops. The huge energy requirements to make solar panels, lithium batteries, more windmills including mining and transport of materials to massive factories, many in China, has already been discussed.

If nuclear was resorted to it would mean a far greater nuclear waste problem, already a big issue. The huge cost of building nuclear power plants compared to coal and gas would mean they would have to be forced and funded by governments, another big cost for taxpayers at the cost of other government services. And eventually a terrible legacy for future generations due to aging defective dangerous nuclear plants costing billions and taking many decades to decommission, to say nothing of the certainty of more catastrophic accidents like Three Mile Island, Chernobyl, and Fukushima with global radiation pollution.

The western democracies would inevitably decline in wealth, power, and the people's living standards compared to the autocracies.

Loss of Freedoms

When societies become subservient to groups within them who passionately believe that they are right, that reasonable ordinary people are wrong or ignorant, and that they can dominate society to force their view, then a tyranny results with the loss of many freedoms. History is full of many examples ranging from the many religious zealots who slaughtered, tortured and burnt at the stake those who did not believe, to racial fanatics including the nazis, and Maoist idealists who murdered millions in Cambodia. These examples are clearly extreme compared with the current crop of carbon dioxide demonizing nett zero enthusiasts. But they are gathering steam, at least in the western democracies, some are intolerant of "climate deniers" and are already using strategies to get their way that are a long way from gentlemanly, reasoned, and tolerant argument including threatening employment, access to investment funds, freedom of travel, other societal pressures as they endeavour to force their plans on everybody. And, as seen many times in the past, these movements, even though they are based on false beliefs, generate their own momentum and become unstoppable because people fear opposing them, or see a selfish individual gain of some kind. The only defence against this is for ordinary people is to become informed about the truth and refuse to permit domination of their freedoms, be it by religious intolerance, believers in the divine right of kings, nazi racists, Maoist zealots, or – those who assert that carbon dioxide, a basic essential component of our world, must be eradicated irrespective of the dire consequences to billions of people. And the ineffectiveness of doing so in terms of influencing the climate.

Science is full of theories and hypotheses which turned out to be false. Yet before they were shown to be so they were promoted vigorously by proponents and all kinds of futile practices followed which are now laughed at or cried over. There are many examples in medicine including believe in bloodletting by leeches to treat fevers; major surgery for peptic ulcers based on the theory that excess stomach acid was the cause when the real cause was *Helicobacter pylori*, a bacteria easily treated by antibiotics; that hand washing and hygiene was not important in hospitals until Semmelweis showed in Vienna the link to sepsis and death; and many others.

The Brightest Future:

Coal, diesel, petrol, fuel oil and aviation fuels have been the bedrock of the great leap in living standards for billions of humans since their use started 200 years ago. And they remain the only practical option for the future. Of course they must be carefully and responsibly used avoiding as far as possible damage to the environment including wildlife. For coal that means the maximum use of all the methods discussed in chapter 14 to reduce pollution, forced by government regulation. And for all the other fossil fuels it means clean efficient use in modern machinery wherever possible. Their ongoing use is essential for the well-being and security of free peoples in the western democracies.

Gas is missing from this list because, as discussed in chapter 14, there are limited supplies which will run out in a few decades. Whereas the other fuels have known reserves for hundreds of years or more.

It is incumbent on reasonable people everywhere to carefully inform themselves about the Earth's surface temperature and make sure those with power and responsibility follow the truth.

Hence this book.

References and Further Reading for Chapter 15 – The Forced Shift to Renewables – Some Consequences

www.tradingeconomics.com/commodity/carbon

www.smh.com.au/environment/climate-change/environment-borger-tax-us-a-warning-to-australia-to-cut-emissions-or-lose-exports-20210715-p58a2u-html

www.ec.europa.eu/eurostat/staistics_explained/index.php?title=Environmental_tax_statistics

www.worldbank.org/en/news/press-release/2022/05/24/global-carbon-pricing-generates-record-84-billion-in-revenue

www.imf.org.en/Publications/fandd/issues/2019/12/the-case-for-carbon-taxation-and-putting-a-price-on-pollution-parry

www.statista.com/statistics/483590/price-of-implemented-carbon-pricing-instruments-worldwide-by-select-country

www.cfib-fcei.ca/en/media/less-than-1-of-the-22b-in-federal-carbon-tax-revenues-have-been-returned-to-small-business#

www.adb.org>institutional-document.pdf

www.tandonline.com/doi/full/10.1080.09640568.2022.2081136

www.mdpi.com/1996-1073/14/7/1986

www.en.wikipedia.org/wiki/Renewable_energy_commercialization

www.en.wikipedia.org/wiki/Energy_subsidy

www.en.wikipedia.org/wiki/Renewable_energy_certificate_(United_states)

www.imf.org/topics/climate-change/energy-subsidies

www.en.wikipedia.org/wiki/Energy_in_russia

www.carboncopy.info>subsidies-for-fossil-fuels-nine-times-higher-then-renewable-energy-study/

www.climatescorecard.org/2023/05/japan-spent-9.5-billion-in-fossil-fuel-subsidies-in-2021/

www.en.wikipedia.org/wiki/Energy_subsidies_in-the_united_states

www.irena.org/Energy-transition/Technology/Geothermal-energy

www.iea.org/data/-and-statistics/charts/government-spending-for-clean-energy-investment-support-and-short-term-consumer-energy-affordability-measures-94-2022

www.en.wikipedia.org/wiki/Renewable_energy_payments

www.world-nuclear.org/information-library/economic-aspects/energy-subsidies-aspx

www.weforum.org/agenda/2023/03/what-do-green-subsidies-mean-for-the-future-of-climate-and-trade-099a016307/

www.globalaustralia.gov.au/news-and-resources/news-items/australian-budget-commits-a25bn-clean-energy

www.afr.com/politics/renewable-energy-subsidies

www.taxpayers.org.au/end-energy-subsidies

www.australianinstitute.org.au/post/australian-fossil-fuel-subsidies-surge-to-11.6billion-in-2021-22

www.stopthesethings.com/tag/cost-of-australias-renewable-subsidies

www.energyinformationaustralia.com.au/research/loose-with-the-truth

www.trade.gov/country-commercial-guides/france-energy-eng#

www.bloomberg.com/news/articles/2021-01-12/germany-paid-record-38billion-for-green-power-growth-in-2020#xj4y7vzkg

www.energypolicytracker.org/country/finland/

www.gov.uk/government/statistics/environment.taxes-bulletin/

www.iea.org/articles/how-much-will-renewable-energy-benefit-from-global-stimulus-package

www.news.climate.columbia.edu/2019/09/23/energy-subsidies-renewables-fossil-fuels

www.beehive.govt.nz>release>government-climate-change

www.cia.gov>docs>DOC_0000381439 (Viet Cong supplies).

www.bbc.com>news>world>asia-41842285 (Taliban supplies).

www.en.wikipedia.org/wiki/list_of_aircraft_Japan_during_world_war_II#

www.military-history-fandom.com/wiki/German_aircraft_production_during_world_war_II

www.warfarehistory.network.com/articles/the-t-34-tank (Soviet tank output WWII).

www.en.wikipedia.org/wiki/Tanks_in_the_Japanese_army

www.quora.com/How_many_German_tanks_were_produced_in_WW2

www.en.wikipedia.org/wiki/British_armoured_fighting_vehicle_production_during_World_war_II#

www.iea.org/countries/korea-democratid-peoples-republic-of (North Korea energy supply).

www.iea.org/countries/iran (Iran energy supplies).

www.en.wikipedia.org/wiki/list_of_countries_by_steel_production

Detailed Summary of Contents and List of Figures

Chapter 1 Heat Energy

Temperature, heat energy, moving heat energy, conduction, convection, radiation, Blackbody radiation, Kelvin temperature, radiation absorption.

Figure 1: Visible Light

Figure 2: Diagram of red light waves

Figure 3: Actual number for each 10^{power}

Figure 4: Electromagnetic waves

Figure 5: Heat waves in the infrared

Figure 6: Sun's radiation wavelengths

Figure 7: Earth's energy (heat) radiation in the infrared

Chapter 2 The Earth's Atmosphere

Nitrogen, Oxygen, water vapour, carbon dioxide, methane, ozone, the layers of the atmosphere, troposphere, tropopause, stratosphere, stratospheric ozone, mesosphere, thermosphere. Exosphere.

Figure 8: Radiation absorption by Oxygen molecules O_2

Figure 9: Radiation absorption by water vapour molecules H_2O

Figure 10: Radiation absorption by Carbon Dioxide molecules CO_2

Figure 11: Radiation absorption by Methane molecules CH_4

Figure 12: The layers of the atmosphere

References and Further Reading for Chapters 1 and 2

Chapter 3 The Earth's Surface Temperature

The energy output of the Sun, cooler and warmer periods historically, ice ages, distance from the Sun, rotation of the Earth, Axis tilt, Precession of tilt, orbit variations, the Sun's radiation reaching Earth, measuring energy, kilowatt hours Sun's energy on Earth, cloud effect, Albedo of Earth's surface, human energy versus Sun's, other energy sources, geothermal, the layers of the Earth, Geothermal energy of the Earth, tidal energy, Infrared radiation from the Earth, Energy loss from the Earth.

Figure 13: Earth's axis tilt.

Figure 14: Axis tilt effect on seasons.

Figure 15: Sun on Earth at the equinox

Figure 16: Earth's orbit and seasons

Figure 17: Incoming Sun radiation to Earth

Figure 18: Effect of the Earth spherical shape on Sun radiation at surface

Figure 19: Sun radiation at surface and latitude

Figure 20: Sun radiation at surface and daily cycle

Figure 21: Total daily Sun radiation at Earth's surface in seasons

Figure 22: The layers of the Earth

Figure 23: Sun and Moon tides

Figure 24: Infrared radiation from Earth

References and Further Reading for Chapter 3

Chapter 4 Water

Water, water on other planets, moons, asteroids, evaporation, condensation, atmospheric pressure, partial pressures, atmospheric gases, vapour pressure and temperature, sublimation and deposition of ice, phase changes, latent heat of evaporation, and condensation, latent heat of melting and fusion of ice, Earth's heat loss by water phase changes, atmospheric water vapour changing to water and ice, cloud formation, rain, hail, snow formation, effect of warmer and cooler times on atmospheric heat loss, negative feedbacks, temperature, clouds, heat radiation from land, ice crystal clouds, heat loss from land, thermal window, slowing heat loss by water vapour, greenhouse effect, heat transfer to poles, Coriolis effect, trade winds, Hadley Cell, Ferrel Cell, Polar Cell, katabatic winds, anticyclones, depressions, roaring forties, other ways water vapour rises to lose heat to space, violent upward air movements, thunderstorms, cyclones, typhoons, hurricanes.

Figure 25: Heat energy moving up the atmosphere to space.

Figure 7: (repeated): Earth's infrared radiation.

Figure 9 (repeated): Radiation absorption by water molecules.

Figure 18 (repeated): Effect of latitude on Sun's radiation at surface.

Figure 26: Hadley, Ferrel and Polar Cells.

References and Further Reading for Chapter 4 – Water

Chapter 5 The Oceans

Temperature of the oceans, major ocean currents, the great ocean conveyor belt, climate oscillations, El Nino – la Nina and others, energy to raise sea temperature

Figure 27: The temperature of the oceans at various depths.

Figure 28: showing the relative amounts of heat energy needed to change the temperature of the whole ocean, just the top 100 meters and the whole troposphere by 1°C,

References and Further reading for Chapter 5 – The Oceans

Chapter 6 Ice and Snow

Ice location on Earth, formation, Antarctic ice sheets, Greenland ice sheet, Ice sheet changes, geothermal effects, Sea ice, energy needed to melt Arctic and Antarctic ice, Snow, Glaciers, regelation, studies of changes,

References and Further Reading for Chapter 6 – Ice and Snow

Chapter 7 Sea Level

Land subsidence and uplift - studies, Australian sea level data, The Isle of the Dead, Forth Dennison, Satellite sea level methods, conclusions about sea level, Pacific coral atolls, sea level oscillation.

References and Further Reading for Chapter 7 – Sea Level

Chapter 8 Carbon Dioxide

Properties and characteristics, CO2 in the Solar System, early Earth history, Photosynthesis, the great oxygen event, Carbon dioxide and the oceans, temperature of the oceans, sea pressure and depth, saltiness, partial pressure CO2, Henry's Law, sea temperature controlling atmospheric CO2 which follows it – studies, the nature of CO2 dissolved in the oceans, Carbonate limestone rock formation, transport of carbon dioxide from the sea to the air, and the air to the sea, Carbon dioxide mixing and distribution in the atmosphere, transfers from, atmosphere to land, photosynthesis, rain, Carbon dioxide transfers to the atmosphere, volcanic, wildfires, land plants decaying, animal respiration, human production of carbon dioxide, Carbon isotopes C^{13} and C^{14}, human production – summary, additional carbon dioxide production by humans, wood burning, summary of carbon dioxide movements, methods of measuring carbon dioxide, Satellite methods, Historical methods, Professor Beck's paper, Ice core studies, other methods for ancient CO2 levels.

Figure 29: a Carbon Dioxide molecule.

Figure 30: the wavelengths of visible light absorbed by chlorophyll plant cells.

References and Further Reading for Chapter 8 – Carbon Dioxide

Chapter 9 Carbon Dioxide Heating of the Atmosphere

Heat loss from CO2 molecules, history of carbon dioxide and water vapour heat energy absorption discoveries.

John Tyndall

Svante Arrhenius

Knut Angstrom and Herr J Koch

Guy Callendar,

Edward O Hulburt

Some prominent modern scientists convinced that human CO2 production is a dangerous cause of global warming – "Alarmists",

Gilbert N Plass,

Charles Keeling

James Hansen

Hubert Lamb

Jerry Mahlman

Michael Mann

Raymond Bradley

Malcolm Hughes

Gavin Schmidt

Susan Solomon

Kate Marvel

Mauro Facchini

Vincent - Henri Peuch

Jim Salinger

Kevin Trenberth

Some prominent scientists who do not consider human CO2 production causes dangerous global warming – "sceptics", or "deniers" or "naturalists".

Sir Fred Hoyle

Joseph D'Aleo

Sallie Balilunas

Willie Soon

Robert Carter

Nicola Scafetta

Vincent Gray

Roy Spencer

Richard Lindzen

William Gray

Siegfried Fred Singer

Ian Plimer

Judith Curry

Guus Berkhouf

Figure 10 (repeated): Radiation absorption by carbon dioxide molecules

Figure 7 (repeated): Earth's energy (heat) radiation showing the wavelengths in the infrared

Figure 9 (repeated): radiation absorption by water molecules

References and Further Reading for Chapter 9 – Carbon Dioxide Heating in the Atmosphere

Chapter 10 Methane

Properties and characteristics, methane formation, Methane measurement issues, Natural sources of methane compared with human sources.

Figure 11 (repeated): Radiation absorption by methane molecules

References and Further Reading on Chapter 10 – Methane

Chapter 11 The Earth's Surface Temperature

The Hadean, Archaen, Proterozoic, Paleozoic, Mesozoic and Cenozoic Periods

Figure 31: Average global temperature for the past 570 million years

Figure 32: Average global temperature for the past 65 million years

Figure 33: Average global temperature for the past 11,000 years

Figure 34: Average global temperatures since 1850

Figure 35: Tropical mid – troposphere temperatures from 1970's compared with IPCC predictions

Figure 36: Temperature variations from 1981

Some Actual Surface Temperatures for Australian Locations

Figure 37: Moruya Heads Pilot Station actual temperature from 1876 to 2022

Figure 38: Actual surface temperature for Cape Bruny Lighthouse, Tasmania 1923 to 2021

Figure 39: Streaky Bay, South Australia 1926 to 2021

Figure 40: Cape Naturaliste Lighthouse, Western Australia 1904 to 2021

Figure 41: Broome Airport 1940 to 2021

Figure 42: Victoria River Downs, Northern Territory 1965 to 2021

Figure 43: Darwin Airport, 1942 to 2022

Figure 44: Palmerville cattle station, North Queensland, 1896 to 2021

Figure 45: Rockhampton 1940 to 2021

Figure 46: Sandy Cape Lighthouse 1907 to 2019

Figure 47: Newcastle Nobby's Head Signal Station 1862 to 2021

Figure 48: Alice Springs Airport

Figure 49: Moruya Heads Pilot Station 1980 to 2021 January and July

Figure 50: Cape Bruny January and July 1980 to 2020

Figure 51: Cape Naturaliste January and July 1980 to 2021

Figure 52: Darwin Airport January and July 1980 to 2022

Figure 53: Newcastle Nobby's Head Signal Station January and July 1980 to 2021

Some Surface Temperature Records from Canada

Figure 54: Alberni Robertson Creek, Vancouver Island, 1960 to 2022, January and July.

Figure 55: Bagotville Air Force Base, Quebec, January and July, 1945 to 2022.

Figure 56: Goose Bay, Newfoundland, January and July, 1941 to 2022

Some Temperature Records from the USA

Figure 57: Santa Rosa, California, January and July 1955 to 2022.

Figure 58: Wichita, Kansas, January and July 1955 to 2022.

Figure 59: Charleston International Airport, January and July 1945 to 2022.

Figure 60: La Guardia Airport, New York, January and July 1949 to 2022.

Some Temperature Records from the United Kingdom

Figure 61: Cambridge, January and July 1952 to 2021.

Figure 62: Aberdeen Airport, January and July 1972 to 2021.

Some Temperature Records from Europe

Figure 63: Deelen Air Base, The Netherlands, January and July 1974 to 2022.

Figure 64: Rugge Airport, Norway, January and July 1973 to 2022.

Figure 65: Albacete, Spain, January and July, 1973 to 2022.

Two Pacific Records

Figure 66: Vunisea, Fiji, January and July, 1973 to 2021.

Figure 67: Christchurch, New Zealand, January and July 1952 to 2022.

The Daily Temperature Cycle

Figure 68: Christchurch January 1985 daily maximums and minimums.

Figure 69: Albacete, Spain, July 1984 daily maximums and minimums.

References and Further Reading for Chapter 11 – The Earth's Surface Temperature

Chapter 12 Conclusions

The Earth's "Sweet Spot" in relation to the Sun, surface liquid water, properties of water molecules.

Carbon dioxide concentration follows temperature changes, ocean storage CO2, CO2 is a trace gas in the atmosphere, CO2 has a limited fingerprint heat absorption compared to water vapour, CO2 increases result in more plant life.

Carbon dioxide alarmism is dominant – why ?

Scientific information presented in a misleading way:

Figure 70: Keeling Curve presentations.

Figure 71: The "Hockey Stick" graph.

Figure 31 (repeated): Average global temperature for the last 570 million years, Graphs, "adjustments", diagrams, omission inconvenient aspects, colour use, reports finalized by bureaucrats and politicians not scientists.

Why not just accept CO2 Alarmism ?

Futility at controlling temperature, Humans need the energy from fossil fuels.

Chapter 13 Alternatives to Fossil Fuels

Introduction to Wind and Solar, batteries, pumped hydro, grid connection problems.

Wind: land and sea required, Bird deaths, Noise, ugliness, limited suitable land, sea turbines, noise effects. Storm effects, limited working life, huge construction costs, undersea cable problems, intermittent.

Solar: solar thermal, solar photovoltaic, cells, modules, panels Solar panel manufacture and disposal, energy required, importance of China, Area required, Floating panels – problems, working life, disposal, recycling.

Solar panel performance issues: efficiency, intensity and duration of sunlight effects, cloud, optimistic calculations of performance.

Figure 19 (repeated): Effect of lower Sun in the sky, Inverter DC to AC losses, transmission line losses

Figure 20 (repeated): Showing effect of the Earth's daily rotation – day and night.

Figure 21 (repeated): seasonal effects on Sun radiation on different places, on Earth, average Sun energy per day at different latitudes, decreased in morning and afternoon compared to midday, latitude effects, season effects, weather, cloudy effects. Sunlight hours for different places, intermittent, Chinese manufacture, government subsidies.

Figure 72: A simple lead acid battery

Batteries: history, how a lead acid battery works.

Figure 73: A simple lithium ion battery

Limited life and recycling lead acid batteries, lithium ion batteries, how lithium battery works, cells wired in series to increase battery voltage, history, John Goodenough, advantages lithium batteries, more power, quick recharging etc. Massive increase production 70 X by 2050 – "nett zero". Manufacture problems, huge energy input for mining and manufacture, Chinese use of coal and diesel for. Limited life, disposal, recycling issues. Landfill dumping, toxicity, difficulties recycling lithium batteries

Electric Vehicles: advantages, global production increased, government incentives, petrol and diesel still dominant globally, most EV's small

Reasons for dominance of petrol and diesel vehicles over electric:

Cost, Norway's situation.

Recharging Problems: types of chargers, long queues recharging, problems charging SUV's and trucks, cost and availability of fast chargers, equipment failures fast chargers, mobile phone APPs required – problems, huge increased demand for electricity for EV's compared to if petrol or diesel, Third World difficulties, truck problems

Range problems, basic science – energy available from petrol and diesel versus batteries, weight lithium batteries in electric vehicles.

Cold problems.

Fire and explosion:, lithium batteries up to twice explosive power TNT, battery fire spreading to house, garage etc danger, battery in floor damage, thermal runaway, fire, accidents – fire and explosion, seawater corrosion, late fires in garages after flooding, damage.

Aviation, aircraft fires, cargo fires, passenger cabin fires, "Av sox" bags, rubbish dump fires, grid storage

Electrocution, electric vehicles 300 volt DC, accidents – rapid acceleration, quiet.

Uncertain future electric vehicles, grid storage batteries, home storage, lead acid or lithium.

Government incentives, large community storage batteries, examples in some countries

Comparison with pumped hydro the largest energy storage world-wide.

Analysis of requirements for "nett zero" for battery storage, impossible practically. duration supply grid storage batteries, flow batteries, Vanadium redox batteries, examples. Iron redox flow batteries.

Hydrogen

Hydrogen production, hydrogen energy output versus energy input to produce, Hydrogen storage, Hydrogen use in vehicles, aircraft, Hydrogen delivery and fuel stations, Hydrogen engine problems, replacing fossil fuels with hydrogen.

Biofuels

Principle biofuels, biofuel controversies and costs, energy to make biofuels, land area required, investor nervousness.

Wood, use in power stations, wood pellet production, global wood pellet economics, controversies regarding wood pellet CO_2 production.

Geothermal

Costs and countries using geothermal, problems with.

Tidal:

History, world tidal power stations, costs and abandoned projects, problems.

Waves: problems.

Nuclear:

History, current world nuclear power stations, power output of Uranium^{235}, structure of nuclear reactors, Nuclear waste, Nuclear waste disposal, accidents, core meltdown, Small modular reactors, problems, Nuclear power stations under construction globally, cost of construction.

References and Further Reading for Chapter 13 – Alternatives to Fossil Fuels

Chapter 14 Gas and Coal – A Modern View

Natural gas, advantages, types of gas plant, Plants world-wide, gas reserves – shortages, costs

Coal, global coal power stations, costs of coal power and comparisons, Coal geology and history, geological formation, types of coal, quality differences, burning of coal and heat production, products of coal burning including pollutants, removal of pollutants – fly ash, circulating fluidized bed burning, baghouses, electrostatic precipitators, cyclone precipitators, scrubbers.

Government regulations, Sulphur dioxide removal, selective catalytic reduction, flue gas desulfurization, activated charcoal, Mercury, Mercury control and regulations, nitric oxides, low nitrogen burners, selective catalytic reduction, overfire air, new methods pollution control, HELE plants.

Human mortality and illness from coal power stations historically, modern plant comparison, stable grid electricity, instability from wind and solar.

Carbon capture and storage problems, abandoned CCS examples.

Coal use essential for modern living especially Asia, coal power list for major countries, tough regulations in many for pollution, trend away from coal in Europa and North America – problems, global reserves of coal

References and Further Reading for Chapter14 – Gas and Coal – A Modern View

Chapter 15   Some Consequences of the Forced Shift to Renewables

Government subsidies and taxes, strategic military considerations, loss of freedom

www.ingramcontent.com/pod-product-compliance
Lightning Source LLC
Chambersburg PA
CBHW062311220526
45479CB00004B/1138